11 기

Noble gas geochemistry

Noble gas geochemistry

MINORU OZIMA

Geophysical Institute, Faculty of Science, University of Tokyo

FRANK A. PODOSEK

Department of Earth and Planetary Sciences and McDonnell Center for the Space Sciences, Washington University, St Louis

CAMBRIDGE UNIVERSITY PRESS

Cambridge

London   New York   New Rochelle

Melbourne   Sydney

Published by the Press Syndicate of the University of Cambridge
The Pitt Building, Trumpington Street, Cambridge CB2 1RP
32 East 57th Street, New York, NY 10022, USA
296 Beaconsfield Parade, Middle Park, Melbourne 3206, Australia

First published 1983

Printed by Nene Litho, Wellingborough, Northants
Bound by Woolnough Bookbinding, Wellingborough, Northants

Library of Congress catalogue card number: 83-5226

British Library cataloguing in publication data

Ozima, Minoru
Noble gas geochemistry.

1. Gases, Rare
I. Title II. Podosek, Frank A.
546'.75   QD162

ISBN 0 521 23939 7

Contents

Preface   ix

1   General introduction   1
1.1   Retrospect and prospect   1
1.2   Geochemical characteristics of the noble gases   3
1.3   Constants and conventions   5
1.4   Nomenclature   7

2   Air   11

3   Nuclear chemistry   15

4   Physical chemistry   23
4.1   Introduction   23
4.2   Adsorption   24
4.3   Solution   35
4.4   Crystal–melt partitioning   39
4.5   Trapping   41
4.6   Clathrates   45
4.7   Diffusion   48
4.8   Stepwise heating   52
4.9   Isotopic fractionation   60

5   Cosmochemistry   68
5.1   Introduction   68
5.2   Elemental abundance patterns   70
5.3   Planetary atmospheres   82
5.4   Isotopic heterogeneity   85
5.5   Light gas isotopery   90
5.6   Heavy gas isotopery   98
5.7   Origin of planetary gas   116

6   Component structures   127
6.1   Introduction   127

6.2   The radiogenic component   127
6.3   The spallation component?   133
6.4   Argon   134
6.5   Neon   135
6.6   Helium   140
6.7   Xenon   143
6.8   Krypton   145
6.9   Excess ^{129}Xe (and ^{136}Xe?)   147
6.10 Anomalies?   149

7   Water   153
7.1   Introduction   153
7.2   Solubility data   155
7.3   Seawater   160
7.4   Meteoric water   166
7.5   Juvenile ^{4}He   168
7.6   Juvenile ^{3}He   169

8   Sedimentary rocks   177
8.1   Introduction   177
8.2   Elemental abundances   177
8.3   Isotopery   184
8.4   Origin of trapped gases   185
8.5   The sedimentary xenon inventory   189
8.6   Paleoatmospheric noble gases?   192

9   Igneous rocks   196
9.1   Introduction   196
9.2   Elemental abundances   196
9.3   Isotopery   207
9.4   Excess Ar and He   208
9.5   Air contamination?   211
9.6   Volcanics   215
9.7   Xenoliths   217
9.8   Diamonds   221
9.9   Plutonic rocks   222
9.10 Cyclosilicates   224
9.11 Noble gas locations within rocks   228

10   Emanation   230
10.1 Introduction   230
10.2 Volcanic emanation   231
10.3 Natural gases   235

10.4	Radon emanation	238
10.5	Earthquake prediction	239
10.6	Stress emanation	246
11	**The mantle**	250
11.1	Introduction	250
11.2	Noble gases and mantle structure	251
11.3	Argon	253
11.4	Helium	257
11.5	The ^{4}He/^{40}Ar ratio	261
11.6	Neon	265
11.7	Xenon and krypton	266
12	**Inventories**	269
12.1	The atmospheric inventory	269
12.2	A secondary atmosphere	273
12.3	Sources	273
12.4	Sinks?	277
12.5	The atmospheric helium budget	280
12.6	The xenon inventory	283
12.7	Inventory of radiogenic gases	285
12.8	Is the earth degassed?	288
12.9	A primary atmosphere?	290
12.10	Venus and Mars	291
13	**Atmospheric evolution**	299
13.1	Introduction	299
13.2	Catastrophic degassing?	300
13.3	Continuous degassing mechanisms	303
13.4	Model structures	306
13.5	Two-reservoir degassing	308
13.6	Argon	315
13.7	Helium	320
13.8	Xenon	324
13.9	Conclusions	335
	References	339
	Index	359

Preface

In addition to being the collective name for the elements in the rightmost column of the periodic table of the elements, the term 'noble gases', also 'rare gases' or occasionally 'inert gases', is a convenient label for the branch of the scientific enterprise concerned with studying the occurrence and distribution of these elements in nature, particularly in the earth and the terrestrial planets. This research area is usually considered part of geochemistry, although other labels would serve as well.

The most familiar and most widely practiced area of geological research involving noble gases is, of course, geochronology, especially K–Ar dating. There are many good books about geochronology and it is not our intention to try to add another. Noble gas geochemistry, the subject of this book, will mean here what it usually suggests in the geological and planetary science community: the study of the natural occurrence of noble gases and what may be learned thereby, other than determination of the ages of rocks.

In the last two decades the discipline of noble gas geochemistry has prospered, at least by the measure of getting its own sessions at scientific conferences and attracting practitioners in sufficient numbers that, regrettably, they no longer all know each other or are even familiar with each other's work. In spite of its fruits, however, noble gas geochemistry often seems to non-practitioners to have much the air of the secret society and its dark art. Among the deplorable consequences of this circumstance are too many cases where good science and important scientific results obtained from noble gases are widely ignored and, on the other hand, too many cases of uncritical acceptance of noble gas results and interpretations which are ambiguous, misleading out of context, fatuous, or outright nonsense.

Among the reasons which could be advanced to explain why noble gas geochemistry has not been very well integrated into the larger scientific community of which it is part is, basically, unfamiliarity. Many of the traditional academic disciplines are involved in the study, and there are no courses in noble gas geochemistry in graduate school. There is also no book which covers the field, and we hope that this one will fill that need. The general style is that of the review paper: the intent is for professional level and currency and, within the limitations of scope, comprehensiveness, with sufficient background and introduction to encourage access by the student of the geosciences. A major aim of this book is that it not only be useful to our colleagues in the field but that it help those in other specializations form a critical appreciation of 'noble gases'.

Our intent is to take the 'geo-' in geochemistry literally, i.e. our focus is on terrestrial noble gases. Still, even more than in most specializations, it is impossible to develop an appreciation of terrestrial noble gases without a corresponding appreciation of the features which have emerged from the study of noble gases in extraterrestrial materials, notably meteorites. Accordingly we have included a chapter on noble gas 'cosmochemistry'. This is background material, however, and not an attempt to treat this subject as thoroughly as we hope to have treated the terrestrial subjects; such an attempt would require a considerable widening of the scope of this book.

The writing of this book was planned and initiated while one author, on sabbatical leave, was a guest in the other's home institution. This was made possible by support from the Japan Society for the Promotion of Sciences, and we are very glad to express our appreciation to the JSPS. We would also like to thank the McDonnell Center for the Space Sciences at Washington University for a variety of kinds of support which greatly facilitated the preparation of this book.

M. Ozima
F. A. Podosek

1 General introduction

1.1 Retrospect and prospect

In view of their scarcity and failure to form chemical compounds, it is not surprising that noble gases remained unknown until relatively late in the history of chemistry. The first known experimental indication of their existence was a persistent gaseous residue after chemical removal of nitrogen and oxygen from air, as noted by Cavendish in 1784; the residue was small, however, 'not more than 1/120th part of the whole', presumably attributed to experimental error, and in any case subsequently ignored. The first definitive identification came when several observers found a previously unknown line in the spectrum of the solar chromosphere during the 1868 eclipse; this was quickly recognized to belong to a new element, not yet known on earth, which was named 'helium' ($\eta\lambda\iota o\nu$: sun), but of course no chemical characterization was possible.

The actual 'discovery' of the noble gases came principally from the work of Rayleigh and Ramsey in the late nineteenth century. In 1892 Rayleigh reported that nitrogen prepared from ammonia was consistently less dense (by 0.5%) than 'nitrogen' prepared from air (by removal of oxygen, carbon dioxide, and water). Both Rayleigh and Ramsey, working in collaboration, followed up this experimental clue; pursuing the possibility that the density difference reflected admixture of a heavier gas in air, they, like Cavendish, found a residue when chemically reactive species were removed from air. This residue was clearly denser than air and spectroscopic examination (by Crookes) gave a spectrum unlike that of any known element. Rayleigh and Ramsay announced the discovery of a new element in a joint paper in 1895. The discovery attracted interest beyond that due a new element because they also found it to be chemically unreactive, an unprecedented property that led to the name 'argon' (($\alpha\rho\gamma o\nu$, negative of $\epsilon\rho\gamma o\nu$): not working, idle, inert, lazy). They further found argon to be a monatomic gas of atomic weight 40. This was a puzzling result since the concept of the periodic table was then well established, and it was surprising to find an element whose weight was that of calcium, with no apparent place for it in the sequence potassium, calcium, scandium.

Further discoveries proceeded rapidly. In seeking possible compounds of argon, Ramsay suspected that 'nitrogen' previously reported to be liberated in acid treatment of a uranium mineral was actually argon rather than nitrogen; he prepared gas in this way (in 1895) and indeed identified argon in its spectrum. He also noted another line in this spectrum; in examining this gas Lockyer and Crookes quickly identified it as the helium line previously found in the solar spectrum. Helium was quickly found to be monatomic, of atomic weight 4, and, like argon, chemically inert. By 1897 Ramsay was able to conclude that helium and argon represented a new column in the periodic table and predicted that at least one additional element, of intermediate atomic weight, remained to be found. In 1898 Ramsay and coworkers conducted fractional distillation experiments on liquid air which led to identification of three new elements: neon (νέον: new), krypton (κρύπτον: hidden), and xenon (ξένον: alien). The identification included spectroscopic characterization, measurement of atomic weight, and confirmation of chemical inactivity. The noble gas column of the periodic table was completed in 1900–1903 by identification of gaseous emanation of thorium and radium as a noble gas, radon (radius: ray).

Since their discovery nearly a century ago, the unique properties of the noble gases have been the subject of much research in theoretical chemistry and physics. These gases have also found many applications as tools for scientific research and many commercial and industrial (even medical!) applications as well. These are well known and our further discussion will focus on the role of noble gases in geochemistry.

The first such application was in geochronology. Even while they were studying the phenomenon of radioactivity of uranium and thorium, Rutherford and Soddy noted its potential value as a geochronometer. The first radiometric age measurement, by the uranium–helium method, was reported by Rutherford at the International Congress of Arts and Science in St Louis in 1904. Uranium-helium dating was subsequently pursued as a major methodology in geochronology, and later supplemented by (and largely supplanted by) the potassium-argon method (Wasserburg & Hayden, 1955), particularly the ^{40}Ar-^{39}Ar variant (Merrihue & Turner, 1966). Geochronological study of terrestrial, meteoritic, and lunar samples continues to be pursued very actively, and to account for much of the research in noble gas geochemistry.

Appreciation of the relevance of noble gases to broader aspects of geochemistry was somewhat later in developing. A noteworthy step was von Weizsäcker's (1937) recognition that atmospheric argon was radiogenic (cf. Section 1.2); the corollary inference of transport of gas from the solid earth to the atmosphere suggests the possibility of quantitative evaluation of the degassing history of the earth, a subject which has received much attention and in which radiogenic noble gases play the principal role. Subsequently, Brown (1952) used noble gas abundances to argue that the earth's atmosphere as a whole was secondary rather than primary, i.e. that the present atmosphere was formed by outgassing of the interior (cf. Chapter 12). The discovery of He in seawater (Chapter 7) by Clarke *et al.* (1969) showed that this degassing process is incomplete and it continues today. Considerable recent work has been done on developing the use of noble gases as geochemical tracers and indices of tectonic activity; as one example, noble gas emanation is of considerable potential value in earthquake prediction (Chapter 10). More generally, characterization of both primordial and radiogenic noble gas abundances in the mantle is a major present concern in noble gas geochemistry because of their profound implications for both the formation of the earth and its thermal, tectonic, and chemical evolution. Such developments and prospects are the principal subject of this book.

As with other scientific disciplines, advances in noble gas geochemistry can be seen to follow not only from prior progress in research but from exploitation of technological and methodological developments. Early work measured gases by volume, using the handling techniques developed by Ramsay and Rayleigh and progressively refined to the point that uranium–helium ages could be determined for ordinary surface rocks rather than only for unusual and scarce uranium and thorium ores (cf. Dubey and Holmes, 1929). The present character of the discipline, however, is set by the research tool which is its *sine qua non*, the mass spectrometer. The first demonstration of multiple isotopic forms of a stable element (neon!) was made in the 'parabolic' mass analyzer by Thomson (1912) and the first focusing mass 'spectrographs' suitable for isotopic abundance measurements were constructed by Dempster (1918) and Aston (1919). During the 1930s and 1940s the basic instrumental configuration for isotope ratio mass spectrometry evolved to the form still most commonly used today: a mono-energetic ion source (with ions of gaseous samples produced by electron bombardment), mass separation by single-focusing magnetic sector analysis, and electrical (rather than photographic) ion detection, usually with an electron multiplier for small samples. A particularly significant development for noble gases was Reynolds' (1956) introduction of techniques and methodologies for static rather than dynamic analysis; this permitted the accurate isotopic analysis of very small quantities of gas which characterizes noble gas geochemistry today and justly may be said to have begun its modern era.

1.2   Geochemical characteristics of the noble gases

Although this whole book is about noble gas geochemistry, there are two important general characteristics that should be noted here. These are everyday knowledge for workers in the field but deserve explicit attention from the nonpractitioner.

The first is that the noble gases are, obviously, noble, i.e. chemically inert. This does not mean that they completely fail to interact with anything (cf. Chapter 4), but the interactions are of the van der Waals type, much weaker than normal chemical reactions. In practical terms, this means that the interactions are less complicated and we may be more optimistic that they can be understood in terms of fairly simple ideas. (That the relevant chemistry is still not very well understood reflects more the lack of empirical data than intrinsic complexity.) It also means that the noble gases constitute a reasonably coherent group, so that geochemical parameters can often be seen to vary more or less smoothly from light to heavy gases. This coherence is the reason why it makes sense to consider their geochemistry as a group, as well as why, experimentally, they are naturally studied as a group.

The second feature is that the noble gases are scarce, whence their common alias, the 'rare gases'. This is a consequence of their inertness. The noble gases are not actually rare in the cosmic sense (Chapter 5), nor in the solar system as a whole, i.e. the sun and presumably the major planets. At some point in its evolution, however, the solar system, or at least that part most familiar to us, passed through a stage in which some of its constituent elements were mostly in solid phases while others, including the most abundant, were in a gas phase. The outstanding chemical characteristic of the terrestrial planets (including meteorites) is that they are made from the solids, to the virtually complete exclusion of the gases. Because of their chemical inertness, the noble gases were overwhelmingly partitioned into the gas phase and so are depleted in the earth and other accessible samples, often to a very striking degree, and in general more than any other elements.

Thus, the noble gases are trace elements *par excellence.* As an example, a not unreasonable value for Xe concentration in a rock is some 10^{-11} cm^3 STP/g (about 3×10^8 atoms/g), or 0.000 06 ppb. It is nevertheless quite feasible to perform an adequate analysis on a gram sample of such a rock, in the sense of a sample to blank ratio in excess of 10^2, 5–10% uncertainty in absolute abundance, and 1% or less uncertainty in relative abundances of the major isotopes. Detection limits are much lower than this, and for the scarcer isotopes the blank and thus the quantity necessary for analysis are two to three orders of magnitude lower. It is worth noting that the reason why such an experiment is possible is the same reason why noble gases are so scarce in the first place: their preference for a gas phase and the ease with which they can be separated from more reactive species.

An important corollary of this scarceness is that there are many effects which are not intrinsically characteristic of the noble gases but are best studied through them. These are effects in which small amounts of material are added to some reservoir. In many cases such additions are undetectable against the background of material already there, but, when the background is low, as it usually is for noble gases, the effects become observable and thereby interesting. An important subclass of such effects is that of nuclear transmutations, many of which are totally unobservable if the daughter nuclide is a 'normal' element but which make prominent changes if the daughter is a scarce noble gas. Nuclear chemistry (Chapter 3) is thus a considerable part of noble gas geochemistry.

Many examples of the utility of scarcity will be found in this book, so one will suffice here. The nuclide ^{40}K is naturally radioactive and thus useful for geochronology. Some 90% of the decays produce ^{40}Ca, but constitute an imperceptible perturbation on normally present Ca in all but a few highly unusual samples, and K–Ca dating is only marginally useful in geochemistry. The 10% of the decays that produce ^{40}Ar generally overwhelm any Ar initially present and are readily observable experimentally, and K–Ar dating is probably the most extensively practiced form of geochronology. By noble gas standards, the earth is awash in radiogenic ^{40}Ar. Indeed, on the basis of atmospheric ^{40}Ar overabundance relative to other Ar isotopes, von Weizsäcker (1937) predicted the decay of ^{40}K to ^{40}Ar even before K was known to be radioactive.

Finally, we should note that while Rn is chemically a noble gas, geochemically it might as well not be. Rn has no stable isotopes and is found in nature only because it is part of U and Th decay schemes (Chapter 3). Because of its extremely low abundance even by noble gas standards, it is observed by different techniques (based on its radioactive decay) than the other noble gases. Most important for our purposes, the geochemical factors that determine the distribution of Rn seldom have any close relation to the factors that control the other noble gases. Accordingly, we will discuss Rn only when it arises in context (cf. Chapter 10), but will otherwise make no attempt to present a comprehensive description of the geochemistry of Rn.

1.3   Constants and conventions
A number of physical constants and conversion factors frequently useful in noble gas geochemistry calculations are collected in Table 1.1. These values have been used for all the calculations in this book.

In subsequent data tabulations in this book, primary experimental data are stated with their attendant uncertainties when it is important to do so. Most numerical data are stated without uncertainties, however, as it is more common that experimental uncertainties are minor in comparison with those originating in the failure of natural systems to conform to simple models or in the fuzziness of quantitative interpretations. In general, numerical values are stated with more significant figures than their precision justifies. This reflects not ignorance but the utility of many such data as starting points for calculations in which the excess figures help alleviate two annoying problems: the accumulation of rounding errors in calculations based only on significant figures and the vexation of obtaining different values of the same quantity when calculated by different paths.

Presumably in continuity with early investigations in which gases were actually measured by volume, in modern geochemical literature quantites of noble gases are generally reported in terms of cm^3 STP. This has the advantage that cm^3 STP are proportional to number of atoms, which is the way gases are measured in the mass spectrometers responsible for most of the modern data. Numbers of atoms, rather than, say, weight per cent or partial pressures, are fortunately also the units generally most convenient for calculations and inter-pretations. Reporting of gas quantities in moles, or actual numbers of atoms, would have the same advantage, but is not customary. Unfortunately, none of these units (nor others that might be chosen) has the desirable feature of being comparable to a sufficiently extensive set of observed data that its use would permit abundance data to be remembered as integers rather than powers of ten.

A widespread convention in geochemistry, particularly for isotopic data, is use of δ values, in which numerical values are expressed as per mil ($^o/_{oo}$) deviations from some standard;

$$\delta_x \equiv (x/x_0 - 1) \times 1000 \qquad (1.1)$$

Table 1.1. Physical constants and conversion factors

Constants	
Avogadro's number	$N_A = 6.022\,17 \times 10^{23}$ molecules/mole
Gas constant	$R = 8.314\,34$ J K^{-1} mole^{-1}
	$= 1.987\,17$ cal K^{-1} mole^{-1}
	$= 0.082\,056$ l atm^{-1} K^{-1} mole^{-1}
at 0 $^\circ$C	$RT = 22.414$ l atm^{-1} mole^{-1}
at 25 $^\circ$C	$RT = 24.456$ l atm^{-1} mole^{-1}
Boltzmann's constant	$k = 1.380\,62 \times 10^{-16}$ erg/K
Ice point 0 $^\circ$C	$= 273.15$ K
Electron charge	$e = 1.602\,19 \times 10^{-19}$ coulomb
Planck's constant	$h = 6.625\,6 \times 10^{-34}$ J sec
Conversions	
1 atm ($\equiv$ 760 torr)	$= 1.013\,25 \times 10^6$ dyne/cm^2
1 cm^3 STP	$= 4.461\,5 \times 10^{-5}$ mole
	$= 2.686\,8 \times 10^{19}$ molecules
	$= 1.091\,5$ cm^3 NTP
1 cal (definition)	$= 4.184$ J
1 eV/molecule	$= 23.06$ kcal/mole
1 amu (^{12}C $= 12$)	$= 1.660\,53 \times 10^{-24}$ g

where x is the quantity of interest and x_0 is the standard value. Other normalizations, e.g. 10^2 for per cent (%) deviations, or 10^4 for highly precise data (unfortunately, noble gas data cannot yet justify using the 10^4) are also used, but without as consistent notation as δ for $^0/_{00}$ deviations. This convention has the advantage that small deviations from the standard value are more readily apparent. It has the more important advantage that if the standardization is based on a physical standard measured under the same circumstances as the sample, δ_x can be determined from the ratio of instrumental responses without the uncertainties involved in correcting either x or x_0 for instrumental isotopic discrimination. In high-precision 'stable isotope' (O, C, N, S, etc.) geochemistry, δ values are thus the norm as primary experimental data, so that variations of isotopic ratios from the standards are known to better precision than are the absolute ratios.

In noble gas mass spectrometry, comparisons using δ values are frequent, but primary experimental isotopic data are generally absolute ratios rather than δ values. To a large extent, the standardization is accomplished by calibrating instrumental performance by analysis of a widely accessible standard (air), but for practical reasons the calibrations are generally less frequent and employ less of a match of sample and standard conditions than is the rule in stable isotope geochemistry. Fortunately, this entails little difficulty in practice, since isotopic effects are frequently so large that minor discrepancies are unimportant or, for all practical purposes, nonexistent. For a few cases of effects near the border of observation, however, a more thorough adoption of δ-value normalization could be of benefit.

1.4   Nomenclature

Most specialized disciplines have their own specialized jargons, each a lexicon and syntax developed partly to make communication more efficient and partly by historical accident. For convenience, this section lists a number of words and phrases used elsewhere in this book and in the pertinent literature which require special attention. Some such terms will be obscure outside the circle of communicants, some have special meanings or connotations beyond their normal dictionary definitions, and some need particular attention because they are not used consistently in the literature.

Component. In isotopic geochemistry, and especially in noble gas geochemistry, this term is widely used to designate any compositionally well-defined and uniform reservoir. It is particularly useful in description of component resolution, i.e. the attempt to interpret analytical data as a superposition of previously known components, the reservoirs of which might have been sampled by the specimen under consideration. A component may have a single and well-understood origin, e.g. the collection of Xe isotopes produced by spontaneous fission of ^{238}U, but it need not. Thus, Xe in air is an ill-understood mixture of Xe contributed by a number of sources (Section 6.7); it is nevertheless useful to consider air Xe itself as a component, since it is a uniform reservoir which can be and is sampled by materials of interest. The term 'component' usually denotes a reservoir of isotopes of a single element. In some cases it may include more than one element, but it is then less useful. Thus, fission Xe from ^{238}U will be produced in propor- tion to ^{4}He, and together the Xe and He will constitute a component (Section 6.2), but the Xe and He can be very easily separated (much more so than, say, ^{136}Xe and ^{134}Xe) and the notion of 'U-derived' gas is less useful in component resolution than is 'U-derived' Xe.

Nuclear component. A nuclear component is one generated by nuclear transmu- tation. For example, Xe produced by fission of ^{238}U is a nuclear component; air Xe, in contrast, would generally not be considered a nuclear component, even though it contains a contribution from ^{238}U fission. Ultimately, all the isotopes have been produced in processes controlled by their nuclear properties, and so everything is a nuclear component, but it is still useful to make a distinction. In particular, it is common to designate as a 'primordial' component all the nuclides present in the solar system when it was isolated from the rest of the galaxy (cf. Chapter 5), in contrast to a variety of nuclear components generated within the solar system since that time. Subdivision of this category is common, e.g. a radiogenic component is one generated by decay of natural radionuclides.

Trapped components and in situ components. Noble gases in solid samples are grouped into two classes, designated 'trapped' and '*in situ*'. The *in situ* label designates a nuclear component which is still in the same locations in which it was generated. Gases which are not *in situ* are trapped. The distinction is opera- tional (although it is sometimes difficult to determine experimentally) and only partly genetic. An *in situ* component is necessarily a nuclear component, but the origin of a trapped component need not be specified. Usually, one thinks of a trapped component as originating outside the specimen in question, so that the specimen sampled some external reservoir; the gas in the reservoir may itself be primordial, a nuclear component, or a mixture. The origin is not necessarily external, however. As an example, a rock may contain some ^{36}Ar of external origin (trapped) and ^{40}Ar generated by decay of ^{40}K (*in situ*). If the rock is melted in a closed system and then solidifies again, the subsequently produced radiogenic ^{40}Ar will be an *in situ* component, but the ^{40}Ar produced before the melting will no longer be in the same microscopic locations in which it was produced, and so will be a trapped component. As specified, the distinction is operational and rests on the possibility, at least in principle, of separating trapped and *in situ* components by mechanical or chemical disassembly of the rock or by different diffusional behavior manifested in stepwise heating. (Whence the distinction between trapped and *in situ* is not made for liquid or gaseous samples.) Thus, in the example, the first-generation ^{40}Ar, although radiogenic, will have become homogenized with the ^{36}Ar and no longer separable from it by any means short of a Maxwell's demon (or mass-dependent isotopic fractionation). The second-generation ^{40}Ar will be separable from these but not, say, from ^{39}Ar produced by ^{39}K (n,p) in a neutron irradiation; together these will constitute a single *in situ* component, since they are sited in the same places, i.e. wherever the K is. The total ^{40}Ar is thus split between two components, one trapped and one *in situ*, and these components can be separated. This situation is not hypothetical: 'excess ^{40}Ar', not uncommon in oceanic rocks, is radiogenic but trapped, not *in situ*.

Solar and planetary. In addition to their normal usage in designating the sun and a planet, respectively, the terms 'solar' and 'planetary' are used to designate specific patterns of noble gas elemental abundances. These are discussed in Section 5.2.

Atmosphere. Air is the gas phase at the surface of the earth. In this book, as is common in the literature, 'atmosphere' will be used in a more general sense (Section 12.1) to designate all the volatiles in surface reservoirs of the earth, including air but also water and sedimentary rocks.

Juvenile. It is generally believed that the earth's atmosphere was formed by degassing of volatiles from the interior (Section 12.2). Volatiles which have remained in the earth's interior and which have never been part of the atmosphere, and which are observed to be entering the atmosphere for the first time, are designated as juvenile. 'Atmospheric' is an antonym to 'juvenile'.

Fractionation and discrimination. Both these terms are used in consonance with everyday meaning but also with special connotations. Fractionation connotes the process of selecting one element in preference to another or, more commonly in this book, one isotope in preference to another of the same element (Section 4.9). For the latter usage in particular, it is supposed that in natural isotopic fractionation processes the only selection criterion is isotopic mass. In nature, isotopic fractionation effects are usually small and measured in per mil (in contrast to elemental fractionation, often measured in powers of ten) and in elements of three or more isotopes can be recognized by their smooth functional (often nearly linear) dependence on differences in isotopic mass. In the technical literature, the term 'discrmination' is usually reserved for the specific case of instrumental rather than natural isotopic fractionation, i.e. that which arises as part of the measurement process. In contrast to solid-source mass spectrometers, noble gas mass spectrometers usually have rather constant discrimination, and the normal procedure is to calibrate this discrimination by analysis of standards and apply the corresponding correction to sample data. When small fractionation effects are inferred, it is sometimes questionable whether they are actually natural fractionations or perhaps simply artifacts (improper correction for instrumental discrimination). To our knowledge no geochemical noble gas investigations have used the double-spike technique which could distinguish between natural fractionation and instrumental discrimination, and even for geochronological applications use of a double-spike (cf. Macedo *et al.*, 1977) is very rare.

Distribution coefficients. In describing the partitioning of a trace element among coexisting phases it is frequently convenient to use a 'distribution coefficient', for a given element defined as a concentration ratio C_2/C_1. Here C is concentration and subscripts identify phases; often the normalizing phase is some convenient reservoir, such as a silicate melt, with which several other phases may equilibrate. For noble gases it is often most convenient to normalize to a gas phase. If the concentrations are expressed in the same units the distribution coefficient is dimensionless. It is conventional to cite noble gas concentrations in condensed phases in $cm^3 STP/g$, however, and to describe the gas phase by partial pressures, so most of our numerical citations of distribution coefficients will be in $cm^3 STP\ g^{-1}\ atm^{-1}$.

2   Air

Air is a major reservoir for terrestrial noble gases. It is possible that air is *the* major reservoir for the earth, accounting for most of the terrestrial noble gas inventory, but this should not be assumed lightly. In any case, air is certainly the most conspicuous and accessible reservoir of noble gases, and its characterization is of fundamental importance in noble gas geochemistry. The geochemical considerations are taken up elsewhere in this book; this chapter focuses on presentation of basic data. It is noteworthy that these data are important not only in the direct sense of providing information about the earth and its atmosphere but also in the sense of serving as a standard for noble gas geochemistry: most data are produced by mass spectrometers whose performance, both sensitivity and mass discrimination, is calibrated by analysis of samples of air or of secondary standards whose parameters are ultimately determined by comparison with air. At least as far as noble gases (except Rn – see Chapter 10) are concerned, it is assumed that air is compositionally uniform (not counting high altitudes and nuclear power plants, which are unimportant for total inventory). Actually, studies directed to this point are scarce, but there seems no reason to challenge this assumption (cf. Mamyrin *et al.*, 1970).

The generally accepted data for elemental abundances and isotopic compositions are given in Tables 2.1 and 2.2, respectively. For a number of applications, it is more convenient to consider abundances of specific isotopes than of elements; representative values are shown in Table 2.3. In particular, the air 'concentrations' in the second line of Table 2.3 are often used as a normalization for observed concentrations in samples. If the atmosphere does account for nearly the total terrestrial inventory, then indeed these values are near the average concentration of noble gases in the materials which accreted to form the earth. Use of these data for normalization does not constitute endorsement of this proposition, however, and whether or not they represent the terrestrial inventory they are a convenient data set with the elemental ratios of air and absolute abundances of the same order of magnitude as many samples.

The primary experimentally-determined data in Tables 2.1 and 2.2 are the isotopic ratios and volume fractions in air and the total mass of air; these are tabulated with their reported uncertainties. Quantities derived from these are shown without error limits in Tables 2.1-2.3; for reasons cited earlier they are stated to more significant figures than are justified by the precision of the primary data from which they are calculated.

It seems likely that application of presently available information and technology could result in some improvement in the elemental abundance data (Table 2.1). At least from the geochemical viewpoint, however, relatively little advantage would result. High-precision absolute abundances, as needed for example in seawater (Chapter 7) or K-Ar studies, are generally isotopic dilution data calibrated volumetrically rather than by comparison with air. Differences between air and other samples are seldom significant below 10%, more commonly not below a factor of two; the major exception is water (Chapter 7).

The opposite is true, however, for isotopic data (Table 2.2). Isotopic comparisons are of major concern in noble gas geochemistry, both terrestrial and extraterrestrial, and experience has shown that as error limits have shrunk in response to technological improvement, the arguments have followed the error limits down to progressively finer levels. In many instances involving comparison

Table 2.1. Elemental composition of dry air

Gas	Molecular weight ($^{12}C = 12$)	Volume fraction[a]	Total inventory	
			grams	cm^3 STP
Dry air[c]	28.9644	1	5.119×10^{21} ± 0.008	3.961×10^{24}
Major gases[b]				
N_2	28.0134	0.78084 ± 0.00004	3.866×10^{21}	3.093×10^{24}
O_2	31.9988	0.20948 ± 0.00002	1.185×10^{21}	8.298×10^{23}
CO_2	44.0099	$(3.15 \pm 0.10) \times 10^{-4}$	2.450×10^{18}	1.248×10^{21}
Noble gases				
He	4.0026	$(5.24 \pm 0.05) \times 10^{-6}$	3.707×10^{15}	2.076×10^{19}
Ne	20.179	$(1.818 \pm 0.004) \times 10^{-5}$	6.484×10^{16}	7.202×10^{19}
Ar	39.948	$(9.34 \pm 0.01) \times 10^{-3}$	6.594×10^{19}	3.700×10^{22}
Kr	83.80	$(1.14 \pm 0.01) \times 10^{-6}$	1.688×10^{16}	4.516×10^{18}
Xe	131.30	$(8.7 \pm 0.1) \times 10^{-8}$	2.019×10^{15}	3.446×10^{17}

[a] Data from *US Standard Atmosphere* (1962) and Mirtov (1961), as tabulated by Verniani (1966).

[b] The next most abundant gas is CH_4 (volume fraction $\sim 2 \times 10^{-6}$), and H_2, N_2O, and CO are more abundant than Xe; a more extensive tabulation is conveniently offered by Walker (1977).

[c] Verniani (1966); based on 0.33% H_2O in total air mass $(5.136 \pm 0.007) \times 10^{21}$ g.

Table 2.2. Isotopic compositional data for noble gases in air

Isotope	Isotopic ratios[h]	Alternative normalization	Per cent abundance (atomic)
Helium[a]			
3	1.399 ± 13	1	0.000 140
4	10^6	714 800	~100
Neon[b]			
20	100	9.800	90.50
21	0.296 ± 2	0.039 0	0.268
22	10.20 ± 8	1	9.23
Argon[c]			
36	$0.337 8 \pm 6$	1	0.336 4
38	$0.063 5 \pm 1$	0.188 0	0.063 2
40	100	295.5^g	99.60
Krypton[d,e,f]			
78	$0.608 7 \pm 20$	1.994	0.346 9
80	$3.959 9 \pm 20$	12.973	2.257 1
82	20.217 ± 4	66.23	11.523
83	20.136 ± 21	65.97	11.477
84	100	327.6	57.00
86	30.524 ± 25	100	17.398
Xenon[d,e]			
124	$0.353 7 \pm 11$	2.337	0.095 1
126	$0.330 0 \pm 17$	2.180	0.088 7
128	7.136 ± 9	47.15	1.919
129	98.32 ± 12	649.6	26.44
130	15.136 ± 12	100	4.070
131	78.90 ± 11	521.3	21.22
132	100	660.7	26.89
134	38.79 ± 6	256.3	10.430
136	32.94 ± 4	217.6	8.857

[a] Mamyrin *et al.* (1970).
[b] Eberhardt *et al.* (1965).
[c] Nier (1950a).
[d] Basford *et al.* (1973).
[e] Nier (1950b).
[f] Nief (1960).
[g] Nier's (1950a) primary data were given, as shown, as ratios normalized to ^{40}Ar; the equivalent is ^{40}Ar/^{36}Ar $= 296.0 \pm 0.5$. The figure 295.5 came into widespread use, however, on the basis of his (rounded) percentage abundances, and Steiger & Jäger (1977) have recommended its adoption as a convention for geochronology.
[h] Uncertainties refer to the last digits given.

of air and sample compositions, or even only the use of air in instrumental calibration, the uncertainties in air isotopic composition are not inconsequential.

For quite some time the 'industry standard' for air isotopic composition was the pioneering work of Nier (1950a,b) and Aldrich & Nier (1948). For Ar, Nier (1950a) calibrated absolute instrumental discrimination by analysis of a spike mixture of separated isotopes whose composition was determined volumetrically. Absolute discrimination calibration for Ne, Kr, and Xe (Nier, 1950b) also used this spike: analysis of Ar^{2+} for discrimination in the Ne mass range, and analysis of Kr^{2+} and Xe^{3+} in the Ar mass range to determine one isotope ratio which was then used to determine discrimination for Kr^+ and Xe^+. Nier (1950a) is still the accepted standard for Ar. The more recent redeterminations of He and Ne also utilize volumetrically-controlled mixtures of separated isotopes. For Kr and Xe, more precise data show persistent deviations from Nier's (1950b) ratios, so that better air compositions are available, but no improved discrimination calibration has been made. Various laboratories (e.g. Podosek *et al.*, 1971; Basford *et al.*, 1973; Bernatowicz *et al.*, 1979) have thus adopted compositions based on detector signal ratios corrected for discrimination so as to agree with Nier's (1950b) data, either in a specific ratio or to minimize overall mass-dependent trends. Such data are thus uncertain by an overall mass-dependent fractionation, although this should not be too severe, probably not more than about 0.1%/amu. The He, Ne, and Ar data in Table 2.2 are generally used for comparisons and calibration; no widely adopted alternative to Nier's (1950b) Kr and Xe compositions has yet emerged.

Table 2.3. Specific isotope noble gas abundances in air[a]

	^4He	^{20}Ne	^{36}Ar	^{84}Kr	^{130}Xe
Absolute abundances					
cm³ STP[b]	2.076×10^{19}	6.518×10^{19}	1.245×10^{20}	2.574×10^{18}	1.403×10^{16}
cm³ STP/g[c]	3.473×10^{-9}	1.091×10^{-8}	2.083×10^{-8}	4.307×10^{-10}	2.347×10^{-12}
Relative abundances					
	1	3.14	6.00	0.124	0.000 676
	0.319	1	1.91	0.039 5	0.000 215
	0.167	0.524	1	0.020 7	0.000 113
	8.07	25.3	48.4	1	0.005 45
	1 480	4 645	8 874	183	1

[a] Calculated from data in Tables 2.1 and 2.2.
[b] Total air inventory.
[c] Total air inventory divided by mass of earth (5.976×10^{27} g).

3 Nuclear chemistry

The noble gases are identified as such on the basis of their chemical rather than their nuclear properties. Except by accident (e.g. the unique role of the α particle in nuclear physics) there is nothing special about the nuclear properties of the noble gases. For convenience of familiarization and reference, the place of the noble gases and neighboring elements in the family of nuclides is illustrated in Fig. 3.1.

Since the synthesis of isotopic species is guided by their nuclear properties, there is also nothing special about noble gas abundances, at least on the cosmic scale. Cosmic abundances are discussed in Chapter 5, abundances in specific samples throughout this book. The noble gas isotopic abundances indicated in Fig. 3.1 are those of air; it is noteworthy that the natural range of isotopic abundances of the noble gases is greater than that of most other elements.

Aside from nucleosynthetic considerations, there are in fact relatively few cases of geochemical importance attached to the nuclear properties of the noble

Fig. 3.1. Isotopes of the noble gases and neighboring elements. The format is that of the *Chart of the Nuclides* (Walker *et al.*, 1977). Proton number is shown on ordinate, neutron number on abscissa. Stable and extant long-lived isotopes are shown as solid boxes with per cent natural abundances. Unstable nuclides are shown as dashed boxes with half-lives in parentheses indicating decay. Dashed arrows indicate fission fragment decay paths. (a)–(f) See individual captions.

Fig. 3.1(a). Isotopes of He and neighboring elements.

Fig. 3.1(b). Isotopes of Ne and neighboring elements.

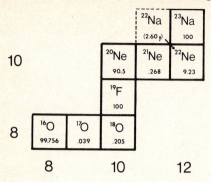

Fig. 3.1(c). Isotopes of Ar and neighboring elements.

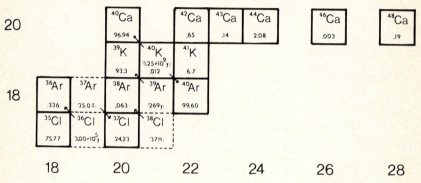

Fig. 3.1(d). Isotopes of Kr and neighboring elements.

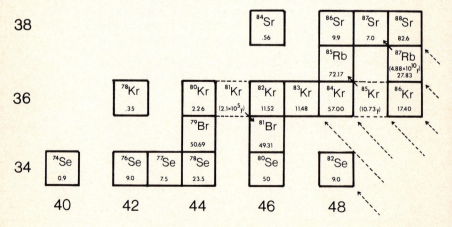

Fig. 3.1(e). Isotopes of Xe and neighboring elements.

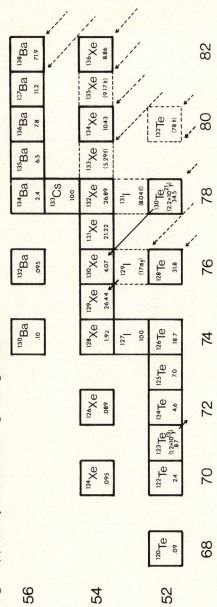

gases themselves. Most of these concern decay schemes and lifetimes of radioactive species, and in the terrestrial environment only the special case Rn is
important. Rn has no stable isotopes (Fig. 3.1(f)) and occurs naturally only
because it is produced in U and Th decay.

In contrast, there are many cases of keen geochemical interest in nuclear
processes in which the final product is a noble gas isotope, as follows from their
characteristic scarcity (Section 1.2). Foremost among these are the decay
schemes of natural radionuclides, which are widely exploited in geochronology
and make often sizeable contributions to the total noble gas inventory of the
earth (Chapters 6 and 12). Parameters for important radionuclides are given in
Tables 3.1 and 3.2. Table 3.3 covers the 'other' category. Tables 3.1–3.3 are
intended to be complete in the sense of geochemical importance (i.e. observa-
bility), known or suggested, but oversights are likely and little benefit would
follow from making the tables exhaustive. Thus, other nuclides are known to
be α active, other transuranics produce Xe and Kr in fission, and scores of
reactions could be added to Table 3.3; important as these may be in other
contexts they are excluded here on grounds of geochemical irrelevance (so far!).
In Table 3.3, only one of each 'kind' of reaction is listed; there are often geochemically related reactions, e.g. ^{7}Li $(\alpha, {}^3\text{He})$ ^{8}Be produces ^{3}He, spallation
produces ^{4}He (in greater abundance than ^{3}He), and ^{17}O (α, n) produces ^{20}Ne,
but these are all less important than the tabulated reactions.

The term 'spallation', used in Table 3.3 and elsewhere, strictly speaking
means nuclear fragmentation caused by impact of high-energy particles. In the
natural environment, the high-energy particles are cosmic rays and their
secondaries. In this book we will adopt the customary geochemical usage of

Fig. 3.1(f). Isotopes of Rn and neighboring elements.

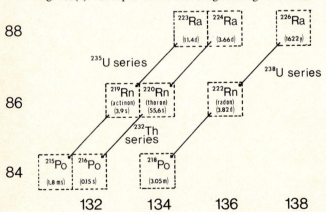

Table 3.1. *Prominent natural radionuclides with noble gas daughters*

Parent [b]	Half-life [a] (10^9 yr)	Daughter	Yield [a] (atom/atom)	Atomic weight of parent	Per cent abundance [a] (atomic)	Noble gas production (cm³ STP/g) [d] Rate (yr⁻¹)	Accumulation
Extant radionuclides							
^{232}Th	14.010	^{4}He	6	232.038	100	2.868×10^{-8}	5.796×10^2
^{238}U	4.468	^{4}He	8	238.029	99.280	11.602×10^{-8}	7.479×10^2
^{235}U	0.7038	^{4}He	7	235.044	0.720	0.467×10^{-8}	4.746
^{147}Sm	106	^{4}He	1	150.4	15.0	0.014×10^{-8}	2.235×10^1
^{40}K	1.251	^{40}Ar	0.1048	39.098	0.01167	3.887×10^{-12}	7.012×10^{-3}
^{238}U	4.468	^{136}Xe	3.50×10^{-8}	238.029	99.280	4.980×10^{-16}	3.210×10^{-6}
Extinct radionuclides							
^{129}I	0.017	^{129}Xe	1	129.904	c	—	$1.739 \times 10^{2\,e}$
^{244}Pu	0.082	^{136}Xe	7.00×10^{-5}	(244.06)	c	—	$6.429 \times 10^{-3\,e}$

[a] Decay constants for ^{232}Th, ^{238}U, ^{235}U, and ^{40}K and isotopic data for U and K are those recommended for adoption for geochronology by Steiger & Jäger (1977); fission data from Table 3.2; other data as tabulated by Walker *et al.* (1977).

[b] This table includes all natural radionuclides known or suggested to account for an observable fraction of the terrestrial noble gas inventory. ^{147}Sm illustrates marginal significance: for 'cosmic' ratio (Cameron, 1973) Sm/Th = 3.3 (by weight), ^{147}Sm produces ^{4}He at 1.6% the rate of ^{232}Th. Several other α-active radionuclides are extant and ^{130}Te decays to ^{130}Xe (Table 3.3), but these are negligible sources.

[c] Likely representative atomic abundances at time of meteorite formation 4.55×10^9 yr ago: ^{129}I/^{127}I = 1.0×10^{-4} (Hohenberg *et al.*, 1967), ^{244}Pu/^{238}U = 0.0068 (Hudson *et al.*, 1983).

[d] Rate is noble gas production (in cm³ STP) per present gram of parent element per year. Accumulation (in cm³ STP) per present gram of parent element in time T ending at present is $(e^{\lambda T} - 1)$ times tabulated value.

[e] Total accumulation (in cm³ STP) of ^{129}I or ^{136}Xe per initial gram of ^{129}I or ^{244}Pu respectively. Note that for initial abundance as in note c, ^{136}Xe accumulation from ^{244}Pu is 27 times that from ^{238}U (in 4.55×10^9 yr).

Table 3.2. *Production of noble gas isotopes in fission*

Parent	Total half-life (10^6 yr)	Branching ratio	Thermal cross-section (Barns)	% yield[a]		^{136}Xe $\equiv 1$ composition[b]				^{86}Kr $\equiv 1$ composition[b]		References
				^{136}Xe	^{86}Kr	^{129}Xe	^{131}Xe	^{132}Xe	^{134}Xe	^{83}Kr	^{84}Kr	
Spontaneous fission												
^{238}U	4468	5.45×10^{-7}	—	6.3 ±0.4	0.95 ±0.06	<0.002	0.076 ±0.003	0.595 ±0.017	0.832 ±0.012	0.03 ±0.01	0.13 ±0.02	1, 2, 3, 7
^{244}Pu	82	1.25×10^{-3}	—	5.6 ±0.6	0.11 ±0.03	0.048 ±0.055	0.246 ±0.020	0.885 ±0.030	0.939 ±0.008	—	—	4, 5, 6
Neutron-induced fission												
^{235}U[c]	704	—	580	6.47 ±0.07	2.04 ±0.02	0.1	0.453 ±0.013	0.677 ±0.020	1.246 ±0.036	0.27 ±0.01	0.50 ±0.01	7, 8, 9

[a] An extensive compilation of data and review are given by Hyde (1964).
[b] Isobaric yields. Some precursors have appreciable half-lives, notably ^{129}I (17 Ma), ^{131}I (8.0 d), ^{132}Te (78 h); all others <1 d.
[c] Isobaric yields are 6.6% at mass 133, 6.5% at mass 135, 1.3% at mass 85. Yield of ^{136}Xe may be appreciably enhanced by ^{135}Xe (n, γ) ($\sigma = 3.6 \times 10^6$ b). Minor amounts of mass 3 and 4 isobars are produced in ternary fission (cf. Halpern, 1971).

References: 1, Wetherill (1953); 2, Young & Thode (1960); 3, Segre (1952); 4, Fields *et al.* (1966); 5, Alexander *et al.* (1971); 6, Lewis (1975); 7, Lederer & Shirley (1978); 8, Farrer & Tomlinson (1962); 9, Farrer *et al.* (1962).

'spallation' to designate not only the fragmentation processes but also, with some liberty, more specific and lower-energy reactions such as (p, n) and neutron capture reactions which arise in exposure to cosmic rays.

Tables 3.1–3.3 are limited to cases important in terrestrial geochemistry. The field of interest would be broadened by expansion of scope to include cosmo-chemistry (Chapter 5). Additional short-lived radionuclides have been suggested to produce observable amounts of noble gases in meteorites. Spallation produces every stable noble gas isotope, often dominating the inventory in meteorites and lunar samples, and unstable ones as well (^{81}Kr, ^{37}Ar, ^{39}Ar); both varieties are useful in studying exposure to cosmic rays.

It is also implicit above that only natural nuclear processes are considered. Artificial processes broaden the field considerably. Nuclear weapons tests and commercial fission reactors involve many reactions. Aside from local effects, e.g. use of Xe isotopes to detect leakage from reactors, the only global-scale effect in noble gas geochemistry is apparently enhancement of ^{3}He (from bomb-produced ^{3}H) in surface water (Section 7.6). Production of fission Xe and Kr is apparently negligible on a global scale (Bernatowicz & Podosek, 1978). Special note should be taken at this point of the Oklo phenomenon, the only known instance of a *natural* fission chain reaction, which provides a unique

Table 3.3. *Selected natural terrestrial nuclear processes[a] generating noble gases*

Isotope	Reaction	References
^{3}He	^{6}Li (n, α) ^{3}H	Morrison & Pine (1955)
^{3}He	spallation[b]	Craig & Lal (1961)
^{3}He	spallation ?[c]	Bernatowicz & Podosek (1978)
^{21}Ne	^{18}O (n, α) ^{21}Ne	Wetherill (1954)
^{21}Ne	spallation ?[c]	Heymann *et al.* (1976)
^{22}Ne	^{19}F (α, n) ^{22}Na ?	Wetherill (1954)
^{38}Ar	^{35}Cl (α, p) ^{38}Ar ?	Wetherill (1954)
^{128}Xe	^{127}I (n, γ) ^{128}I	Srinivasan *et al.* (1971)
^{129}Xe	^{130}Te (μ^-, η) ^{129}Sb[e]	Takagi *et al.* (1974)
^{130}Xe	^{130}Te ($\beta\beta$)	Inghram & Reynolds (1950)
^{131}Xe	^{130}Ba (n, γ) ^{131}Ba	Srinivasan (1976)
^{132}Xe	^{131}Xe (n, γ) ^{132}Xe	Drozd *et al.* (1974)
^{136}Xe	^{232}Th (sf)[d]	Wetherill (1953)
^{136}Xe	^{239}Pu (nf)[d]	Drozd *et al.* (1974)

[a] Other than as noted in Tables 3.1 and 3.2.
[b] In present atmosphere.
[c] In preaccretionary earth materials.
[d] Fission also produces other heavy isotopes of Xe and Kr. Other fission processes (Table 3.2) are more common.
[e] The secondary neutrons give rise to detectable ^{130}Te (n, γ) $\rightarrow$ ^{131}Xe.

exercise in nuclear geochemistry not only for the noble gases (Drozd *et al.*, 1974) but other elements as well.

Several geochemical methodologies also involve artificial nuclear processes, notably neutron-irradiation (^{40}Ar–^{39}Ar, ^{129}Xe–^{128}Xe, fission Xe dating, and trace element analysis), in which a further variety of noble gas nuclear effects are important. Examples include ^{235}U neutron-induced fission, and neutron capture by Cl, Br, and I to produce Ar, Kr, and Xe, respectively. Occasionally, the noble gases are even important as targets as well as products: in some cases neutron capture by ^{129}Xe and ^{131}Xe causes perceptible changes in Xe composition and neutron capture by ^{135}Xe (3.6×10^6 barns!) can appreciably enhance the fission yield of ^{136}Xe.

4 Physical chemistry

4.1 Introduction

The noble gases are also well known as the inert gases, reflecting their characteristic lack of chemical interaction with other elements. In the extreme case, a substance whose atoms fail to interact with other substances except by elastic collisions would always be an ideal monatomic gas. In general, the noble gases approach this extreme more closely than other elements. Nevertheless, of course, the noble gases do not fail completely to undergo interactions, and such interactions as do occur are responsible for governing their geochemical distributions.

A number of the basic parameters characterizing the noble gases as elements are presented in Table 4.1. This chapter will treat those aspects of noble gas interactions with other substances which are important geochemically. A much broader and more extensive treatment of the fundamental physical and chemical characteristics of the noble gases can be found in Cook (1961).

It is now well known that despite their name the noble gases (at least Rn, Xe, and Kr) do in fact participate in interactions normally considered 'chemical', notably with F but also with other elements, and the Xe–F bond strength is a substantial 30 kcal/mole. Noble gas chemistry is accordingly a subject of considerable theoretical interest. Nevertheless, it is extremely unlikely that conditions resulting in the formation of noble gas compounds would be encountered outside the laboratory, so noble gas chemistry will not be important in geochemistry and will not be discussed here. Treatments of noble gas chemistry are presented by Hyman (1963), Claasen (1966), and Holloway (1966).

In general, noble gases interact with other elements (and themselves) chiefly through van der Waals forces, also known as dispersion forces, which arise in electronic polarization effects. Although noble gas atoms are spherically symmetric and have no permanent dipole or higher multipole moments, they can interact even with similar atoms because fluctuations in the instantaneous dipole moment of one atom can induce moments in other atoms. The principal term in such resonant fluctuations is the dipole–dipole interaction, whose potential V has an inverse sixth power dependence on interatomic separation r:

$$V(r) = -C/r^6 \tag{4.1}$$

An exact calculation of the constant C is in general not possible; a well-known approximation is that of London (1930), who treated atoms as harmonic oscillators of single frequency ν and obtained

$$C = \tfrac{3}{2}\alpha_1\alpha_2 h\nu_1\nu_2/(\nu_1 + \nu_2) \tag{4.2}$$

for the interaction of atoms designated by subscripts 1 and 2. The characteristic frequencies ν must be determined by empirical fit to this formula but usually $h\nu$ is close to the ionization energy. The major variations in C are due to the electronic polarizability α (the ratio of induced dipole moment to inducing field).

At sufficiently close approach, overlap of electronic wave functions leads to a repulsive potential with a sharp spatial dependence, usually modeled as e^{-ar} or r^{-n} with n about 12. Together the repulsive and attractive forces create a potential well with a minimum at interatomic spacing of a few Å.

Interaction of a noble gas atom with condensed matter is considerably more complicated, and is usually approximated simply by summing or integrating potentials pairwise. Such treatments are necessarily crude but nevertheless allow an appraisal of the general features of an interaction and often provide realistic numerical values as well. Young & Crowell (1962), for example, review theoretical treatments of noble gas adsorption along these lines; predicted potentials for adsorption on various forms of carbon, to consider one example, range from a few hundred cal/mole for He to a few kcal/mole for Xe, in reasonable agreement with observed heats of adsorption.

While the details of van der Waals interactions in a given situation may be complex, a generalization worth noting is that the strength of the interaction will usually be roughly proportional to polarizability (cf. eq. (4.2)). Noble gas polarizabilities (Table 4.1) increase regularly with the number of electrons, and it is this feature which accounts for the general tendency of a heavier noble gas to be somewhat less noble than a lighter noble gas.

Other parameters which often are or may be important in interactions of noble gases with condensed matter or otherwise influence their distribution are atomic size and ionization potential (Table 4.1). These also follow a regular progression according to atomic number.

4.2 Adsorption

Adsorption is a common mechanism by which the noble gases may interact with other materials, and it is frequently cited as a possibly relevant factor in noble gas geochemistry. Adsorptive phenomena in general are a convenient way of studying intermolecular interactions and the properties of

Table 4.1. Physical and chemical properties of the noble gases

	Units	He	Ne	Ar	Kr	Xe	Rn
Atomic number	–	2	10	18	36	54	86
Atomic weight	amu	4.002 60	20.179	39.948	83.80	131.30	–
Triple point	K	–	24.6	83.8	116.0	161.3	202
Normal boiling point	K	4.2	27.1	87.3	119.8	165.0	211
Critical point	K	5.3	44.5	151.9	209.4	289.7	378
Triple point pressure	torr	–	324	516	548	612	–
Heat of fusion at triple point	cal/mole	–	80	281	391	549	–
Heat of vaporization at NBP	cal/mole	19	414	1 558	2 158	3 020	–
Van der Waals const. a	l^2-atm	0.034 12	0.2107	1.345	2.318	4.194	–
Van der Waals const. b	l	0.023 7	0.017 09	0.032 19	0.039 78	0.051 05	–
Atomic radius (crystal)	10^{-8} cm	1.8	1.6	1.9	2.0	2.2	–
Atomic radius ('univalent')	10^{-8} cm	0.93	1.12	1.58	1.69	1.90	–
Ionization energy	eV	24.48	21.56	15.76	14.00	12.13	10.75
Ionization cross-section (80 eV electrons)	10^{-17} cm^2	3.1	6.8	36	51	75	–
Polarizability	10^{-24} cm^3	0.201	0.390	1.62	2.46	3.99	–

surfaces and the subject has accordingly been afforded extensive theoretical development and empirical investigation. Although many empirical studies have included noble gases, data describing adsorptive interactions between noble gases and materials of geochemical importance are rather scarce. Nevertheless, adsorption is a relatively nonspecific interaction and much can be inferred on the basis of generalizations. Noble gas adsorption tends to be a rather weak interaction at temperatures of principal geochemical interest, but even so it can be of substantial importance, and there may be exceptions to this rule. There are many parallels between adsorption and solution (Section 4.3) and in many cases it can be difficult to distinguish between the two in practice (Section 4.4).

The term adsorption designates the situation in which the molecules of a gas (the sorbate) are concentrated at the surface of a solid (the sorbent) with which the gas is in contact. The concentration results from an attractive potential experienced by the sorbate at the sorbent surface, and is usually viewed as a temporary residence of sorbate molecules 'on' the sorbent surface. The gas–solid interaction is the usual picture for adsorption, but it is not the only one possible: the sorbate could be a dilute constituent in a condensed phase (e.g. gases dissolved in water) and/or the sorbent could be a liquid.

In many treatments a distinction is made between 'physical adsorption' and 'chemisorption'. Physical adsorption usually designates the case in which the attractive force involved is the van der Waals or 'dispersion' force arising in mutual electric dipole induction. This interaction is relatively weak (a potential of a few kcal/mole or less, limited by the strong electronic repulsion which prohibits closer approach) but is also ubiquitous and relatively nonspecific, depending principally on atomic number and radius rather than detailed electronic configuration. Chemisorption designates cases of electron sharing or transfer normally considered 'chemical' interaction. The energies involved are higher (tens of kcal/mole) and the interaction much more sensitive to the specific identities of sorbent and sorbate. The distinction between physical adsorption and chemisorption is not sharp: a continuum of energies is possible. It is sometimes said that physical adsorption is reversible while chemisorption is irreversible. This is not literally true, of course: rather, at the higher energies involved in chemisorption, desorption lifetimes can be long compared to realistic laboratory timescales. Clearly, it is expected that noble gas adsorption will be in the physical adsorption rather than the chemical adsorption regime.

In the following discussion it will be understood that in algebraic expressions and equations all physical quantities are designated in a consistent set of units, e.g. pressure in $dyne/cm^2$, energy in ergs, etc.; in this case quantities of elements will be dimensionless (number of atoms). Numerical citations in text and tables, however, will be in more familiar laboratory units, e.g. pressure in atm or torr, energy in cal, and gas quantities in cm^3 STP.

In equilibrium, the quantity N of a given sorbate which is adsorbed on a given sorbent depends on its partial pressure (fugacity) P in the gas phase and on the temperature T. A basic phenomenological description is specification of the functional dependence between N, P, and T. Both experimental observations and theoretical or thermodynamic descriptions are often cast in univariant functional descriptions: the relation between N and P at constant T (an isotherm), between N and T at constant P (an isobar), or the relation between P and T at constant N (an isostere).

Table 4.2. *Selected data for Xe and Kr adsorption*

Sorbent	Temperature (°C)	Henry constant (cm^3 STP g^{-1} atm^{-1})		Heat of adsorption[f] (kcal/mole)		Specific area[a] (m^2/g)
		Xe	Kr	Xe	Kr	
Martinez shale[b]	0	3.3	0.27	5.5	3.9	18
	25	1.3	0.15	–	–	–
'Volcanic ash shale'[b]	0	1.2	0.24	5.9	–	52
	25	0.5	–	–	–	–
'Volcanic ash shale'[c]	0	5.6	0.43	4.8	4.0	–
	25	2.5	0.21	–	–	–
	−77	220	10	–	–	–
Kibushi shale[c]	0	1.2	0.43	5.3	3.1	–
	25	0.5	1.1	–	–	–
	−77	69	24	–	–	–
Wyoming montmorillonite[d]	−77	38	7.6	–	–	36
Julian limonite[d]	−77	76	3.0	–	–	16
Vacaville basalt[d]	−77	30	7.6	–	–	4
Columbia River basalt[c]	0	0.7	0.28	4.5	3.3	–
	−77	24	3.9	–	–	–
Uncompaghre quartzite[c]	0	5×10^{-5}	3×10^{-4}	6.7	4.1	–
Allende meteorite[e]	−160	10^4	200	–	–	3
Activated charcoal[c]	0	5 835	292	7.5	4.9	–

[a]   BET areas (cf. Fig. 4.1).
[b]   Fanale & Cannon (1971a); Henry constants estimated from graphical data presentation and heats estimated from Henry constants and eq. (4.12).
[c]   Podosek *et al.* (1981).
[d]   Fanale *et al.* (1978).
[e]   Fanale & Cannon (1972).
[f]   All heats are negative.

An important special case of an isotherm is a linear dependence of N on P (Henry's Law):

$$N = \mathcal{H}P \tag{4.3}$$

The Henry constant $\mathcal{H}$ is a function of T but not P. (In some theoretical treatments the Henry constant is the ratio of fugacity to quantity adsorbed, i.e. the inverse of the sense used here.) It is generally expected that at sufficiently low pressures adsorption will be governed by Henry's Law. It is possible to construct theoretical models for adsorption in which an isotherm does not reduce to Henry's Law, eq. (4.3), even in the limit $P \to 0$, but it is not clear that such situations obtain in practice and doubtful that they are important in noble gas geochemistry.

The noble gases are all monatomic as gases, have unreactive closed shell electronic configurations, and can be expected to be present at only very low partial pressures in any natural situation of geochemical interest. We will accordingly assume that Henry's Law is applicable in all cases of noble gas geochemistry; semiquantitative justifications of this generalization can be made (see below), and in any event there is certainly no evidence to the contrary.

A simple but instructive kinetic theory model due to Langmuir (1918) is often used as a starting point for an atomic scale view of adsorption. It is assumed that the sorbent surface, of total area A, is comprised of N_s sites each capable of binding only one sorbate atom (or molecule). If, at a given instant, a fraction θ of these sites are occupied by sorbate atoms, and the mean life against desorption is τ, and τ is independent of θ (i.e. no interaction among sorbate atoms), the flux ϕ of desorbing atoms is

$$\phi = \theta N_s / A\tau \tag{4.4}$$

If the gas phase is ideal, the flux of impacting atoms on the surface is $P(2\pi mkT)^{-1/2}$, where m is atomic mass and k is Boltzmann's constant. If α is the probability that an atom incident on an unoccupied site will be adsorbed, the flux of adsorbing atoms is

$$\phi = \alpha P(1 - \theta)(2\pi mkT)^{-1/2} \tag{4.5}$$

In equilibrium the two fluxes are equal, whence

$$N = N_s\theta = N_s bP/(1 + bP) \tag{4.6}$$

for

$$b = \alpha\tau a(2\pi mkT)^{-1/2} \tag{4.7}$$

where $a = A/N_s$ is the area of an individual adsorption site. Eq. (4.6) is generally designated the Langmuir isotherm.

At sufficiently low pressures ($\theta \ll 1$), the Langmuir isotherm reduces to Henry's Law with

$$\mathcal{H} = \alpha A \tau (2\pi mkT)^{-1/2} \tag{4.8}$$

At sufficiently high pressures ($\theta \to 1$), the adsorbed quantity N approaches the saturation limit N_s.

This model is not only incomplete (e.g. it makes no prediction of α or τ), it is clearly oversimplified, and it is well known that the Langmuir isotherm is not a very good description of most sorbent–sorbate interactions, particularly at high pressures (charcoal is an important exception). It is nevertheless quite valuable as a starting point for more sophisticated generalizations and as a framework by which to form a qualitative evaluation of numerical data.

Thus, for example, eq. (4.8) provides a basis for estimating the mean time τ that an adsorbed atom resides on the sorbent surface before it is desorbed. (Assumption of fixed attachment sites is not necessary in the derivation of (4.8), as suggested by the absence of explicit dependence on N_s or a.) As will be seen later, values of geochemical importance are of the order $\mathcal{H} = 1$ cm^3 STP g^{-1} atm^{-1} and $A = 10^5$ cm^2/g. For these figures (for Xe at 0 °C and $\alpha = 1$) the lifetime τ is 2×10^{-9} sec. For comparison, the time required for a gas atom to travel across an atomic distance scale (cf. Table 4.1) at thermal speeds is of the order of 2×10^{-12} sec. In this example the atom is thus adsorbed for a 'long' time. We can also infer that unless τ *is* substantially greater than a thermal flight time across the few Å of surface attractive potential the concept of adsorption will have little meaning.

The Langmuir model also provides a convenient basis for estimating when Henry's Law or saturation effects can be expected. If an individual attachment site has an area $a \approx 2 \times 10^{-15}$ cm^2, the order of atomic cross-sectional area (cf. Table 4.1), then $N_s \approx 5 \times 10^{14}$ atoms/cm^2 = 2×10^{-5} cm^3 STP/cm^2. Surface concentrations approaching this order of magnitude can be expected to exhibit saturation behavior. Conversely, much lower concentrations indicate $\theta \ll 1$ and lead to the expectation of Henry's Law behavior. Possible adsorption effects important in noble gas geochemistry always involve much lower concentration than this illustrative value, which is one reason why Henry's Law violation is not expected.

The Langmuir model permits an explicit evaluation of surface areas from experimental isotherm data. Rearrangement of eq. (4.6) gives

$$P/N = (1 + bP)/N_s b \tag{4.9}$$

The slope of this linear relationship between P and P/N is $1/N_s$, which permits calculation of A for an assumed value of $a = A/N_s$. (An approach to saturation is needed to extend beyond the Henry's Law regime $P/N = $ constant.)

Since the Langmuir isotherm is not an adequate description for most systems, eq. (4.9) is not much used for area measurement. There are a number of other isotherm formulations which utilize adsorption in surface area measurements, however (cf. Young & Crowell, 1962, for example). The best known and most widely used is the BET (Brunauer, Emmett, & Teller, 1938) theory, a generalization of the Langmuir model to multilayer adsorption. Assuming that for the second and succeeding molecular layers the 'adsorption' potential is the same as in bulk condensation the isotherm analogous to eq. (4.6) is

$$N = N_s c x (1-x)^{-1} [1 + (c-1)x]^{-1} \tag{4.10}$$

where $x = P/P_0$ (P_0 is the vapor pressure of the sorbate) and c is a dimensionless constant peculiar to the sorbate–sorbent pair (and a function of T). Rearranging to a form analogous to eq. (4.9) gives

$$x/N(1-x) = [1 + (c-1)x]/N_s c \tag{4.11}$$

This is a convenient linear relationship against which experimental data can be compared (cf. Fig. 4.1). If experimental data conform to this prediction, the slope and intercept of this linear relation yield both c and N_s, and thus A.

There are substantial theoretical objections to the BET model. Nevertheless, experimental data often conform to eq. (4.11) admirably (Fig. 4.1) and the BET model is used extensively for the determination of areas of microscopically complicated surfaces.

Many materials of geochemical importance, notably clays and shales, turn out to have large BET areas (Table 4.2). Specific areas are often of the order of $10 \, m^2/g$. For comparison, perfect spheres (of density $3 \, g/cm^3$) would need radius $0.1 \, \mu m$ to have specific areas of $10 \, m^2/g$. The high areas correspond to a large degree of microscopic surface involution and pores and channels which greatly enhance adsorption capabilities.

One additional isotherm which has been used in parameterizing noble gas adsorption on natural materials (Fig. 4.2) is the Freundlich isotherm:

$$N = \text{constant} \times P^{1/n} \tag{4.12}$$

where $n \geq 1$ is a constant (the special case $n = 1$ is Henry's Law). It is possible to derive this isotherm theoretically for special models, but it is usually regarded as a 'semiempirical' form to which to fit data. It does not reduce to Henry's Law as $P \to 0$ nor does it predict saturation (Langmuir isotherm) or bulk condensation (BET isotherm) at sufficiently high P, so its application as an interpolation/extrapolation formula is generally considered restricted to an 'intermediate' pressure range.

None of the isotherm models considered above make explicit predictions for absolute values of the relevant parameters nor of their temperature dependence. Useful information on temperature dependence can be obtained by application of thermodynamics, usually via a heat of adsorption. A common parameter is the isosteric heat of adsorption:

$$\Delta H = (RT^2/P)(\partial P/\partial T)_\theta \tag{4.13}$$

which is also the calorimetric heat evolved in a reversible isothermal adsorption. We will restrict further consideration only to cases in which Henry's Law applies. In such case eq. (4.13) becomes

$$\Delta H = -RT^2\, d(\ln \mathcal{H})/dT \tag{4.14}$$

Fig. 4.1. BET plot for adsorption of N_2 on shale samples at $-195\,^\circ$C. Linear data arrays indicate conformity to eq. (4.11), by which indicated specific surface areas were calculated. Reproduced from Fanale & Cannon (1971a).

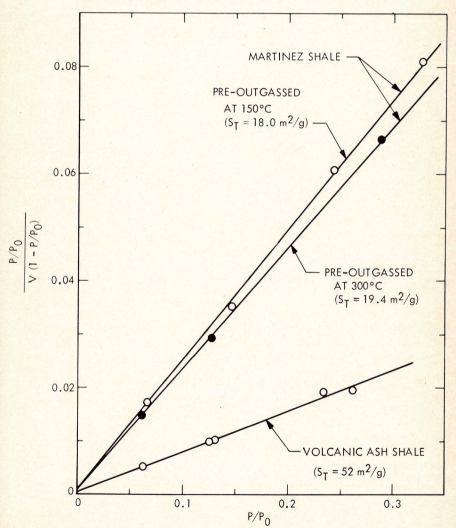

If ΔH is known eq. (4.14) can be integrated to obtain the temperature dependence of $\mathcal{H}$. In general, ΔH is not constant because gas and adsorbed phases have different heat capacities, but often in practice the variation in ΔH over a limited temperature range is not severe.

Explicit predictions for adsorption parameters can be obtained by application of statistical mechanics to detailed models for the adsorbed phase. Many such models have been developed (cf. Young & Crowell, 1962; Ross & Olivier, 1964),

Fig. 4.2. Freundlich isotherm (eq. 4.12) plots for Xe and Kr adsorption on volcanic ash shale. Reproduced from Fanale & Cannon (1971a).

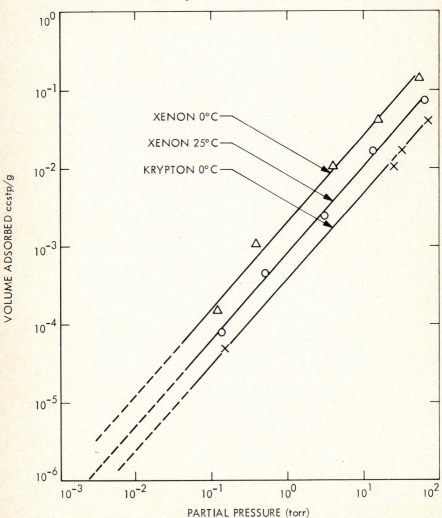

but a model suitable for a given application cannot in general be chosen *a priori*. A simple example, used by Podosek *et al.* (1981) for fitting of very low pressure (Henry's Law region) Xe and Kr adsorption on shales (Fig. 4.3) is

$$\mathcal{H} = Ah(2\pi mkT)^{-1/2}(kT)^{-1}e^{-\Delta E/RT}(1 - e^{-h\nu/kT})^{-1} \tag{4.15}$$

This formula represents the model in which the adsorbed phase has perfect lateral freedom (two-dimensional ideal gas) on the sorbent surface and vibrates harmonically (fundamental frequency ν) normal to the surface; the sorbent surface is assumed to have a uniform sorbate potential $\Delta E < 0$ (including the zero-point oscillator energy) relative to the gas. By eq. (4.8) the lifetime corresponding to eq. (4.15) in the special case $\alpha = 1$ and $h\nu/kT \gg 1$ is

$$\tau = (h/kT)e^{-\Delta E/RT} \tag{4.16}$$

Representative adsorption data of geochemical interest are given in Table 4.2. Only Xe and Kr data are listed; the lighter gases are much poorer sorbates, sufficiently so that it is possible to question whether adsorption has any geochemical relevance for them. It should be noted that it is very difficult to measure even Xe and Kr adsorption on the materials and at the T and P of greatest geochemical interest. In general, measurements are made at high P or low T or both, so that considerable extrapolation by models of questionable reliability is involved. The accuracy of the surface temperature Henry constants in Table 4.2 is thus probably somewhere between a factor of two and an order of magnitude. For only one sample, the 'volcanic ash shale', is there an opportunity to compare different methods. The Fanale & Cannon (1971a) measurements (of Fig. 4.2) were made at the designated temperatures but at pressures of around 1 torr, a factor of 10^3 or more higher than the atmospheric pressures of primary concern. The Podosek *et al.* (1981) measurements were made at characteristic pressures of 10^{-9} torr and less, but require temperature extrapolation of a factor of 10^3 or more in $\mathcal{H}$. The agreement (Table 4.2) seems about as good as can be expected.

Adsorption is sometimes invoked not only as a factor in geochemistry but also as a laboratory nuisance. Analyzed samples are often found to have a superficial or 'loosely bound' component ascribed to air contamination which is frequently described as 'adsorbed' on the sample. Without further qualification (e.g. as in section 8.4) this makes little sense. An air contamination effect certainly exists, as can be inferred clearly when 'intrinsic' sample gas is isotopically distinct (cf. section 4.4), but whether adsorption is responsible or even involved is questionable. In all such noble gas analyses a necessary step is storage in laboratory vacuum before gas extraction. By definition, adsorbed gas is desorbed under vacuum. The relevant factor is the timescale required to reach equilibrium, and in general desorption timescales are short. If eq. (4.15) is even approximately correct, an adsorption energy even as high as 6 kcal/mole (cf. Table 4.2) gives a lifetime against desorption $\tau < 10^{-8}$ sec at 25 °C. A lifetime of order 10^5 sec at 25 °C, minimum for laboratory vacuum exposure, requires 24 kcal/mole, a reasonable energy for chemisorption but seemingly well out of reach for a noble gas.

The data in Table 4.2 are for natural samples. It is notoriously difficult to prepare and maintain a 'clean' solid surface, since even in a rather good laboratory vacuum any freshly created surface quickly becomes 'contaminated' with adsorbed species. Any naturally occurring solid material must be considered to

Fig. 4.3. Henry constants for Xe and Kr adsorption on 'volcanic ash shale' (cf. Figs. 4.1 and 4.2). The solid lines are fits according to eq. (4.13); small symbols with bars (lower left) indicate extrapolation to 0 °C and 25 °C. Reproduced from Podosek *et al.* (1981).

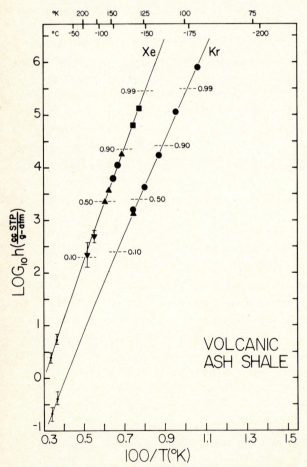

have a surface extensively populated by adsorbed atoms and molecules rather than a 'pristine' surface. Such surfaces are what are examined in most laboratory experiments, e.g. those reported in Table 4.2, and, of course, it is just such surfaces which are geochemically relevant for noble gas adsorption. It is interesting to note, however, that in other situations, noble gas adsorption can be rather a stronger effect. Thus, for example, Bernatowicz *et al.* (1983) examined Xe adsorption on a vacuum-crushed lunar rock and concluded that a small part of the freshly created surface had an adsorption potential as high as 14 kcal/mole but that in a few days at 10^{-8} torr this surface was rendered inaccessible to Xe by other chemical species which were better competitors for the sorbent surfaces.

4.3   Solution

When a condensed phase (the solvent), solid or liquid, equilibrates with a gas phase (the solute), some concentration of the gaseous species will be dispersed in the solid or liquid, i.e. some gas will be dissolved. Solution is the most general way in which a noble gas will interact with other materials. It should be noted, however, that the term 'solution' implies a more or less uniform microscopic-scale admixture of solvent and solute molecules or complexes of molecules, an assumption that is presumably reasonable for liquid solvents but perhaps not for solids, and difficult to test experimentally.

There is a wealth of both theoretical treatment and empirical data for the phenomenon of solution, including empirical data for noble gas solution. As with adsorption, however, data for solution in geochemically important materials are sparse. A prominent exception, the only one, is water, extensive data for which are presented in Chapter 7. Most of the available data for other materials of principal geochemical interest are summarized in Table 4.3.

In equilibrium, the quantity N of a given solute which is dissolved in a given solvent depends on its gas phase partial pressure (fugacity) P and on the temperature T, and a basic phenomenological description of the equilibrium is specification of the functional relationship between N, P, and T. At sufficiently low pressures it is expected that the pressure dependence is linear (Henry's Law):

$$N = \mathcal{H}P \tag{4.17}$$

where the Henry constant $\mathcal{H}$ is a function of T but not P. (In some theoretical treatments the Henry constant is the ratio P/N, i.e. the inverse of the above definition.) In this and subsequent algebraic expressions, it will be understood that physical quantities are in a consistent set of mechanical units, e.g. pressure in dyne/cm^2, energy in erg, etc., and quantities of elements are dimensionless (number of atoms), while numerical citations in text and tables are in more convenient units, e.g. pressure in atm and gas quantities in cm^3 STP.

The basic description above (and much of the further treatment below) is intentionally similar to that given for adsorption in Section 4.2. This stresses the point made there that in terms of macroscopic thermodynamics there are great similarities between the two phenomena.

Henry's Law behavior can be expected as long as the solute concentration is low, in the sense of not approaching a saturation limit or becoming more than a trace constituent of the solution. For liquid solutes we thus expect that Henry's Law will be valid as long as the mole fraction of solute is low. Experimental verification of Henry's Law behavior for noble gases in silicate melts has been reported by Kirsten (1968) and by Hayatsu & Waboso (1982), for example. This should be the case in any situation of geochemical interest: in a material of density 3 g/cm^3 and average atomic weight 24, for example, an atomic fraction of 10^{-6} corresponds to a gas concentration of 3×10^{-3} cm^3 STP/g, much higher than any concentration likely to be encountered in nature. For solid solutes, however, gas atoms are likely to be preferentially 'dissolved' in special accommodation sites such as vacancies or other lattice defects. These are likely to be nonuniform in their interactions with solute atoms and the number of accommodation sites of a given kind might be quite low in comparison with the number of solvent atoms. In such cases the range of validity of Henry's Law solution might be very limited, perhaps even nonexistent, and solubility might be highly variable from sample to sample of the same solvent and sensitive to the particular history of a given specimen. Nevertheless, there are neither theoretical treatments nor empirical data which indicate that nonlinearity actually is important for noble gases in nature, and natural levels of dissolved gases are typically so low (atom fraction no more than $\sim 10^{-8}$) that nonlinearity would not be expected. In the rest of this section, we will accordingly assume that Henry's Law is adequate for the description of noble gas solution effects in geochemistry and we will not pursue description of any more complicated pressure dependence.

The Henry constant (per unit mass of solvent) can be expressed

$$\mathcal{H} = (\rho k T)^{-1} C_1/C_g \qquad (4.18)$$

where ρ is solvent density, C_g is the concentration of gas phase solute, and C_1 is the equilibrium concentration of solute in the solution. Equal concentrations in gas and solution thus correspond to $\mathcal{H} = (\rho k T)^{-1}$, for example $\mathcal{H} = 1$ cm^3 STP g^{-1} atm^{-1} for $\rho = 1$ g/cm^3 at $T = 0$ °C, and $\mathcal{H} = 0.06$ cm^3 STP g^{-1} atm^{-1} for $\rho = 3$ g/cm^3 at $T = 1200$ °C. Comparing with the data in Table 4.3 we see that noble gas solubility is relatively low even in water and quite low in likely magmas, in the absolute sense that the atoms are more dilute in the solution than in the gas phase. Even the silicate melt solubilities are not negligibly low in other important contexts, however: for the tabulated values equilibrium with

Table 4.3. *Data for noble gas solubility*

Solvent	Temperature (°C)	Henry constant for solution (cm³ STP g⁻¹ atm⁻¹)					Heat of solution (kcal/mole)					Ref.
		He[a]	Ne	Ar	Kr	Xe	He	Ne	Ar	Kr	Xe	
Silicate Melts												
Enstatite	1 500	0.000 12	0.000 07	0.000 02	(0.000 013)	(0.000 006)	—	—	—	—	—	1
Basalt	1 200	—	—	0.000 025	0.000 013	0.000 009	—	—	—	—	—	2
Tholeiitic basalt	1 200	—	0.000 19	0.000 023	0.000 008	—	—	—	+20.5	—	—	3
Alkali-olivine basalt	1 200	—	0.000 22	0.000 046	0.000 016	—	—	—	—	—	—	3
Basaltic andesite	1 200	—	0.000 26	0.000 090	0.000 021	—	—	+4.2	+16.7	+19.3	—	3
Solids												
Magnetite[b]	227	0.042	0.016	3.6	1.3	0.88	−2.42	−2.20	−15.3	−13.0	−12.5	4
Serpentine	340	—	0.000 37	0.000 46	0.000 76	0.001 2	—	—	—	—	—	5
Water												
Fresh water	0	0.009 5	0.0127	0.0536	0.111	0.226	−0.93	−1.87	−4.05	−5.08	−6.23	6

[a] Jambon & Shelby (1980) report permeation-technique (Section 4.7) measurements of He solubility in obsidian glasses in the range 0.002–0.003 cm³ STP g⁻¹ atm⁻¹ for temperatures approximately 200–300 °C.

[b] These magnetite solubility data are well known but are listed here only for illustration and reference; these data have been discredited by Yang et al. (1982) and at present must be considered invalid.

References: 1, Kirsten (1968) (He, Ne, Ar values measured, Kr, Xe values by extrapolation according to eq. 4.19); 2, Fisher (1970); 3, Hayatsu & Waboso (1982); 4, Lancet & Anders (1973); 5, Zaikowski & Schaeffer (1979); 6, Tables 7.1 and 7.3 and eq. (4.17).

noble gas pressures at air values will produce gas concentrations of the general order of magnitude observed in igneous rocks, and it is often hypothesized that in some cases trapped gases in igneous rocks were indeed acquired in just this way (see Chapter 9).

The temperature dependence of solubility is related to a heat of solution in the same way as previously described for adsorption (Section 4.2). If Henry's Law is obeyed the heat of solution is

$$\Delta H = RT^2 \, d \, (\ln \mathcal{H})/dT \tag{4.19}$$

If the heat of solution is known, eq. (4.19) can be integrated to express the temperature dependence of the Henry constant $\mathcal{H}$ (Fig. 4.4). The heat will vary somewhat with temperature, however, because of different heat capacities of gaseous and dissolved phase, and this variation must be taken into account in extrapolation over a large temperature range. In contrast to the case for adsorption (Table 4.2), it should be noted that heats of solution (Table 4.3) are generally positive (greater solubility at higher temperatures). Also, heats of solution are of a greater magnitude than heats of adsorption, so that the temperature dependence is characteristically stronger for solution than for adsorption.

The degree of adsorption of the various noble gases on a given sorbent exhibits a monotonic progression with atomic number, i.e. the gases follow a sequence, a generalization which evidently applies to noble gas solution in liquids as well. Solution in solids may not be so simple, however, particularly

Fig. 4.4. Arrhenius plot for temperature variation of Henry constant for solution of noble gases in silicate melts (cf. Table 4.3). Reproduced from Hayatsu & Waboso (1982).

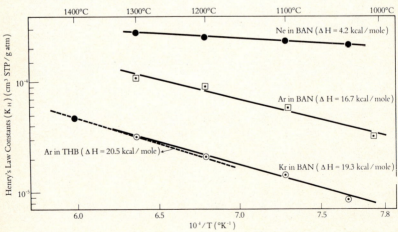

if the noble gases atoms must be accomodated primarily in 'holes' of fixed size, e.g. structural channels such as those in cyclosilicates (cf. Section 9.10) or accidental anion vacancies. In such cases atomic radius may be critical and a sharp drop in solubility may occur for atoms too big to fit in the holes.

It is interesting to consider a simple but specific model for noble gas solution in liquids which was proposed by Blander *et al.* (1959) (also cf. Grimes *et al.*, 1958; Uhlig, 1937). The basic premise is that the principal energy involved in gas–solvent interaction is the surface energy of a cavity in the solvent in which the solute atom resides. The process of going into solution can be conceived to occur in three separate steps: (i) expansion of the gas from concentration C_g to C_1, (ii) insertion of the gas atoms, as point particles, into the solvent, and (iii) expansion of the gas atoms to a finite radius r. If the only interaction is the creation of the solvent surface area, the molar Gibbs free energy change for the total process is

$$\Delta G = -RT \ln C_g/C_1 + 4\pi r^2 \sigma N_A \tag{4.20}$$

where N_A is Avogadro's number. Here σ is the 'surface' tension, described as the 'microscopic' surface tension and perhaps not very different from the ordinary macroscopic surface tension. For solution in equilibrium $\Delta G = 0$, whence

$$C_1/C_g = e^{-4\pi r^2 \sigma/kT} \tag{4.21}$$

The Henry constant can then be calculated by eq. (4.18). This model is clearly oversimplified and even if the basic premise of no interaction between solvent and solute is nearly satisfied, there is considerable latitude in the choice of appropriate r and σ. Nevertheless, Blander *et al.* (1959) found that its predictions for solubility and heat of solution of noble gases in fluoride salt melts agreed reasonably well (within an order of magnitude in Henry constant) with experimental observations. A useful feature of this model is that it has only one parameter, the surface tension, which characterizes the solvent, so that a single experimental measurement defines a model which can be used for a prediction of both temperature behavior and the solubility of other gases. This model is often applied, in the absence of adequate experimental data, to the description of noble gas solution in silicate melts. In the few cases where it can be tested it accounts reasonably well for the different solubilities of the different noble gases in a given melt (Kirsten, 1968; Hayatsu & Waboso, 1982). We might expect that the Blander *et al.* model predicts minimum solubilities for gases in liquids, since any attractive van der Waals interaction between solute and solvent would lead to higher solubility.

4.4   Crystal–melt partitioning

The crystal–melt distribution coefficient, $K_D = C_s/C_1$, where C_s is concentration in a solid and C_1 concentration in coexisting liquid, is a key parameter in trace element studies of igneous systems. A noble gas crystal–melt distribution coefficient is the ratio of the gas solubilities considered above. As seen in Table 4.3, solubilities have now been reported for a variety of melt compositions, but solubility data are still very scarce for solids in general and there are no data at all for the major minerals of common igneous rocks.

There are a number of trace elements commonly described as 'incompatible' because their distribution coefficients (for typical mantle mineralogy) are low, often as low as 10^{-2} or 10^{-3} or less. Prominent examples are K, Rb, U, Th, and the rare earths. The principal reason for their incompatibility is evidently their large ionic radii. Concentrations of these elements in mantle-derived rocks are relatively low and the generalization emerges that they have been expelled from the mantle (at least the upper mantle) and concentrated in the crust, especially continental crust.

It is generally believed that the noble gases should behave like incompatible elements, only more so. Their atomic radii are also relatively large (Table 4.1). Moreover, they have only very weak interactions with other elements; in igneous conditions they will not be ionized or share electrons in covalent bonds. It is not expected that they could substitute for regular lattice atoms except by accident, i.e. if there is a structural vacancy. In analogy with the lithophile incompatibles, their abundances are low in mantle-derived rocks and they are greatly concentrated in a 'crust' (the atmosphere). On the other hand, the mantle clearly contains at least some juvenile gas which has never been part of the atmosphere and there is substantial uncertainty about what significance is to be attached to low concentrations in mantle materials and even exactly how low these concentrations actually are (see Chapters 9 and 11). There is precious little actual observational data by which to judge this expectation of very low crystal–melt distribution coefficients for the noble gases. This lack hampers theoretical treatments of mantle degassing and atmospheric evolution and essentially prohibits any quantitative evaluation of the behavior of noble gases in igneous systems.

One approach to estimating distribution coefficients is examination of natural samples. Batiza *et al.* (1979) reported noble gas data for primitive and evolved volcanic rocks putatively from the same magma. From petrologic estimates of the amount of crystallization involved between the former and the latter they calculated crystal–melt distribution coefficients assuming closed-system behavior. Calculated coefficients were about 10^{-1} for Ne but greater than unity for the heavier gases (in the range 2–8). From this unexpected result Batiza *et al.* concluded that the system was not closed, i.e. that gas was lost from the more evolved rocks or from the magma, and that the calculated distribution coefficients were not real.

The straightforward approach of measuring crystal-melt distribution coefficients in the laboratory is a formidable experimental problem, and only recently have any data been reported. Hiyagon & Ozima (1982) grew crystals from a melt and then separated quenched crystals and glass and compared their gas contents. Their results are given in Table 4.4. The inferred distribution coefficients are less than unity but are surprisingly large in comparison with the usual expectation. In view of the substantial experimental difficulties and ambiguities involved these results must be considered provisional and subject to considerable uncertainty; we cite them nevertheless because there have been no other data of this kind reported at all.

If ultramafic xenoliths are the solid residue of partial melts represented by volcanic rocks, their relative gas concentrations should be governed by crystal-melt partitioning and should permit estimation of the distribution coefficient. In practice, forming an estimate on such grounds is rather difficult because the measured concentrations, especially of the xenoliths, span such a wide range, because there is little basis for comparing coexisting liquid and solid, and because of the possibilities of posteruptive gas loss and/or atmospheric contamination (Chapter 9). Nevertheless, there does seem to be a trend for lower concentrations in xenoliths than in lavas for the lighter gases He and Ne (Figs. 9.2 and 9.3) but a lesser bias, if any at all, for the heavier gases Ar, Kr, and Xe (Figs. 9.4-9.6). Albeit rather inconclusively, this again suggests that at least for the heavier gases the crystal-melt partition coefficient does not conform to the expectation that it should be extremely small.

4.5 Trapping

The term 'trapped', as applied to noble gases, has come to have a well-defined denotation as designating noble gases which were not produced *in situ* (Section 1.4). Its original connotation, however, is that gases somehow got 'into'

Table 4.4. Crystal–melt distribution coefficients[a]

Crystal	Ne	Ar	Kr	Xe
Clinopyroxene[b]	$(0.6)^f$	0.5^d	0.5^d	$(0.2)^f$
Olivine[c]	—	0.14^e	—	—

[a] Data from Hiyagon & Ozima (1982). Melt composition is basaltic.
[b] Crystals grown at pressure of 15 kbar, temperature 1175–1250 °C.
[c] Crystals grown at pressure of 1 bar, temperature 1300 °C.
[d] Average of two determinations.
[e] Average of five determinations (0.08–0.18).
[f] Relatively poor experimental quality.

a sample, by a process or processes which it is unnecessary and often impossible to specify, and have subsequently been unable to get back out of it. The term is still used with this connotation.

Strictly speaking, there are only two ways in which gases can be 'in' a sample. If gas atoms are mixed with host atoms on a microscopic scale, the gas is dissolved; if gas atoms are 'on' the surfaces which bound the sample, the gas is adsorbed. There are quite a few complications and variations on these basic themes, however, as will be discussed below. From a macroscopic viewpoint, and from the viewpoint of laboratory practice, it is sometimes difficult to distinguish between solution and adsorption, particularly for samples inferred to have extensive and complicated 'internal' surfaces (embayments, pores, cracks, etc.). It can be and has been argued, for example, that in some experiments purporting to measure noble gas solubility the relevant phenomenon is actually adsorption, and vice versa. When the structure of the sample in question is sufficiently complicated, e.g. in clays, activated charcoal, or poorly crystalline or amorphous materials, it is not clear that a distinction can or should be made between a trace element's presence 'in' (solution) or 'on' (adsorption) its host. From a macroscopic viewpoint, however, it is also unnecessary to make the distinction in such cases. As pointed out in Sections 4.2 and 4.3, the thermo-dynamic descriptions of solution and adsorption are quite similar. The Henry constants which describe both phenomena (Tables 4.2 and 4.3) are examples of distribution coefficients and have a meaning which is independent of the micro-scopic interpretation of how the gas atoms interact with the host.

An additional connotation of the term 'trapped' gases deserves explicit note here. The gases must be observed experimentally before they can be identified as trapped gases. The experimental procedure always involves loading the samples into a laboratory vacuum system before analysis of gases subsequently (hours or months later) released by heating or other means; usually, the samples are also heated to approximately 100 °C *in vacuo* to remove 'air contamination' or 'loosely bound' gases or other volatiles which would interfere with the analysis. Solution and adsorption are both equilibrium effects for which the gases con-tained in the sample are functionally dependent on external partial pressure, and when this pressure is effectively zero, as in the laboratory vacuum, so too are the equilibrium gas contents. The same remarks apply to the immediately prior history of the samples: field residence and laboratory stroage in which the external medium is air and equilibrium gas contents should be dictated by the partial pressures in air.

Trapped gases are thus those gases contained within samples which, for kinetic reasons, are not in equilibrium with their present or immediately prior external pressures: at some epoch in the past they were incorporated in the samples, in equilibrium with that past environment or otherwise, and have survived their later history, usually on a geological timescale, without equilibration in response to a changing environment. If the gases are dissolved, there is no particular problem with maintaining this disequilibrium even over geological timescales: at ambient temperatures on the surface of the earth (and in interplanetary space) noble gas diffusion coefficients in many materials are quite low and the time-scales for equilibration by diffusive redistribution are quite long (see Section 4.7). Indeed, it can be argued that noble gas atoms can be so efficiently encaged in a crystalline lattice that they move only few or even no lattice spacings in the age of the solar system. Adsorption timescales, in contrast, are quite short, and adsorption/desorption are often considered essentially instantaneous on a lab-oratory timescale.

Trapped noble gases are thus usually considered to be of the dissolved rather than the adsorbed variety. It must be remembered that the relevant definition is operational, however: gases which appear in the mass spectrometer in the usual analysis were trapped no matter what their original microscale distribution in the sample. Zaikowski & Schaeffer (1979), for example, report that noble gases dissolved in phyllosilicates may equilibrate with air - both lose and gain gas - in laboratory (and museum) conditions and timescales. Conversely, Podosek *et al.* (1981) suggest that in materials of sufficiently convoluted surfaces adsorbed

Table 4.5. *Distribution coefficients for trapping of Ar*

Material	Distribution coefficient $(cm^3 STP g^{-1} atm^{-1})$	Ref.
Condensation from vapor		
Magnetite	0.000 02	1
Mg	0.000 09	1
CdTe	0.000 02	1
Zn	0.000 01	1
C	0.3	2
Pyroxene	<0.000 6	2
Other processes		
Soot[a]	0.02	3
Kerogen[b]	0.035	3
Aqueous polymer[c]	0.011	3

[a] Colloidal fraction of soot collected on quartz rod in open propane flame.
[b] Solvent/acid insoluble fraction of material synthesized in modified Miller–Urey experiment (electric discharge).
[c] Melanodoidin precipitated from aqueous solution of ribose and histidine.

References: 1, Honda *et al.* (1979); 2, Kothari *et al.* (1979); 3, Frick *et al.* (1979).

gases may not escape the sample in the laboratory vacuum (see Section 8.4). Whether the term 'trapped' *should* be applied to gases thus contained in natural samples is a semantic issue we will not pursue here.

While trapped gases are either dissolved or adsorbed, there are nevertheless a variety of 'ways' in which they may be trapped. One such possibility is that gases may be contained in small regions, fluid or crystalline, wholly enclosed in a crystal of another material (cf. Section 9.11). The gases in this case are not really *in* the enclosing material, but this circumstance may not be evident in the laboratory. Gas atoms can be implanted if they strike a solid at sufficiently high velocity to penetrate more than a few lattice spacings; solar wind gases so trapped in meteorites and lunar samples are common (Section 5.2). Gases present on surfaces or in nearby materials, or even in a gas phase, might be 'fixed' (dissolved?) in transient high pressures and temperatures resulting from high-velocity impact shock, or even in static compression at only moderately high pressures (cf. Honda *et al.*, 1979).

It is often suggested that noble gases can be trapped during the growth of crystals, especially during condensation from vapor but also during crystallization from a liquid. Gases may be present on or near growth surfaces because of solution, adsorption, energetic implantation, or some other mechanism, and then covered over – trapped – by the continuing growth. The term 'occlusion' seems to be used for a variety of meanings in the literature, but it frequently suggests this process.

It is clearly very difficult to make a quantitative analysis for these ways in which trapped noble gases may arise. Equilibration with local conditions may dominate or it may be irrelevant. It should be noted that the magnetite and serpentine solubilities in Table 4.3 are based on experiments in which these materials were synthesized in the presence of noble gases, whence it can be questioned whether the resultant data actually reflect equilibrium solubility.

There are a number of recent studies of the extent to which noble gases are trapped during the laboratory synthesis of solids which are very difficult to describe in terms of any simple model. These include direct (and decidely nonequilibrium) condensation from vapor (Honda *et al.*, 1979; Kothari *et al.*, 1979) and also precipitation from solution and from an electric discharge (Frick *et al.*, 1979). In general, an adequate description of the processes involved or the conditions of formation is not possible; even a description of the resultant materials is difficult. Nevertheless, noble gases are trapped. The results can be described in terms of distribution coefficients (Table 4.5), but in most cases the variation of trapped gas content with ambient gas pressure was not examined, so the 'distribution coefficients' are only ratios of trapped gas to pressure in a single experiment. The distribution coefficients are in some cases quite high, and in some cases produce quite strong elemental fractionation. In at least one case, a very impressive isotopic fractionation was produced (Fig. 4.5). While the mechanisms for these processes are unknown and clearly complex, so too are the mechanisms involved in production of trapped gases in many natural samples, and these data are perhaps relevant to the interpretation of the observations.

A comparison of the various ways in which noble gases can be put into various materials is made in Table 4.6. The tabulated values are the gas concentrations which correspond to the distribution coefficients in Tables 4.2–4.4 and partial pressures in air. Some of these are also illustrated in Fig. 8.6.

4.6 Clathrates

Clathrates are generally described as 'cage' compounds in which one substance, the host, forms a lattice with large structural 'holes' in which molecules of a second substance, the guest, reside. Often the presence of the guest is essential in that the host will not form that crystal structure without the guest. The struc-

Fig. 4.5. Isotopic fractionation patterns in Xe and Kr trapped in electric discharge kerogen (cf. Tables 4.5 and 4.6). Ordinate is per mil deviation (Section 1.3) of trapped gas from ambient gas (air composition) in which the kerogen was formed. Reproduced from Frick *et al.* (1979).

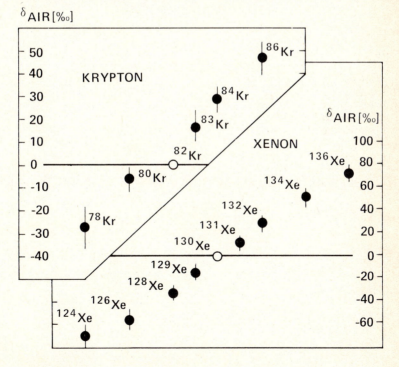

ture of a unit cell may be quite complicated and involve many molecules, and so the proportion of guest to host is variable; the ratio of guest to host will still be one of relatively small integers, however, i.e. the guest is not a trace element.

In the ideal case, guest and host will occur in definite stoichiometric proportions and thus, at least from a thermodynamic perspective, the clathrate is a specific substance, i.e. a chemical compound. In normal usage the distinction between such a 'clathrate' and a 'chemical compound' is presumably that in a clathrate the forces involved are van der Waals forces rather than transfer or sharing of electrons.

If the gas phase activity of the host is controlled by the presence of a pure condensed phase, solid or liquid, the equilibrium between host and guest in a stoichiometric clathrate can be described in terms of the gas phase pressure of the guest. This is, in effect, a vapor pressure for the guest: at higher pressures the guest will condense to form clathrate and at lower pressures the clathrate will decompose. Temperature variation of this pressure will follow the Clapeyron equation which, with the usual assumptions (ideal gas behavior of the vapor and

Table 4.6. *Concentrations for incorporation of noble gases by various processes at gas partial pressures in air*

| Sample | Gas concentrations (cm^3 STP/g) | | | | |
	^{4}He ($\times 10^{-8}$)	^{20}Ne ($\times 10^{-8}$)	^{36}Ar ($\times 10^{-8}$)	^{84}Kr ($\times 10^{-10}$)	^{130}Xe ($\times 10^{-12}$)
Solution[a]					
Water (0 °C)	5.0	21	169	721	802
Serpentine (340 °C)	–	0.61	1.4	4.9	4.2
Enstatite melt (1 500 °C)	0.063	0.12	0.063	0.085	0.021
Adsorption[b]					
Martinez shale (0 °C)	–	–	–	1 755	11 685
Kibushi shale (0 °C)	–	–	–	2 795	4 250
Other ?[c]					
Soot	–	0.24	75	918	6 030
Kerogen (discharge)	0.12	1.0	110	184	1 600

[a] For distribution coefficients in Table 4.3.
[b] For distribution coefficients in Table 4.2.
[c] Data from Frick *et al.* (1979); cf. Table 4.5.

negligible volume of the condensed phase), reduces to the Clausius-Clapeyron equation:

$$d(\ln P)/dT = \Delta H/RT^2 \tag{4.22}$$

(cf. eqs. 4.14 and 4.19). Here ΔH is the heat of 'evaporation' of the guest, specifically the enthalpy change for the process in which clathrate decomposes to gaseous guest and condensed-phase host. Over a limited temperature range in which ΔH is approximately constant this can be integrated to

$$P = P_0 e^{-\Delta H/RT} \tag{4.23}$$

where P_0 is a constant of integration. This can be used as an approximation to vapor pressure as a function of temperature.

More generally, clathrates often tend to be nonstoichiometric: the clathrate is a distinct phase but its composition is not definite, i.e. the degree to which the structural holes are filled depends on the gas pressure of the guest. In such cases the clathrate can presumably be viewed as an extreme case of selective solution or adsorption, according to whether the holes are viewed as interior to the host structure or as its surface (cf. Sections 4.2, 4.3, and 4.5). In such cases, however, the solute/sorbate (guest) is not a trace constituent and Henry's Law will not apply.

Noble gases are well known to form clathrates with a number of hosts, notably organic compounds such as hydroquinone. An important special case, the only one of even potential significance in geochemistry, is ice–noble gas clathrate. For ideal stoichiometric behavior the parameters for vapor pressure according to eq. (4.23) calculated from fits by Sill & Wilkening (1978) to data of Barrer & Edge (1967), are given in Table 4.7. The heats are higher than the heats of vaporization of the pure noble gases and the decomposition temperatures are higher than the normal boiling points (Table 4.1). In the presence of sufficient H_2O, pure noble gases would condense as ice clathrates before they would condense as pure liquids. The heats of clathrate formation are lower, however, than the corresponding heats for aqueous solution (Table 4.3).

Noble gas clathrates will not now form on the earth, as can be seen from the 'air pressure' decomposition temperatures in Table 4.7. They might, however, form in cooler regions of the primitive solar nebula. Sill & Wilkening (1978) note that for pressures in a plausible model nebula pure ice clathrates of Ar, Kr, and Xe could form at 40 K, 45 K, and 62 K, respectively.

As with other compounds, solution effects can elevate the 'condensation' temperatures of clathrate guest species. Sill & Wilkening calculate that in a gas of solar composition the major clathrate, and the first to form, will be ice–methane, and that noble gases can substitute for the methane at temperatures higher than decomposition temperatures for noble gas clathrates. They calculate, for example, that in a total nebular pressure of 2×10^{-3} atm (high in comparison with most model pressures currently considered), ice–methane clathrate at 80 K will have dissolved 99% of the available Xe (and substantially smaller amounts of the other noble gases).

Sill & Wilkening propose that clathrates formed in the cold outer parts of the solar system and then transported to the inner solar system (e.g. in comets) might help account for the atmospheres of the terrestrial planets. They contend that infall of 1 ppm of ice–methane clathrate with noble gases dissolved as above could account for the present inventories of Ar, Kr, and Xe in the terrestrial atmosphere.

4.7   Diffusion

Diffusion of noble gases into and out of magmas, minerals, and rocks is a pervasive process which alters not only noble gas concentrations but also the ratio of one element to another and, to a lesser extent, the isotopic composition of a given element. An appreciation of the basic processes involved is thus fundamental to deriving any useful conclusions from observations of elemental compositions of noble gases in natural samples. Since the subject of diffusion is one of the most extensively developed areas of mathematical physics, this section gives only a brief description and a presentation of relevant data for the noble gases.

Quantitative treatments of diffusion almost invariably begin with the 'diffusion equation'

$$\frac{\partial C}{\partial t} = D\nabla^2 C \tag{4.24}$$

where C is concentration and t is time; D is the diffusion coefficient. This

Table 4.7. *Decomposition vapor pressure data for ice–noble gas clathrates*

	Ne[b]	Ar	Kr	Xe
P_0 (10^4 atm)[a]	2.82	2.09	1.77	6.11
ΔH (kcal/mole)[a]	0.92	2.96	3.90	5.78
Temperature (K) for $P = 1$ atm[c]	45	150	201	264
Temperature (K) for $P =$ air pressure[c,d]	22	102	84	107

[a] Parameters for eq. (4.23), equivalent to fits by Sill & Wilkening (1978) to experimental data of Barrer & Edge (1967).
[b] Data by extrapolation of properties of heavier gases.
[c] From P_0 and ΔH in first two rows.
[d] For 1 atm air pressure with composition given in Table 2.1.

equation assumes an isotropic medium in which diffusive flux is proportional
to concentration gradient. An additional term must be included if the concentration is nonconservative, i.e. if there are sources (e.g. production by radioactive decay) or sinks. Solutions of the diffusion equation for a variety of
initial and boundary conditions are available in standard texts, e.g. Carslaw
& Jaeger (1959) or Crank (1975); an illustrative example is given in the next
section.

It is also generally assumed that the diffusion coefficient D is independent of
concentration and has a temperature dependence given by an Arrhenius equation:

$$D = D_0 e^{-Q/RT} \tag{4.25}$$

where D_0 is a constant and Q is the 'activation energy'. Selected data for noble
gas diffusion coefficients are given in Table 4.8 and Fig. 4.6.

Solutions of the diffusion equation inevitably involve the dimensionless
parameter Dt/a^2 in such a way that diffusive redistribution becomes significant
as this parameter approaches unity (cf. eq. 4.27). Here a is some characteristic
dimension of the diffusive region. In the case where the medium is homogeneous,
without a microscopic substructure, e.g. a glass or a liquid, a is the macroscopic

Table 4.8. Selected data for noble gas diffusion in rocks and minerals

Sample	Gas	Temperature range[a] (°C)	E (kcal/mole)	D_0 (cm²/sec)	D_0/a^2 (sec⁻¹)
Sanidine[b]	Ar	500–800	18	—	—
Microcline[b]	Ar	300–700	52	—	—
Biotite[c]	Ar	<800	35	—	—
	Ar	>800	49	—	—
Orthoclase[d]	Ar	500–800	43	0.009 82	—
Phlogopite[e]	Ar	550–1 080	57.9	0.75	—
Volcanic glass[f]	He	125–400	19.1	0.067	—
Obsidian[g]	He	200–300	8.01	7.7×10^{-4}	—
Obsidian[g]	He	200–300	8.59	1.6×10^{-3}	—
Shungite[h]	Ar	200–1 000	10.5	—	0.032
(amorphous	Kr	200–1 000	15.5	—	0.1
carbon)	Xe	200–1 000	9.2	—	0.003 6

[a] Temperature range in which the diffusion experiments were made.
[b] Baadsgaard et al. (1961).
[c] Wescott (1966).
[d] Folland (1974).
[e] Giletti & Tullis (1977).
[f] Kurz & Jenkins (1981).
[g] Jambon & Shelby (1980).
[h] Rison (1980b).

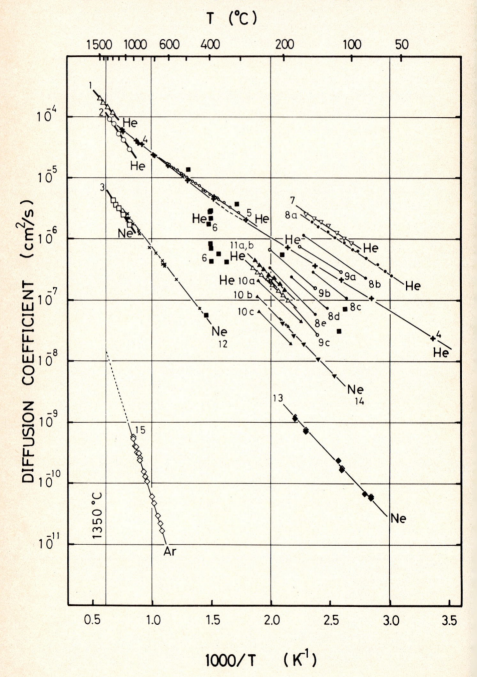

Fig. 4.6. Arrhenius plot showing temperature variation of noble gas diffusion coefficients. Samples 1–3 are glass melts; 4, 5, and 14 are vitreous silica; 6 is commercial glass; 7 and 14 are B_2O_3; 8–10 are mixtures of alkali oxides with B_2O_3, SiO_2, and Al_2O_3; 11 and 12 are obsidians; 13 and 15 are SiO_2. Reproduced from Hiyagon (1981).

dimension. In rocks which consist of numerous mineral grains the relevant a is usually the individual grain size rather than the macroscopic dimension. Often, an experimental study of diffusion in natural samples leads to determination of an average value for D/a^2 (Table 4.8) but if a reliable estimate for a is unavailable a value for D itself cannot be calculated.

The importance of the characteristic size a as well as the diffusion coefficient D in controlling the rate of diffusion is evident. Subdivision of mineral grains into smaller grains by exsolution or alteration can thus markedly increase the speed of diffusion. As an illustrative example, K-feldspar often exsolves into perthite, a fine-grained mixture of albite and orthoclase or microcline, and this perthitization has been invoked to help explain the frequently poor retention of radiogenic ^{40}Ar in K-feldspar (e.g. Folland, 1974).

Diffusive loss of gases from crystalline rock is often viewed as a two-stage process, first 'volume' diffusion from grain interiors to grain boundary surfaces, then 'grain boundary' or 'surface' diffusion for diffusion out of the macroscopic sample. Each process has its characteristic diffusion coefficient and activation energy; in general, volume diffusion has a lower diffusion coefficient and higher activation energy than grain boundary or surface diffusion. As in the example cited above, it is usually considered that volume diffusion from grain interiors is the rate-limiting process, and that once gas reaches grain exteriors its diffusion along grain boundaries is relatively rapid. In some cases, however, the converse may be true. Ozima & Takigami (1980), for example, studied the loss of radiogenic ^{40}Ar by stepwise heating and concluded that while volume diffusion is the limiting process in granites, grain boundary diffusion is the limiting process in oceanic basalts. This conclusion is consistent with the view that in granites K is distributed throughout relatively large grains while in basalts it is already concentrated near or at the boundaries of relatively small grains.

Diffusion of noble gases in geological materials is generally very slow (cf. Fig. 4.6) at relatively low temperatures, so laboratory diffusion experiments are usually conducted at fairly high temperatures and extrapolated to lower temperatures by means of eq. (4.25). Alternatively, a geologic estimate of diffusion coefficient can be made, e.g. by measurement of radiogenic ^{40}Ar in rocks whose age is determined by independent geochronometers. Various authors (e.g. Wescott, 1966) have pointed out that the geologic estimates are generally higher than the laboratory measurement extrapolations to the relatively low temperatures of principal interest ($\lesssim 200\ °C$). This discrepancy might be attributable to faster diffusive loss in nature because of grain size reduction by weathering or exsolution. Much of the discrepancy may also be due to failure of eq. (4.25) when extrapolated over too wide a temperature range. This is particularly plausible when significant structural changes occur in the relevant minerals at temperatures spanned by the extrapolation. Wescott (1966), for example, studied Ar diffusion in biotite and concluded that two different activation energies were involved, 35 kcal/mole below 800 °C and 49 kcal/mole above 800 °C.

A convenient experimental technique for measuring the diffusion coefficient (and solubility) of a noble gas in a solid is that of permeation, the flux of gas through it in response to a pressure differential (cf. Jambon & Shelby, 1980). If a slab of thickness L and area A experiences a partial pressure difference P across its faces, the steady state flux through it will be KPA/L, where the permeation coefficient $K = D\mathcal{H}$ for diffusion coefficient D and solubility (Henry constant) $\mathcal{H}$. The product $D\mathcal{H}$ can be separated by examining the time variation of the flux in response to a change in boundary conditions, e.g. abruptly decreasing P to zero, which permits evaluation of D (and thus $\mathcal{H}$ for known K).

4.8   Stepwise heating

Stepwise degassing is a very powerful and widely used technique in noble gas analysis. The approach is simple: piecemeal extraction. The sample is heated at a given temperature for a given time and the evolved gases are collected and analyzed; then the sample is heated to a higher temperature, gases are again collected and analyzed, and so on. Typical parameters are 100 °C temperature steps and half-hour to hour heatings, but these vary considerably.

Often in terrestrial samples, and almost always in extraterrestrial samples, the total noble gas complement is a superposition of components with different origins, sited in physically different locations in the samples. In general, then, they will have different release patterns in stepwise heating, i.e. gases collected in different temperature steps will have different proportions of the various components. It is this partial separation of components that is the great utility of stepwise heating: the constraints on composition and quantity which can be established thereby, and indeed the demonstration of the very existence of distinguishable components. Examples abound. Much of what has been learned (Sections 5.5 and 5.6) about noble gas components in the early solar system (and almost all of what was known prior to the advent of the selective dissolution approach described in Section 5.2), for example, is based on stepwise heating analysis of meteorites.

An obvious special case is when gases reside in different minerals. Fig. 4.7 shows stepwise heating data for whole rock and separated biotite and feldspar samples of the JG1 granodiorite. Gases are released from biotite some 300 °C earlier than from feldspar, so that the early part of the whole rock pattern is dominated by biotite, the later by feldspar. It should be noted, however, that the possibilities for mineralogical discrimination in stepwise heating are rather limited, and are not very great for more than two minerals: indeed, even in JG1

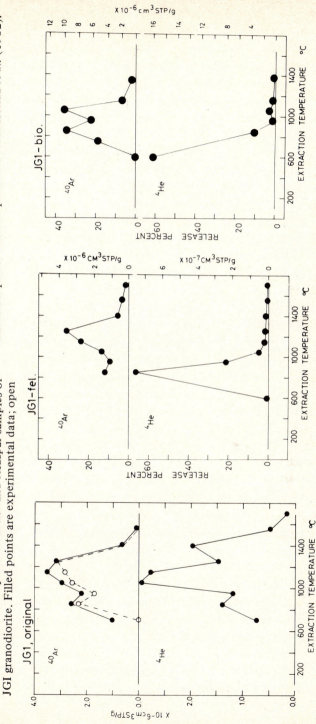

Fig. 4.7. Radiogenic ^{40}Ar and ^4He release in stepwise heating of whole rock and separated biotite and feldspar samples of JGI granodiorite. Filled points are experimental data; open circles in the whole rock pattern are results predicted from the mineral separate data. Reproduced from Honda et al. (1982).

whole rock (Fig. 4.7) the contributions of different minerals might not be evident without the data for the separates. Furthermore, the release pattern in stepwise heating depends not only on the intensive properties of a given mineral, but also rather sensitively on grain size.

It should also be noted that the components separated in stepwise heating are implicitly understood to be components of a single element. Different elements may separate in stepwise heating even if they are in the same place to begin with because they can have greatly different diffusion coefficients. Different isotopes of the same element will be released together, however, so that a single component in the sample remains a single component in stepwise heating. (As discussed in Section 4.9, this is not quite true, but isotopic fractionation will be much smaller than elemental fractionation and follow a regular pattern in mass, and observers are usually alert to the possibility.)

Component separation is often particularly prominent when one or more of the components present is *in situ.* There is no requirement that different components reside in different minerals; indeed, it is often not known at all *where* the components reside. An *in situ* component will be sited wherever its parent element is, and this siting might be rather different from that of some other *in situ* component or a trapped component, as illustrated in Fig. 4.8. Radiogenic gases are common *in situ* components, and stepwise heating is frequently applied to resolving them from other components.

There is a family of quite similar geochronological methodologies which are particularly well suited for demonstrating the features of stepwise heating: the ^{129}I-^{129}Xe method (Jeffrey & Reynolds, 1961), the ^{40}Ar-^{39}Ar method (Sigurgeirsson, 1962; Merrihue & Turner, 1966), the ^{244}Pu-^{136}Xe method (Podosek, 1970b; 1972), and the fission Xe method (Teitsma & Clarke, 1978). All of these are based on natural radionuclides (^{129}I, ^{40}K, ^{244}Pu, ^{238}U) which decay to noble gas daughters (^{129}Xe, ^{40}Ar, fission Xe, fission Xe), and the first essential step in the method is a neutron irradiation which produces a different isotope of the daughter element (^{128}Xe, ^{39}Ar) or a distinct fission Xe component, from a different isotope of the parent element (^{127}I, ^{39}K, ^{235}U, ^{235}U) (there is no extant isotope of Pu, so in the third case it is merely hoped that U is now sited as Pu was when ^{244}Pu was extant). If the natural daughter has accumulated undisturbed since a given time in the past, it will now be found in fixed ratio to its parent and thus to the artificial daughter, and the value of this ratio carries the desired geochronological information. Moreover, the natural daughter and the artificial daughter should both be in the same physical locations (wherever the parent element is) and thus constitute a single component. (A qualification: production of ^{39}K involves a finite recoil range, and the methodology is compromised if the phases which host the parent are very small; cf. Huneke

& Smith, 1976. Fission Xe fragments have even longer recoil ranges, but fission is involved in both the natural decay and the activation, so there is no significant separation.) It is noteworthy that these methods eliminate problems due to sample heterogeneity and can establish a relative chronology on the basis of an isotopic ratio (natural daughter to artificial daughter) rather than a typically less accurate elemental ratio (natural daughter to parent). More important, however, is the second essential step: analysis by stepwise heating to test the basic assumption, i.e. to determine whether such a well-defined component actually exists, and if so, to measure its composition. Of these four methods,

Fig. 4.8. Correlated isotopic variations observed in stepwise heating of lunar breccia 60019. The linear array suggests that only two components, in varying proportions, are present: one is trapped solar wind, presumably with composition near BEOC-12, the other is *in situ* cosmic-ray-induced spallation, relatively rich in ^{124}Xe and ^{126}Xe. Numeral '3' indicates $300\,^{\circ}$C extraction, etc. Reproduced from Bernatowicz *et al.* (1978).

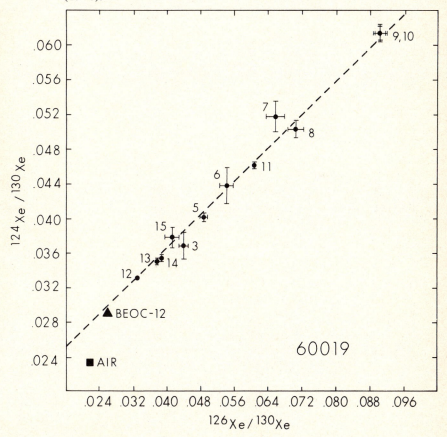

the most familiar and widely applied is the ^{40}Ar–^{39}Ar method, and we will use it to provide specific examples in the discussion below.

The application of the basic methodology is illustrated in Fig. 4.9. The *in situ* I-derived component is a mixture of natural daughter ^{129}Xe and activation daughter ^{128}Xe; the spread in observed compositions reflects variable proportion of this component to trapped Xe. The linear correlation at high temperatures indicates a well-defined ratio of I-derived ^{129}Xe to ^{128}Xe (and thus ^{129}I to ^{127}I), whose value is the correlation slope. It is noteworthy that the lower temperature data deviate from the line in the direction usually interpreted as partial loss of

Fig. 4.9. Correlated isotopic variations observed in stepwise heating of the (neutron-irradiated) meteorite Bjurböle (cf. Fig. 4.8). Trapped Xe is at lower left (M = Murray meteorite, A = air), and variations are caused by addition of *in situ* ^{129}Xe (from decay of ^{129}I) and ^{128}Xe (from neutron irradiation of ^{127}I). The linear variation indicates a constant ratio of the added ^{129}Xe and ^{128}Xe. The 1000 °C datum (and lower temperature data, not shown) deviate from this linear relationship in the direction indicating partial loss of radiogenic ^{129}Xe. Reproduced from Drozd & Podosek (1976).

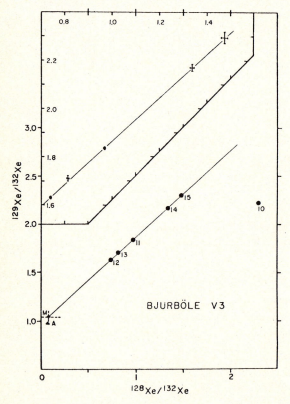

radiogenic ^{129}Xe from the least retentive sites. The most important feature of the methodology is that the desired chronological information can be obtained from the high-temperature data in spite of the low-temperature loss. The same considerations apply to the ^{40}Ar–^{39}Ar method, for which the corresponding diagram is ^{40}Ar/^{36}Ar versus ^{39}Ar/^{36}Ar, but in practice such a diagram is seldom used. The principal objective of examining this correlation is separation of trapped and *in situ* components, and trapped ^{40}Ar is usually negligible in extra-terrestrial samples and assumed to be 296 times ^{36}Ar (the air ratio) in terrestrial samples. The effects of partial ^{40}Ar loss are usually more serious, however.

To proceed with quantitative illustration it is necessary to adopt some specific model for gas release. It is usually assumed that Fick's Law applies and the problem can be treated as volume diffusion (Section 4.7). As a first approximation, it is often assumed that the sample is composed of uniform spheres of radius a. If it is further assumed that at some initial time $t = 0$ the gas concentration is uniform and that the concentration of gas is always nil at the surface of the sphere, then after diffusion to time t the fraction f of the original gas which remains in the sphere is (Carslaw & Jaeger, 1959)

$$f(x) = \frac{6}{\pi^2} \sum_{n=1}^{\infty} \frac{1}{n^2} e^{-n^2 x} \tag{4.26}$$

The extent of diffusion is conveniently parameterized by the dimensionless variable

$$x(t) = \frac{\pi^2}{a^2} \int_0^t D(t') \, dt' \tag{4.27}$$

which scales both the time and sphere size.

To model the effects of partial loss we will suppose that some rock forms at time $t = 0$ and accumulates uniform radiogenic ^{40}Ar (designated component ^{40}Ar$_1$) until some time t_1, when it experiences a partial degassing event. Diffusive loss in this event is specified by some value x_1 (eq. (4.27)), so the ^{40}Ar remaining is ^{40}Ar$_1 f(x_1)$. Subsequently, diffusion is arrested and the rock again accumulates uniform radiogenic ^{40}Ar, which can be designated a second component, ^{40}Ar$_2$, until the present. The rock is then irradiated, producing a ^{39}Ar component, and then heated in the laboratory to an extent which can be described by an (unsubscripted) value x (eq. (4.27)). At any stage x in the laboratory heating the Ar remaining in the rock is then given by

$$^{40}\text{Ar} = {}^{40}\text{Ar}_1 f(x_1 + x) + {}^{40}\text{Ar}_2 f(x) \tag{4.28}$$

$$^{39}\text{Ar} = {}^{39}\text{Ar}_0 f(x) \tag{4.29}$$

where ^{39}Ar$_0$ is produced from all the K in the rock. The instantaneous Ar composition released at any stage x is obtained by differentiation:

$$\frac{{}^{40}\mathrm{Ar}}{{}^{39}\mathrm{Ar}} = \frac{{}^{40}\mathrm{Ar}_1}{{}^{39}\mathrm{Ar}_0} \frac{g(x_1 + x)}{g(x)} + \frac{{}^{40}\mathrm{Ar}_2}{{}^{39}\mathrm{Ar}_0} \tag{4.30}$$

where

$$g(x) = \mathrm{d}f/\mathrm{d}x = -\frac{6}{\pi^2} \sum_{n=1}^{\infty} e^{-n^2 x} \tag{4.31}$$

At the beginning of the experiment, $x = 0$, the measured ${}^{40}\mathrm{Ar}/{}^{39}\mathrm{Ar}$ ratio is ${}^{40}\mathrm{Ar}_2/{}^{39}\mathrm{Ar}_0$, from which the time t_1 of the episodic loss can be determined. Near the end of the release, for large x, the measured ${}^{40}\mathrm{Ar}/{}^{39}\mathrm{Ar}$ ratio asymptotically approaches the sum of ${}^{40}\mathrm{Ar}_2/{}^{39}\mathrm{Ar}_0$ and $e^{-x_1}\,{}^{40}\mathrm{Ar}_1/{}^{39}\mathrm{Ar}_0$; if x_1 is small, this sum will be close to the nominal value which would have been obtained had there been no loss ($x_1 = 0$) at t_1, i.e. the high temperature ratio gives a close estimate of the 'true' age in spite of the partial loss at t_1.

As a numerical example, if the loss at t_1 is characterized by $x_1 = 0.02$, the fractional loss at that time is 15%, a significant error in a conventional whole rock K–Ar age. Nevertheless, the high-temperature ratio of the remaining ${}^{40}\mathrm{Ar}_1$ to ${}^{39}\mathrm{Ar}_0$ will approach a value 98% of what it would have been without the loss; by 50% release of ${}^{39}\mathrm{Ar}$ ($x = 0.3$) it will have reached 95% and by 75% release ($x = 0.9$) it will have reached 97.6%. In this example, the latter half of the release will be characterized by a plateau (nearly constant ${}^{40}\mathrm{Ar}/{}^{39}\mathrm{Ar}$) at very nearly the value correct for the 'true' age. Conversely, there will be only a very narrow region in the initial release which gives the episodic age t_1. For higher values of x_1 there will be a broader low-temperature plateau corresponding to t_1 but, of course, a narrower and lower high-temperature plateau.

More generally, x_1, t_1 and ${}^{40}\mathrm{Ar}_1$ can all be regarded as parameters whose values generate families of curves for the total release pattern, and fits to the whole curve, not just the ends, can be used to select the best values and test the model assumption. It is also possible to generalize the model (at the expense of more assumptions and free parameters) to non-uniform grain sizes and/or activation energies for the diffusion by summing or integrating f values (eq. (4.26)) over the appropriate distributions.

Turner *et al.* (1966) were the first to apply this kind of model analysis to ${}^{40}\mathrm{Ar}$–${}^{39}\mathrm{Ar}$ dating. In general, this approach is reasonably successful in accounting for the observations. Turner *et al.* were able to show, for example, that data for the Bruderheim meteorite (assuming a log-normal grain size distribution) fit a model for an 0.5 Ga event which caused extensive but not complete loss of ${}^{40}\mathrm{Ar}$ accumulated in a much older primary age (Fig. 4.10).

In actual experiments, of course, things are not quite so simple. In the first place, one does not observe the instantaneous release but only its integration in discrete temperature steps. Problems also arise because of the recoil effect mentioned above and various interferences and other uncertainties arising in the neutron irradiation. In practice, discrete events are identified and meaningful ages assigned to them only when the release pattern provides a significant plateau, at least two or three temperature fractions with the same ^{40}Ar/^{39}Ar extending over about 25% or more of the total ^{39}Ar release.

In most applications of stepwise heating, the objective is the separation of components rather than characterization of their locations, and the relevant degassing parameter, when it is of interest at all, is the integral in eq. (4.27). There is thus a tendency for experimenters to be rather sloppy about heating schedules and the absolute temperatures involved. With better control over these parameters, it is possible to gain considerably more information.

Thus, the fraction f_i of gas remaining in a sample after the ith temperature step is determined directly from the experimental data. If the gas concentration was initially uniform and the sample is spheres of radius a, f is given by eq. (4.26). Since $f(x)$ is a monotonic function, each f_i is associated with a specific value x_i. If the temperature T_i was constant, then from eq. (4.27) we obtain

$$(\pi^2/a^2)D_i = \Delta x_i/\Delta t_i \qquad (4.32)$$

Fig. 4.10. Release pattern obtained in a ^{40}Ar–^{39}Ar analysis of the Bruderheim meteorite. The theoretical curves are for a primary age of 4.5 Ga and a 500 Ma thermal event in which a 90% loss of radiogenic ^{40}Ar occurred. Reproduced from Turner *et al.* (1966).

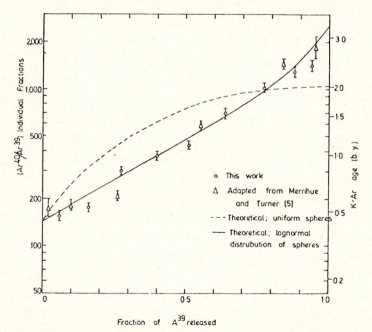

Fraction of A^{39} released

where $\Delta x_i = x_i - x_{i-1}$ and Δt_i is the duration of the ith step, and D_i is the diffusion coefficient at T_i. If $D(T)$ follows the expected functional form (eq. 4.25), display of eq. (4.32) in an Arrhenius diagram will define a linear trajectory and observation of such a result will help justify the assumptions. If a is known the exact value of D is determined; even if a is not known the slope determines the activation energy. If the assumption of uniform grain size is dropped we can generalize by integration of the left-hand side of eq. (4.27) over the distribution in a, and the same remarks apply as long as all grains have the same diffusion coefficient (or at least the same activation energy).

An example is shown in Fig. 4.11. It is noteworthy that the granodiorites and feldspar have activation energies in the range 37–48 kcal/mole, while the submarine volcanics have much lower energies, 12–20 kcal/mole.

Diffusive release of gases for other geometries and boundary conditions, and for nonuniform initial distributions are more complex but still relatively straightforward, and standard treatments such as that of Carshaw & Jaeger (1959) are readily available. Although standard solutions are usually framed in terms of constant D, the variable D in stepwise heating can, in general, be handled by scaled parameterizations analogous to eq. (4.27).

Albarède (1978) has presented a treatment of the inverse problem: the determination of an initial spatial distribution from stepwise heating data. An important premise in his discussion is that there is at least one isotope whose initial distribution is uniform. Stepwise heating data for this isotope then establish values for the f_i, and thus the x_i, described above. These values, in conjunction with the observed f_i for the other isotopes and an application of linear inversion theory, establish constraints on the initial distributions of these other isotopes. Fig. 4.12 illustrates application of Albarède's method (assuming uniform spheres) to a ^{40}Ar-^{39}Ar analysis of lunar rock 15415 by Turner (1972). The isotope assumed initially uniform is ^{37}Ar, a byproduct (from ^{40}Ca) of the neutron irradiation. The initial distribution of ^{39}Ar (i.e. of ^{39}K) is essentially uniform except for a surface enhancement, a feature frequently inferred on other grounds as well. The ^{40}Ar distribution shows the surface enhancement also, and a somewhat deeper depletion consistent with moderate diffusive loss on the moon.

4.9 Isotopic fractionation

When the equilibrium partitioning or transfer rate of a given element between two reservoirs depends on atomic mass, isotopic fractionation may arise. For elements such as O, C, H, S, N, and others, significant isotopic fractionations are common, occasionally reasonably well understood, and often exploited to provide information about the processes responsible for the fractionation or as geochemical tracers. Noble gas fractionations share only the first of these three features.

Isotopic fractionation in equilibrium partitioning between two reservoirs arises in two ways. The density of quantum states, and thus the partition function, usually varies as some low power of mass (e.g. $m^{3/2}$ for the translational partition function of an ideal gas). Such mass dependence may be nearly but not quite equal in the two reservoirs, whence the distribution coefficient will depend on

Fig. 4.11. Arrhenius plot for diffusion coefficients obtained from stepwise heating release of ^{39}Ar (from neutron-irradiation of ^{39}K) from submarine basalts (identified by sample number), granodiorites, and K-feldspar. The ordinate is $\ln (D\Delta t/a^2)$, calculated from eq. (4.32). Reproduced from Ozima & Takigami (1980).

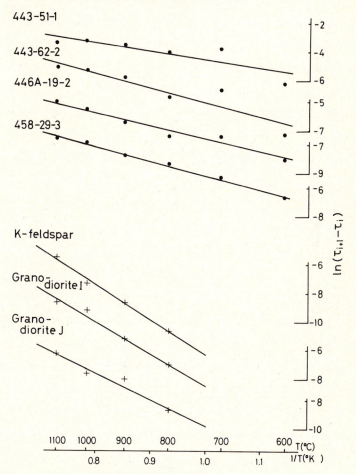

mass. Fractionation so arising will be relatively small, the isotopic partition coefficient for a mass difference δm differing from unity by some small fraction of $\delta m/m$, and only weakly or not at all dependent on temperature. In a manner of speaking, by this means the mass enters the Gibbs free energy through the entropy term. The other way is through the energy term, most commonly

Fig. 4.12. Initial spatial distributions of (a) ^{39}Ar (from neutron irradiation of ^{39}K) and (b) ^{40}Ar (radiogenic) in assumed spherical grains in lunar sample 15415. Both concentrations and radius are scaled to unity. The paired curves indicate the bounds ($\pm$ one standard deviation) of initial distribution inferred from application of linear inverse theory to stepwise heating data of Turner (1972). Reproduced from Albarède (1978).

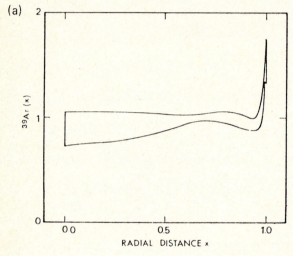

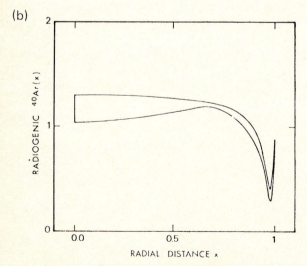

because of the mass dependence of the zero point vibrational energy ($\frac{1}{2}\hbar\omega$, where ω is proportional to $m^{-1/2}$). This effect contributes to the isotopic partition function of a factor in which the term constant x $\delta m/mKT$ is exponentiated, where the proportionality constant depends on the difference in bond strengths in the two reservoirs. While quantitative evaluation is complicated by the occurrence of many vibrational modes in condensed phases, it is clear that the resultant fractionation can be relatively large and will be relatively sensitive to temperature (the principle of isotopic geothermometry), and will be greater at lower temperatures. Actually, we can expect that this effect will not be important for noble gases (although there is very little empirical evidence to support this expectation). The effect arises only if most oscillators are in the ground state ($KT \ll \hbar\omega$). While this is the usual case for most chemical bonds at surface temperatures, the available data indicate that this will not be true for the weak van der Waals 'bonds' which characterize noble gas interactions. It should be noted that vibration is not the only way in which mass can enter the thermodynamic energy term, however. In a nonturbulent atmosphere (e.g. the earth's upper atmosphere), for example, each isotopic species will reach its own equilibrium with the gravitational field, and since the scale height (KT/mg) depends on mass, the isotopic composition will depend on altitude.

It is difficult to make generalizations for isotopic fractionation in adsorption and solution equilibrium partitioning. In considering possible fractionations predicted by various statistical models for adsorption, Podosek *et al.* (1981) noted that there might be a strong enrichment of heavy isotopes in the adsorbed phase if the zero-point vibrational energy term is important, but that otherwise any fractionation should be weak, with both magnitude and sign dependent on the particular statistical model assumed. In particular, for the mobile film model which is the basis of eq. (4.15) the first-order mass dependence of the Henry constant vanishes, in which case fractionation effects would be negligibly small. There have been no experimental studies of noble gas isotopic fractionation in adsorption on geochemically relevant materials, but it may be that adsorption is implicated in noble gas trapping in sedimentary rocks (Section 8.4), in which case some observed fractionations (cf. Fig. 8.5) might be attributable to adsorption. An important parameter controlling the distribution coefficient in solution is evidently atomic radius (Section 4.3); since the radius is independent of isotopic mass there should be no fractionation by this mechanism. It is not clear whether or how the interaction energy between solvent and noble gas solute should depend on isotopic mass. There is, however, one relevant datum: Weiss (1970b) finds that ^{3}He is 1.2% less soluble in water at 0 °C than ^{4}He is. This is a rather small effect for so great a mass difference and so low a temperature, so it is probably reasonable to infer that fractionation by solution will be unobservably small for the heavier gases.

Fractionation also arises in the nonequilibrium transfer of material from one reservoir to another if the transfer rate is mass dependent (and the transfer is incomplete). As for equilibrium partitioning, we can distinguish two basically different cases according to whether the mass enters as a relatively low power or in an exponential. In the first case, the fractionation is typically small, a fraction of $\delta m/m$, and insensitive to temperature. The most common example is diffusion, in which fractionation is usually modeled by assuming that the diffusion coefficient is proportional to kinetic velocity, i.e. to $m^{-1/2}$. (This relationship is intended to encompass different isotopes of a given element and *cannot* be expected to be even approximately correct for different elements.) In the latter case, the exponential is usually the Boltzmann factor $e^{-Q/KT}$, proportional to the probability that a given atom at a given time has an energy at least equal to Q (if $KT \ll Q$). In this case the fractionation can be large and in general will be sensitively temperature dependent. A mass dependence in the 'activation energy' Q can arise in many contexts. If Q is a dissociation energy, for example, it will depend on mass through the vibration frequency, since the ground state is $\frac{1}{2}\hbar\omega$ above the bond potential energy minimum. Lighter isotopes thus react faster (i.e. at room temperature $^{16}O^{16}O$ dissociates 17% faster than $^{18}O^{18}O$). As above, however, the effect is important only if most systems are actually in the ground state, which seems unlikely for noble gases at surface temperatures. Probably the most important case in geochemistry is escape from a gravitational field (Q proportional to mg). The problem is a complicated one involving not only the probability that an exosphere atom has enough energy to escape but also the rate at which the exosphere supply is replenished from below; nevertheless, it is clear that light isotopes will escape faster than heavy ones, and the escape rate can be quite sensitive to mass. Escape from the earth's atmosphere is important for He (Section 12.5), for example, and it is noteworthy that while 'thermal' (Jeans) escape is the dominant loss mechanism for ^{3}He, Jeans escape for the heavier isotope is sufficiently slower that it is *not* the dominant loss term for ^{4}He.

It is well known that even when 'single stage' fractionation is small, fractionation effects may be cumulative in appropriate circumstances. Diffusion provides a convenient example, as illustrated in Fig. 4.10. The fractional mass difference between ^{134}Xe and ^{132}Xe in $\delta m/m = 15^o/_{oo}$. For a mass dependence according to $m^{-1/2}$ the diffusion coefficient for ^{134}Xe will be approximately $\frac{1}{2}\delta m/m = 8^o/_{oo}$ less than for ^{132}Xe. The first gas to diffuse out of material with an initially uniform concentration of both isotopes will be depleted (relative to the initial composition) in the heavier isotope by $\frac{1}{4}\delta m/m = 4^o/_{oo}$. Because of the continuing preferential loss of ^{132}Xe, the $^{134}Xe/^{132}Xe$ in the residual gas becomes larger the greater is the depletion of the original reservoir. At a certain point the escaping gas has the composition of the original reservoir, and beyond that point both the residuum and the escaping gas are enriched in the heavier isotope (Fig. 4.13). Conversely, comparable effects of light isotope enrichment occur for gas diffusing *into* the reservoir of interest. Similar cumulative effects can occur in any process involving fractionation; the equivalent and more familiar effect for equilibrium partitioning between different phases is Rayleigh distillation, for example. A characteristic feature of such cases is that (unless there is reflux or its equivalent) the extreme effects occur only in a very small part of the total quantities involved.

Since isotopic fractionation is mass dependent, there will be a 'smooth' relationship between effects observed in various isotopic ratios. In most cases of interest, the magnitude of fractionation and the fractional mass range are both small in an absolute sense, and to first order the degree of fractionation

Fig. 4.13. Illustration of isotopic fractionation effects in diffusion. The model is that ^{132}Xe and ^{134}Xe are initially uniformly distributed throughout spheres in the ratio ^{134}Xe/^{132}Xe = 0.382 and then allowed to escape by diffusion with the boundary condition that the concentration vanishes on the surface (eq. 4.26). The figure shows the instantaneous composition of the released gas at various stages, assuming that the diffusion coefficient varies as $m^{-1/2}$. 'Single component' locus is for all spheres having the same radius; 'mixed components' is for a distribution of sizes. Reproduced from Funk *et al.* (1967).

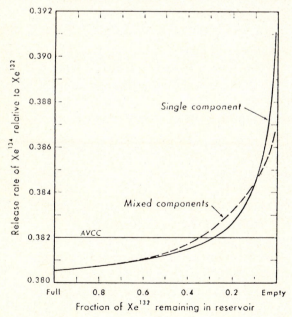

is approximately linear in mass: δ_m is proportional to δm, where δ_m is a fractional measure (e.g. in per mil) of the fractionation effect between isotopes of mass m and m_0 eq. (1.1)) and $\delta m = m - m_0$. In elements of three or more isotopes, this is the signature by which fractionation is distinguished from specific isotope effects: if two or more isotopic ratios are observed to vary in this fashion appropriate for fractionation, it is usually considered that this is not an accident and that fractionation has indeed occurred (rather than two or more independent effects which depend on the nuclear identity of the isotopes in question).

Among the noble gases the element whose isotopes span the greatest relative mass range is He and it is thus expected that He should exhibit the greatest fractionation effects in natural occurrence. This is presumably true, but it is usually difficult to tell. Since He has only two isotopes, the smoothness test for mass dependence cannot be applied, and there are so many possible and potentially large specific isotope effects (Sections 5.5, 6.2, and 6.6) that unambiguous assignment of some modest isotopic variation to fractionation is usually impossible. Ne has three isotopes, so that a smoothness test can be applied. In practice, however, ^{21}Ne is difficult to measure accurately because its abundance is so low, and a specific isotope effect at ^{21}Ne is not uncommon (Section 6.2), so that it is not too easy to judge competing hypotheses for variation of ^{20}Ne/^{22}Ne – fractionation or specific isotope effect (cf. Section 6.6) – simply by appeal to Ne compositional variation alone. In contrast, Ar also has only three isotopes, and specific effects at ^{40}Ar are so common that ^{36}Ar and ^{38}Ar must often be treated as a two-isotope system decoupled from ^{40}Ar. Even so, however, plausible specific isotope effects at ^{36}Ar and ^{38}Ar are so uncommon and so small that except in very unusual samples (Section 6.2) any observed variations in ^{38}Ar/^{36}Ar ratio are usually *assumed* to be due to fractionation. The smoothness test really comes into its own for Kr and Xe, which have so many isotopes that fractionation is obvious, even when it is not anticipated or understood (Figs. 4.5 and .85) and even when there actually *are* complicating specific isotope effects at some isotopes (cf. Fig. 6.2).

It is a characteristic feature of noble gas chemistry that there are many cases where a fractionation signature is so clear that one must conclude that fractionation has occurred even when there is no adequate understanding of *how* it has occurred, e.g. as is the case for 'electric discharge kerogen' (Fig. 4.5). An important example is the apparent very strong fractionation of terrestrial Xe relative to any plausibly identified or even hypothesized source elsewhere in the solar system (cf. Section 6.7).

In concluding this section it is pertinent to take note of a special kind of isotopic fractionation: ubiquitous, often quite severe, and arguably the most important source of fractionation which must be taken into consideration in noble gas geochemistry. This is the fractionation which arises in mass spectrometric analysis: contributory effects can and do arise in gas extraction and transport through the vacuum system, in the ion source (especially when a source magnet is used), in beam transmission, and in ion collection and detection (especially when an electron multiplier is used). As noted in Section 1.3, sample data are corrected for instrumental (and procedural) discrimination which is calibrated by analysis of some standard gas (usually air). This is a roundabout and imperfect near-equivalent to the δ value convention which is the norm in 'stable isotope' geochemistry (O, C, H, S, N, etc.). The reproducibility of instrumental discrimination inferred from repeated calibration analysis is usually quite satisfactory, but seldom is any care taken to try to match operating conditions in samples and calibration analyses. It is thus a matter of faith – undoubtedly quite justified in most cases, but faith nonetheless – that discrimination is the same for sample and calibration. The only unequivocal method for distinguishing natural from laboratory fractionation is double-spike analysis, which to our knowledge has never been applied in noble gas geochemistry.

5 Cosmochemistry

5.1 Introduction

On the basis of several lines of evidence, a picture was developed that at some stage in its history the solar system (rather, the inner solar system, consisting of sun, terrestrial planets, and meteorites, the only parts for which adequate data are available) was compositionally uniform. This was conceived as a quite strict uniformity, of isotopes as well as elements, on all scales from astronomical units down to counting statistics. This unique composition is what is usually designated by the adjective *primordial*. For cases of changing composition due to decay of primordial radionuclides, an epoch must be specified; this is taken to be the time of formation of primitive meteorites (see below), where formation means isotopic closure, the customary definition in geochronology. The existence of such a thing as a primordial composition, as well as the data which describe it, is a very important boundary condition in studies of the solar system, since astrophysical theory is required to produce this composition and to account for its uniformity in the nascent solar system, and planetary science is required to account for the generation of presently-observed elemental and isotopic variations in terms of its subsequent evolution.

It has become evident that the solar system was not ever totally homogeneous, so that there is no unique primordial composition applicable to all parts of the solar system. While the implications of primordial heterogeneity are of great import to solar system science, still it seems, at least at present, that the compositional variations were small in an absolute sense, and that revolutionary revision of theories for the evolution of the solar system after the formation of meteoritic solids is not necessary on this account. Thus, the concept of a primordial composition, from which samples of the solar system subject to analysis must have been derived, is still valid and useful as long as it is remembered that some variation of ancestral composition is allowed.

The tabulation of Cameron (1973) is generally accepted as the best available estimate of solar system primordial composition; selected data are presented in Table 5.1. In the literature, primordial composition is frequently referred to by 'cosmic abundances', the term having been coined in the expectation that the solar system was a sample of a truly cosmic composition. Actually, spatial and temporal variations in even the local cosmos are readily evident (cf. review by Trimble, 1975), so the term is a misnomer, and 'cosmic' abundances really describe only what they are based on: the inner solar system.

In primordial composition, the noble gases are no more nor less rare than other elements. The primordial abundance of any nuclide is determined by summation of various nucleosynthetic processes (cf. Burbidge *et al.*, 1957; Clayton, 1968; Trimble, 1975) which are sensitive to nuclear rather than chemical properties, and the primordial abundances of the noble gases fit smoothly into the pattern of neighboring elements. With a few exceptions, these abundances are preserved in the sun. That noble gas abundances in all other accessible samples are so very much lower (Table 5.2 and Figs. 5.1–5.5) is attributable to chemical fractionation processes occurring within the solar system. Actually, a hedge must now be added: while the Cameron (1973) abundances of He and Ne are based on observations of solar flares, the Ar, Kr, and Xe abundances are based not on the observation of any material but on the *assumption* of the proposition just stated, and so on interpolation from neighboring elements according to nucleosynthetic theory. While the reasoning is circular and its application may introduce inaccuracies, the validity of the principle has not been challenged and its overthrow *would* be revolutionary. In any case, if the Cameron (1973) noble gas abundances were seriously (more than about a factor of two) in error the smooth relationships seen in 'solar' gas abundances (Section 5.2) would have to be considered the fortuitous result of complex hypothetical processes, so there is not much doubt about the nominal abundances.

Table 5.1. 'Cosmic' abundances

Element	Abundance[a]
H	3.18×10^{10}
He	2.21×10^{9}
C	1.18×10^{7}
N	3.74×10^{6}
O	2.15×10^{7}
Ne	3.44×10^{6}
S	5.0×10^{5}
Cl	5.7×10^{3}
Ar	1.172×10^{5}
Kr	46.8
Xe	5.38

[a] Values are given as numbers of atoms per 10^{6} atoms of Si (Cameron, 1973).

Noble gas data for cosmic abundances are restated as absolute abundances in Table 5.2 and relative abundances in Table 5.3. The data included in Table 5.2 are for a hypothetical rock containing 17% Si (by weight). Actually, cosmic proportions would not produce a rock, and even the He alone would be 54 g, so these data are best read as 'cm^3 STP of gas per 0.17 g of Si'. They are a convenient basis for estimating the degree of depletion from cosmic proportions in real samples.

The general field of scientific investigation describable as cosmochemistry is a problem-oriented rather than a technique-oriented discipline and draws on many sources of information. Much of it comes from the study of meteorites. These are samples of terrestrial planetary material (i.e. the 'condensable' elements), but their original location is not rigorously established. Most of them are very old; in the usual sense of isotopic closure, they are the oldest objects in the solar system. Since they also contained, at the time of their formation, short-lived radionuclides considered not to have been produced in the solar system, they also define the age of the solar system as a whole (usually cited as 4.5×10^9 yr, with contention about only the third significant figure). For our present purposes, their most important characteristic is that most (not all) of them, the class designated as chondrites or undifferentiated meteorites, are primitive. The general view is that they are solids which formed in and from an early solar nebula, accreted into planetary bodies of presumably asteroidal or cometary character and size, and there experienced varying degrees of secondary alteration, but often not very severe and in particular not including igneous differentiation. In comparison with all known samples of the earth, its moon, presumably the other terrestrial planets, and the complementary class of differentiated meteorites, the undifferentiated meteorites give a much more easily readable picture of circumstances in the early solar system in which they and the other planets originated.

Overall reviews showing the emergence of these generalizations are readily had, e.g. Mason (1962); Wasson (1974); Wood (1979). Here we will of course concentrate on the noble gases. Representative data for noble gas contents in meteorites are included in Table 5.2.

5.2 Elemental abundance patterns

In extraterrestrial solids there are two prominent classes of *in situ* nuclear processes which produce noble gases (Chapter 3): decay of primordial radionuclides and cosmic-ray spallation. These are interesting and useful in many ways but our immediate concern here is with the complementary component, the trapped gases (cf. Section 1.3). In practice, 'trapped' is operationally synonymous with 'nonradiogenic and nonspallation', since that is mostly how trapped gases are identified.

Since it was first observed that meteorites indeed contain trapped noble gases (Gerling & Levskii, 1956), it has been clear that they occur in a wide range of abundances, none of them even close to their cosmic abundances (Table 5.2 and Figs. 5.1–5.5). In spite of the variety and complexity, however, it is possible to make meaningful generalizations that are useful in considering their origin. Specifically, gases characteristically occur in one of two patterns, designated the 'solar' and 'planetary' patterns (Signer & Suess, 1963; Pepin & Signer, 1965). Isotopic characterizations are discussed in Sections 5.4 and 5.5. Here we will focus on the elemental abundances; the discussion is summarized in Table 5.3.

There are some classes of extraterrestrial materials, notably the moon and some differentiated meteorites, whose representatives are very poor in volatile species generally and essentially devoid of trapped noble gases in particular, a circumstance believed characteristic of their parent planetary bodies. Nevertheless, other representatives of these classes exhibit quite high concentrations of noble gases, characteristically in abundance patterns illustrated in Fig. 5.1. This is the solar pattern, defined by elemental abundance *ratios* nearly the same as the cosmic ratios.

The origin of solar gas is not in any serious doubt: the gas is captured solar wind, manifested primarily in the noble gases because their indigenous abundances are so low. The association with solar wind is based not only on the noble gas abundance ratios but on a variety of other lines of investigation and reasoning (e.g. Suess *et al.*, 1964). For the moon and evidently for at least most meteorites as well, exposure to the solar wind occurs in regolith at the surface of the parent body rather than in independently orbiting grains. For lunar samples, it is commonly found that soils and breccias made from soils are rich in solar gas, which is essentially absent from the parent rocks. Meteorites characteristically exhibit a visible macroscopic dark–light structure, solar gases being abundant in the dark fraction and less abundant or absent in the light fraction, both features reflecting different degrees of exposure at the planetary surface. The differing absolute gas concentrations evident in Fig. 5.1 are attributable simply to different degrees of solar wind acquisition.

Stipulation of the solar gas pattern by elemental abundance ratios nearly equal to the cosmic (i.e. sun) ratios requires some qualification. In Fig. 5.1 there is an obvious, more or less smooth, and nontrivial trend of heavy gas enhancement such that in solar gas the Xe/He ratio is some 10 to 20 (sometimes more) times higher than the cosmic Xe/He ratio. It is possible, but not likely, that the cosmic abundance estimates systematically fail to represent the composition of the sun by this much, or that the solar wind itself systematically differs from the composition of the sun. At least for the light gases, however, neither possibility seems particularly likely to be responsible for this overall trend, since measurements of the actual solar wind (Geiss *et al.*,

Table 5.2. Noble gas abundance data (cm³ STP/g)

Sample	Notes	^{4}He	^{20}Ne	^{36}Ar	^{84}Kr	^{130}Xe
'Cosmic'						
'Cosmic rock' (17% Si)	a	3.00×10^5	4.32×10^2	1.34×10^1	3.62×10^{-3}	3.14×10^{-5}
Lunar soil						
12001 ilmenite (125 μm)	b	1.04×10^{-1}	4.08×10^{-4}	2.00×10^{-5}	1.27×10^{-8}	2.74×10^{-10}
12001 ilmenite (10.9 μm)	b	2.00×10^{0}	7.60×10^{-3}	2.25×10^{-4}	1.57×10^{-7}	3.44×10^{-9}
12001 'bulk' (10.4 μm)	b	1.69×10^{-1}	2.44×10^{-3}	4.80×10^{-4}	3.22×10^{-7}	8.74×10^{-9}
Meteorites						
Ivuna (CI)	c	9.43×10^{-5}	2.78×10^{-7}	8.21×10^{-7}	1.0×10^{-8}	7.3×10^{-10}
Orgueil S2 (CI)	c	1.89×10^{-4}	5.44×10^{-7}	3.00×10^{-6}	2.5×10^{-8}	2.8×10^{-9}
Mokoia (C3V)	c	1.08×10^{-3}	3.88×10^{-6}	2.57×10^{-7}	3.2×10^{-9}	3.1×10^{-10}
Allende (C3V)	d	–	4.2×10^{-8}	1.7×10^{-7}	1.6×10^{-9}	2.4×10^{-10}
Allende 3C1	d, i	1.89×10^{-3}	1.09×10^{-5}	3.72×10^{-5}	3.83×10^{-7}	4.88×10^{-8}
Dimmitt (H3)	e, h	–	2.0×10^{-7}	4.7×10^{-8}	6.05×10^{-10}	1.30×10^{-10}
Dimmitt AmlI	e, i	6.07×10^{-5}	7.00×10^{-7}	1.87×10^{-5}	1.89×10^{-7}	5.35×10^{-8}
Estacado (H6)	e, h	–	–	3.5×10^{-9}	8.3×10^{-11}	1.6×10^{-11}
Estacado AmlI	e, i	–	6.0×10^{-8}	1.41×10^{-6}	9.17×10^{-9}	2.63×10^{-9}
Bruderheim (L5)	f, h	–	–	5.5×10^{-9}	8.2×10^{-11}	1.6×10^{-11}
Pesyanoe (Au)	g	8.7×10^{-3}	2.23×10^{-5}	1.36×10^{-6}	6.00×10^{-10}	1.29×10^{-11}
In situ components						
In situ (typical)	j	1×10^{-5}	3.5×10^{-8}	3.5×10^{-9}	4×10^{-12}	8×10^{-14}
Planetary atmospheres [k]						
Earth	–	3×10^{-9}	1.1×10^{-8}	2.1×10^{-8}	4.4×10^{-10}	2.4×10^{-12}
Mars	–	–	5.6×10^{-11}	1.3×10^{-10}	4.2×10^{-12}	7×10^{-14}
Venus	–	6×10^{-7}	2.3×10^{-7}	1.5×10^{-6}	l	$<8 \times 10^{-10}$

b Eberhardt et al. (1972).
c Mazor et al. (1970).
d Lewis et al. (1975).
e Moniot (1980).
f Average for 'Berkeley standard Bruderheim', unpublished (circulated by J. H. Reynolds).
g Marti (1969).
h Lightest gas tabulated includes modest spallation correction.
i Acid-resistant residue.
j ^{4}He is half of expected radiogenic accumulation from 12 ppb U (with Th/U = 3.7) in 4.5×10^9 yr; others are spallation products in 10^7 yr exposure to cosmic rays, chondritic target abundances (corresponding ^{3}He is 2×10^{-7} cm^3 STP/g). The radiogenic contribution to ^{4}He was subtracted before plotting in Figs. 5.1–5.4.
k Atmospheric inventory divided by mass of planet, data from Table 12.4.
l Two different atmospheric abundances, corresponding to 1.5×10^{-9} and 3.0×10^{-8} cm^3 STP/g, have been reported; see Section 12.10.

Table 5.3. Prominent noble gas elemental compositions

Occurrence	^{4}He/^{20}Ne	^{20}Ne/^{36}Ar	^{36}Ar/^{84}Kr	^{84}Kr/^{136}Xe	References
'Cosmic' (= sun)	694	32	3 700	115	Table 5.2 (Cameron, 1973)
Solar wind	600	37	—	—	Geiss et al. (1972)
'Solar' (= moon)	253	27	1 590	38	Eberhardt et al. (1972) (12001 ilmenite)
'Planetary' (= meteorites)	220	0.28	80	8	Mazor et al. (1970) (Type CII)
Earth atmosphere	—	0.52	48	180	Table 5.2

1972) during the Apollo missions match the cosmic ratios quite nicely (Fig. 5.1). More likely, the characteristic elemental fractionation, relative to cosmic ratios, can be attributed to differential retention effects in the targets, specifically the relative ease of diffusive loss of light gases in comparison with heavier gases and the greater bearing of saturation effects for the more abundant light gases. Some fractionation, at least, obviously arises in the target, as evidenced by compositional variations and dependencies on grain size and mineralogical identity of the target (Eberhardt *et al.*, 1972), and on the whole the elemental fractionation in the solar gas pattern is plausibly understandable in terms of processes acting during and after solar wind incidence.

Fig. 5.1. Illustration of noble gas abundance patterns. For each specific noble gas isotope the ordinate is the logarithm (base 10) of the ratio of the observed concentration to the 'cosmic rock' concentration. The ordinate position of the solar wind composition is arbitrary. The vertical dashed lines illustrate roughly the magnitude of typical *in situ* contributions; at He, the horizontal tick is for ^{4}He, the upward extension for ^{3}He; trapped gases at lower concentrations would be difficult to identify. Solar wind composition from Table 5.3, all other data from Table 5.2. All the patterns shown here are of the 'solar' type (cf. Figs. 5.2–5.4). NB The scales and ordinate ranges are the same in all of Figs. 5.1–5.4.

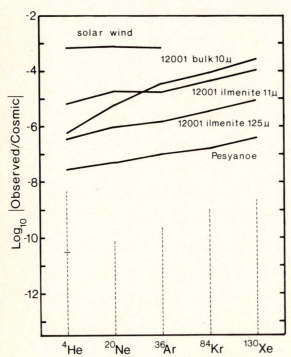

Undifferentiated meteorites (= chondrites) contain trapped noble gases also, sometimes in quite high abundances and usually not in the solar abundance pattern. Generally, gas contents correlate with degree of 'primitiveness' (= absence of postaccretionary processing in the parent body). Notable among the undifferentiated meteorites are the carbonaceous chondrites, generally recognized as the most primitive meteorites known (but also cf. McSween, 1979) and also characteristically the most prominent bearers of trapped noble gases. An extensive data tabulation and general discussion of carbonaceous chondrite noble gases is given by Mazor *et al.* (1970).

Noble gas abundances for some carbonaceous chondrites are illustrated in Fig. 5.2. The abundances define a distinctive signature designated as the *planetary* pattern. As is the case with solar gas, planetary gas is characterized by a set of elemental ratios: some meteorites contain more, some less, but the composition remains sensibly the same in samples in which the absolute abundances range over three or more orders of magnitude. Also as is the case with solar gas, the planetary gas elemental ratios, relative to cosmic abudance ratios, show a smooth (monotonic, anyway) dependence on atomic weight. The elemental fractionation is much greater, however: in planetary gas the Xe/He ratio is some five orders of magnitude higher than the cosmic ratio, in contrast to the one order of magnitude typical of solar gas.

As is also the case with solar gas, there are some variations among compositional patterns identified as planetary. The variations are beyond experimental uncertainty, but still small in comparison with the differences between the two patterns. The differences between solar and planetary patterns, as illustrated by comparison of Figs. 5.1 and 5.2, are quite clear, and it is evident that the two must have basically different origins. There are also characteristic isotopic differences between solar and planetary gas, as discussed in Sections 5.4 and 5.5.

For lack of a reasonable alternative, planetary gas is regarded as a biased sample of primordial gas composition, and the regularity and reproducibility of the pattern indicates that the sampling process was probably relatively simple and, at least, widespread. It is also generally considered that planetary gas was acquired by solid grains from the ambient solar nebula before their accretion into the meteorite parent bodies (cf. solar gas, acquired on the surface of a parent body). Beyond such generalizations, the origin of planetary gas is obscure. Several hypotheses have been advanced, as described in Section 5.7, but none has gained a wide acceptance.

For the heavy gases, compositional variations among different specific examples of planetary gas are relatively modest. For carbonaceous chondrites, Mazor *et al.* (1970) indicate only slight typical variation, about 30%, in Kr/Xe ratio, and, for Ar/Kr, a somewhat larger variation, 35–40%, with a tendency for

Ar/Kr to correlate positively with absolute Ar content, ^{36}Ar/^{84}Kr ranging
from about 80 to 200. They also find an apparent systematic variation in
^{4}He/^{20}Ne according to meteorite class, this ratio ranging from about 220 to
about 350. The largest variations in planetary composition, according to Mazor
et al., are found in the ^{20}Ne/^{36}Ar ratio, ranging downwards from a high of about
0.3 to effectively zero in systematic fashion according to meteorite class.

Determination of planetary trapped abundances for the light gases He and Ne
is generally a more risky business than for the heavier gases: the relative influences
of *in situ* (cf. Figs. 5.1 and 5.2) and possibly superposed solar (see below and
cf. Fig. 5.3) contributions are greater, and the lighter gases are more susceptible
to compositional modification by diffusive loss during processing in the parent
body. On the whole, however, planetary noble gases, particularly the heavier
gases, seem to have survived parent body processing remarkably intact, as
judged from compositional fidelity. Larimer & Anders (1967) have argued that
the observed compositional variations are largely primary rather than secondary,
i.e. that they were established by equilibration with the ambient nebula before

Fig. 5.2. Illustration of noble gas abundance patterns (cf. Fig. 5.1). All
the patterns shown here are of the 'planetary' type (cf. Figs. 5.1–5.4).

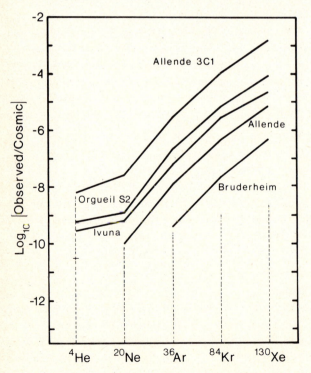

or during accretion rather than in subsequent parent body evolution. Granting this, the general compositional features are still such as to suggest that the same basic physical process, whatever it may have been, is responsible for the planetary trapped gases in the various meteorite classes.

It is observed, not unexpectedly, that some meteorites which contain planetary trapped gas also contain solar gas (cf. Pepin & Signer, 1965). The total noble gas abundance pattern is then, in general, a hybrid resulting from superposition of the two basic patterns, the solar component most visible in the lighter gases and the planetary component most visible in the heavier gases (cf. Fig. 5.3). For ordinary chondrites, such cases typically show the same signs

Fig. 5.3. Noble gas abundance pattern for Mokoia (cf. Figs. 5.1 and 5.2); data from Table 5.2. The shallow dashed line shows expected abundances for observed ^{20}Ne and other gases in nominal 'solar' proportions (Table 5.3). The steeper dashed line shows expected abundances for observed ^{84}Kr and other gases in nominal 'planetary' proportions (Table 5.3); on this scale the 'planetary' ^{36}Ar/^{84}Kr ratio is indistinguishable from the observed ratio. Mokoia is an example of a hybrid pattern: the planetary component dominates the heavier gases Kr and Xe, the solar component dominates the lighter gases He and Ne, and they make contributions of the same order of magnitude to Ar.

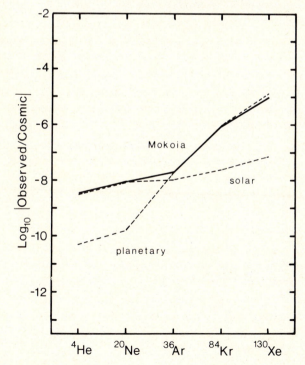

of planetary surface exposure as do the differentiated meteorites and lunar samples devoid of planetary gases, e.g. a dark–light structure. Superposed solar gases also occur in carbonaceous chondrites as illustrated in Fig. 5.3 (Mokoia is an extreme case; more typically in carbonaceous chondrites the relative proportion of solar gas is lower and the Ar is inferred to be mostly planetary rather than solar). In some cases the solar wind exposure presumably may also occur on a planetary surface, but there is less evidence to go on. Jeffery & Anders (1970) find that in Orgueil the gases in silicates carry a planetary signature but that the gases in magnetite are solar, and Mazor *et al.* (1970) suggest that at least in this case solar wind irradiation occurred before accretion.

The planetary elemental abundance pattern (Fig. 5.2) seems to have a kink at Ne, i.e. more He than an extrapolation of the pattern might suggest. More specifically, it is an interesting and rather puzzling feature of planetary gas that its ^{4}He/^{20}Ne ratio is essentially the same as the ^{4}He/^{20}Ne ratio in solar gas, or at least that the systematic difference between the two varieties is smaller than the range of variation within each. It might be suggested on these grounds that He and Ne are very low or absent in planetary gas, and that the observed He and Ne are really solar, e.g. as in Mokoia (Fig. 5.3), but this is apparently not so. It would be a curious coincidence for independent effects to combine in such faithful proportion as characterizes 'pure' planetary gas in a number of samples (Fig. 5.2). Furthermore, solar and planetary gases have distinct isotopic signatures in He and Ne, and can be resolved on that basis without appeal to elemental ratios. Still, an impression arises that in planetary gas He and Ne are somehow apart from the heavier gases, particularly since the relative proportions of He and Ne on the one hand, and of Ar, Kr, and Xe on the other, are less variable than the proportions of the two sets, so that the major compositional variations of planetary gas occur in the ^{20}Ne/^{36}Ar ratio. (This should not be confused with the second major reason for Ne/Ar variation. The observations of Mazor *et al.* (1970) referred to above are for *planetary* gas, and the class-dependent ^{20}Ne/^{36}Ar ratios range mostly *downwards* from the nominal planetary value in Table 5.3. Superposition of solar gas is an independent effect and will drive the ^{20}Ne/^{36}Ar ratio *upwards.*)

Although most of the detailed descriptions of planetary gases are based on carbonaceous chondrites, the same generalizations apply to the more numerous ordinary chondrites as well, as was first clearly evident in the comprehensive general survey by Zähringer (1968). The less intensely metamorphosed ordinary chondrites frequently have noble gas concentrations of the same order of magnitude as carbonaceous chondrites, with the same compositional features, i.e. planetary composition, with some apparent cases of solar gas contributions to Ne (e.g. Dimmitt, Fig. 5.4). The more metamorphosed ordinary chondrites (e.g. Bruderheim, Estacado) contain lower concentrations of trapped gases, but in the same planetary proportions (Figs 5.2 and 5.4). Strictly speaking, this latter generalization applies only to the heavier gases, Ar, Kr, and Xe, since the anticipated planetary He and Ne are usually submerged in the *in situ* components. The matter of trapped He and Ne in such cases is an interesting topic, since most of the compositional structure observed in carbonaceous chondrites relates to the He and Ne abundances, and information about the He and Ne trapped in different planetary-scale regions might prove useful in illuminating the origin of planetary gas. This subject is also interesting as an analog for noble gases in the terrestrial planets (Section 12.10) which apparently have (originally) trapped gases at abundance levels which would be unobservable in meteorites (as noted in Section 6.3, spallation gases in the earth are at least four orders of magnitude less abundant than in typical meteorites).

Some progress on this problem has been made through selective chemical attack experiments on meteorites (Lewis *et al.*, 1975). Attack by HCl and HF leaves a residue (typically about 1%) rich in trapped gases. Oxidative attack on such residues or fractions thereof removes most of the trapped gases. The noble

Fig. 5.4. Noble gas abundance patterns in ordinary chondrites; cf. Figs. 5.1–5.3.

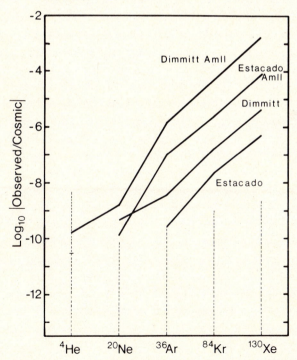

gas carrier, denoted Q (Aristotelian quintessence), has not been isolated and is not very well characterized other than by high noble gas content, resistance to HCl and HF, and lack of resistance to oxidizing agents; otherwise its identity is a matter of dispute, and it may be more than one phase. It is a minor constituent of the bulk meteorite ($<10^{-3}$), has greatly enhanced noble gas contents (at least 10^2 to 10^3 times the bulk), frequently accounts for a significant ($>10\%$) and often for a major ($>50\%$) fraction of total trapped gases; with allowance for possible mechanical losses or partial attack in the initial chemical treatment, it may account for nearly all the trapped gas. Concentration of gases in acid-resistant residues (Allende in Fig. 5.2, Dimmitt and Estacado in Fig. 5.4) is characteristic of carbonaceous chondrites (Lewis *et al.*, 1975; Reynolds *et al.*, 1978), and of ordinary chondrites as well (Alaerts *et al.*, 1977; Moniot, 1980).

Since *in situ* (and solar component) gases are distributed more or less throughout the major phases, the concentration of planetary trapped gases in a minor fraction allows their study with considerably less interference. Trapped Ne can be so observed, for example (Fig. 5.4), in samples in which it would otherwise be overwhelmed by spallation Ne. In general, the noble gases, at least the heavy ones, occur in the same proportions in Q as in the bulk meteorite, i.e. in planetary ratios. Particularly in the limited data yet available for the ordinary chondrites, there is a noticeable tendency for Ne to be less abundant than would be expected on this basis, however. Ne also shows a tendency to be less completely removed by oxidative treatment. It may be that Ne (and He) are preferentially lost or redistributed in processes occurring in the parent body or in the laboratory. This behavior of Ne is also suggestive of the $^{20}Ne/^{36}Ar$ variations observed in bulk meteorites (Mazor *et al.*, 1970), however, so it may be that this reflects systematic compositional variation in planetary trapped gases according to local conditions prior to accretion. Alaerts *et al.* (1977) argue for this case on the basis of correlation of $^{20}Ne/^{36}Ar$ ratio with ^{36}Ar abundance in their data for LL chondrites, while Moniot (1980) finds scant support for their thesis in this data for H chondrites.

In closing this section, a few remarks on nomenclature are in order. The terms 'unfractionated' and 'fractionated', applied to noble gas elemental composition, are often used synonymously with 'solar' and 'planetary', respectively. In their general connotations of reflecting unbiased or biased sampling of a reservoir, use of these terms causes no problems. (That 'solar' is not really an unbiased sampling of the sun's composition is forgiveable in view of the much greater bias in the planetary component.) These terms sometimes suggest a more specific connotation of mass-dependent fractionation, however, which was in fact one of the first candidate mechanisms invoked to produce planetary gas. In this latter connotation this usage is presumptive and probably (but not assuredly) incorrect, since in most current models other properties are invoked, e.g. solubility, sorbability or ionization potential (Section 5.7); these properties also reflect the identity of the gas and vary progressively with atomic mass, but not as a predictable function of mass, and would lead to different predictions, say, for isotopic fractionation. The term 'gas-rich' is usually used in the special sense of rich in solar-type gas, but unfortunately not always, and is sometimes used in a more general sense, e.g. to denote carbonaceous chondrites rich in planetary gas. The terms 'cosmic' and 'solar', as used in this chapter, are both misnomers, each coined in the expectation that they were literally appropriate, but later found to be over-reached. The adjective 'solar' is particularly suited to create confusion, since depending on context it may designate the composition of the sun or a composition relatively close to but significantly and systematically different from that of the sun (Fig. 5.1). Both terms are quite entrenched, however. The term 'planetary' was coined in reflection of the similarity of planetary composition to that of the earth (Fig. 5.5); this too may be overly presumptive, since it is not assured that the earth has planetary noble gases (Section 5.3). Finally, in this chapter and generally (but not quite universally) in the literature, 'trapped' has the specific meaning 'not produced *in situ*' (cf. Section 1.4); in many instances in the literature the term 'primordial' is used with the same meaning. In other contexts, and as used in this book, 'primordial' has a more specific designation, 'present at the time of formation of the solar system' (in some publications, 'primeval' is used to convey this meaning, but the term never caught on). It appears that in most cases trapped noble gases are indeed primordial, but there is at least one important exception: in solar-type gas, ^{3}He is certainly trapped but probably not primordial (Section 5.4).

Finally, the meaning of the terms solar and planetary has undergone change. Originally, these were essentially labels for empirically distinguishable elemental abundance patterns. As more data, both elemental and isotopic, were obtained and subjected to close scrutiny, the usage of the terms has shifted, and presently bears more the flavor of a genetic classification than a phenomenological one. In this sense, solar gas is that which comes from the sun, as distinct from the early solar nebula, and exists in planetary materials because it was embedded in them by virtue of kinetic energy; planetary gas is that which is not solar, and was acquired from the early solar nebula by process or processes yet unknown. Each type is then readily allowed to be a family rather than a monolithic component which may be sampled to varying degrees but without compositional alteration. From this viewpoint it is then a matter of observation that in spite of variations within each family, the differences between the two are sufficiently large and persistent as to be diagnostic of origin.

5.3   Planetary atmospheres

The terrestrial planets, like meteorites, are samples of the 'condensable' elements, and there has been considerable interest in the extent to which meteorite data can be used to predict whole-planet elemental compositions. The analogy between primitive meteorites and planets is by no means exact, and it is clear that the terrestrial planets are not simply chondrites aggregated into bodies large enough to undergo internal differentiation processes. The analogy is nevertheless useful and a good starting point, and is presumably valid in the generalization that the terrestrial planets' compositions are determined by the same processes that determine meteoritic compositions. Thus, it is generally expected that planetary compositions can be understood in terms of only a few kinds of components, each consisting of a group of elements which were not fractionated from each other in the solar nebula prior to planetary accretion. Bulk compositional differences among different classes of meteorites and the larger planetary bodies would then be attributable to accretion of these components in different proportions (e.g. Ganapathy & Anders, 1974; Morgan & Anders, 1980).

It would thus be expected that noble gases in the larger terrestrial planets, as in meteorites, would be a superposition of the planetary and solar gases discussed in Section 5.2. It might further be expected that the solar gas contribution would be low or negligible: solar wind accretion is not now significant and would hardly be expected to be important on a planetary scale for a planet with low surface-to-volume ratio (unless there were substantial solar wind exposure of planetary materials prior to accretion).

In evaluation of this expectation the relevant parameters are total planetary gas contents. In practice, comparison must be based on atmospheric data: these are the only data available at all for other planets, and even for the earth a rigorous assessment of the total noble gas inventory is not possible (Chapter 12). The presumption, which may be regarded as a hypothesis to be examined, is that the planet is fully degassed, in which case the atmospheric inventory is the total inventory, or that even if incomplete the degassing process has not fractionated elemental composition, e.g. complete degassing of part of the planet.

As seen in Fig. 5.5, the earth's atmosphere indeed has a definite planetary signature, at least as far as the middle three gases are concerned. The relative proportions of Ne, Ar, and Kr are indistinguishable from planetary values, the exact Ne/Ar and Ar/Kr ratios well within ranges observed in meteorites. The absolute magnitudes for the atmospheric abundance of these three gases are also well inside the meteorite range (cf. Fig. 5.2); this in itself, however, is not particularly strong evidence that the earth is fully degassed, since these values could easily be imagined to be moved upwards another order of magnitude or

Fig. 5.5. Noble gas abundance patterns in the atmospheres of the terrestrial planets: atmospheric abundances divided by mass of planet (Table 5.2), normalized to 'cosmic rock' concentrations as in Figs. 5.1–5.4. Two Kr values are shown for Venus, corresponding to different atmospheric concentrations reported for Pioneer and Venera missions; the Xe abundance is an upper limit. The solar and planetary patterns illustrate compositions only: ordinate positions are arbitrary. Data from Tables 5.2 and 5.3.

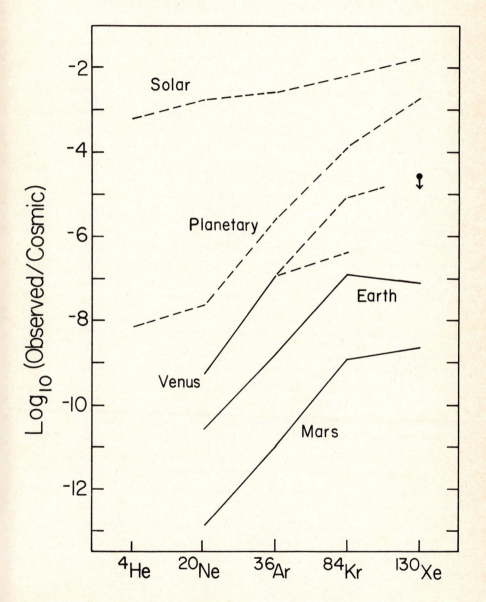

two without stretching the meteorite analogy. On the grounds of the elemental abundances, there is no reason to invoke any addition of solar gas, but the elemental abundance constraints are not so tight as to make a strong argument against a modest or even substantial solar component contribution to Ne (cf. Section 6.5).

It is unfortunately not possible to extend the comparison to He. The atmospheric abundance of He is, to be sure, quite low: specifically, the ^{4}He abundance is $10^{-13.9}$ times the cosmic concentration in Table 5.2. This is essentially irrelevant, however, since He escapes from the atmosphere and atmospheric ^{4}He is mostly radiogenic anyway (Section 12.5). ^{3}He is no more useful in this sense. It is in fact not possible to make quantitative assessments for initial He abundances *except* by the meteorite analogy.

The comparison *can* be extended to Xe, however, and it is evident (Fig. 5.5) that Xe is way out of line. Specifically, on the basis of the Kr/Xe ratio, which is the compositional ratio showing the least variation in meteorites, the atmosphere is deficient in Xe by a factor of 23 from what would be expected for planetary gas. This cannot be accommodated within the range of meteoritic variation. It has often been pointed out that the atmospheric Xe/Kr ratio is closer to the solar ratio than to the planetary ratio. This is true but hardly seems relevant. The ratio is in fact significantly lower than the solar ratio, and is even lower than the cosmic ratio. Neither observation is pertinent unless it is extended to the hypothesis that the earth acquired its noble gases from such reservoirs; in this case the anomaly would be up to two orders of magnitude depletion of Ar and Ne coincidentally ending up in planetary proportions.

The low atmospheric abundance of Xe constitutes a problem and has attracted attention only because of the meteorite analogy, the case for which has been made above. It is only an analogy, but it is a strong one, since there is no analog for such a low Xe/Kr ratio and since it works so well for Kr, Ar, and Ne (with a legitimate excuse for He). There are three possible resolutions to this problem. One is simply to abandon the meteorite analogy, and stipulate that whatever process was responsible for meteoritic gases, a different one must be invoked for the earth. The second is to assume that while atmospheric gases are not planetary, total terrestrial gases are, i.e. that Xe, preferentially, has not been degassed from the solid earth. The third resolution is to assume that terrestrial gases are planetary, and that they have been outgassed entirely or without compositional bias, so that atmospheric gases are also planetary, but that the Xe is mostly in some atmospheric reservoir other than air (cf. Section 12.6). This proposal has been advocated by Fanale & Cannon (1971a), and seems to have gained the widest acceptance. Fanale & Cannon suggested shales, but the idea is readily generalized, and other kinds of sediment, ice, or subduction of atmospheric Xe in oceanic sediments serve the same purpose. Podosek *et al.* (1980), however, have argued that available data for shales offer scant support for this hypothesis.

Isotopic comparisons between terrestrial and extraterrestrial gases (Chapter 6) also bear on this issue. The pertinent points can be summarized briefly: He is again irrelevant, Ne is ambiguous, terrestrial Ar is indistinguishable from planetary Ar, terrestrial Kr is only slightly different from planetary Kr, and terrestrial Xe is strikingly different from planetary (or solar) Xe in a fashion without analog in extraterrestrial samples or known explanation in terrestrial evolution.

Atmospheric noble gas abundances for Mars and Venus (Section 12.10) also fit the meteorite analog quite well, and thereby strengthen it substantially. It is noteworthy that the overall abundances on Venus are significantly higher than on earth and those of Mars significantly lower. These results have attracted considerable interest (e.g. Anders & Owen, 1977; Pollack & Black, 1979), but it is not clear whether these differences reflect total planetary abundances or different degrees of degassing nor how such differences may be related to the mechanism of trapping of planetary gases. The moon has not accumulated a permanent atmosphere and appears to be extremely deficient in volatiles in general: all the noble gases observed in lunar samples can apparently be interpreted in terms of solar wind irradiation, nuclear effects (radioactive decay and cosmic-ray spallation), or terrestrial contamination, and no indigenous lunar primordial noble gases have been unambiguously identified.

The Xe abundance for Mars has attracted particular interest. Within experimental uncertainties the Xe/Kr ratio for Mars apparently might be the planetary value, but the suggestion is clearly that it is more like the terrestrial air ratio. If so, the same alternative interpretations exist for Mars as for the earth. Fanale *et al.* (1978) favor a planetary composition for the Martian atmosphere, with most of the Xe adsorbed on the regolith. The comparison also supports the alternative view, however, that noble gases on earth and Mars are more akin to each other than either is to meteoritic planetary gases (Bernatowicz & Podosek, 1978; Podosek *et al.*, 1981).

5.4   Isotopic heterogeneity

As noted earlier, an important concept in the study of the chemical history of the solar system is that of its primordial composition: a single universal composition from which all parts of the solar system have been derived. The strongest evidence that a primordial composition existed in the first place is isotopic (cf. Reynolds, 1967). In comparison with chemical variation, processes capable of generating isotopic variation are relatively few, simple, and amenable

to quantitative evaluation. A wealth of data supporting isotopic uniformity –
the same isotopic composition of all elements in all accessible samples of the
solar system – led to the view that the contributions of what must have been
a large number of nucleosynthetic processes were thoroughly homogenized
before the formation of the sun and planetary objects now available for analysis.
In this view known isotopic variations were those plausibly attributable to
processes acting within the solar system to modify this once-uniform composition: mass-dependent isotopic fractionation, decay of radionuclides included
in primordial material, and nuclear reactions such as occur in the sun or which
are induced in planetary materials by cosmic rays or energetic particles emitted
in radioactive decay. In most geochemical usage the term 'isotopic anomaly' is
reserved for an isotopic variation which could not be understood in terms of
these processes. The dogma of initial uniformity thus asserted that anomalies
were only apparent and not real: either a previously unrecognized solar system
process, one of those above, was involved, or experimental error was involved
and the effect would vanish with more careful work.

This approach was apparently successful for quite some time but in the past
decade it has become untenable. Real isotopic anomalies exist, i.e. the various
nucleosynthetic components of which the solar system is made were never
completely homogenized into a uniform primordial composition. In an absolute
sense the anomalies so far known are minor: small variations in few elements
and/or few samples. In particular, it does not appear that heterogeneities were
large enough to change bulk chemistry, and so the concept of primordial composition retains its utility as the starting point for chemical evolution. The very
existence of isotopic anomalies, however, has led to significant changes in models
for the formation of the solar system, and promises a wealth of information
regarding its origin and evolution. The study of isotopic anomalies has grown
into a complex field to which intense research activity is dedicated, and our
discussion is intended to be an outline appropriate as background for noble
gas geochemistry rather than a comprehensive review. A number of reviews of
broader scope are available, e.g. Podosek (1978); Clayton (1978); Begemann
(1980); Wasserburg *et al.* (1980).

An important class of samples in the study of isotopic structures is the
so-called 'refractory inclusions'. These are fragments, up to centimeter size but
typically of millimeter size, often found in carbonaceous chondrite meteorites.
The defining feature of these inclusions is that their chemistry is dominated by
refractory elements (Ca, Al, Ti, and characteristic suites of both lithophile and
siderophile trace elements), specifically those elements more refractory than
Si, Mg, and Fe. At appropriate temperature and pressure and in bulk chemistry of
primordial composition, these refractory elements would exist as solids (typically

as oxides or oxide-assemblage minerals, with minor metal) while most of the Si, Mg, and Fe were gaseous. In general terms, this presumably accounts for the origin of the inclusions.

The first effect generally accepted as an isotopic anomaly was found in O in these refractory inclusions (Clayton *et al.*, 1973). For these inclusions in general, different minerals characteristically have different O compositions, and span a range of some $40\%_{00}$ (Fig. 5.6). These variations are nonlinear in isotopic mass, i.e. cannot be regarded as fractionation of a single primordial composition, and must be regarded as representing admixture of two distinct components, specifically components differing primarily in the proportion of ^{16}O relative to ^{17}O and ^{18}O. Since O is an abundant element for which plausible nuclear processes which could generate the variations within the solar system are lacking, these results are interpreted as a true isotopic anomaly, i.e. nonhomogenization of distinct presolar nucleosynthetic components. As far as is now known, O is the only element which displays isotopic heterogeneity on such a microscopic scale,

Fig. 5.6. Oxygen isotope compositions in various classes of meteorites and the earth, displayed as per mil variations from the SMOW standard. The indicated extent of the lines correspond to actual data ranges. The solid lines (slope 1/2) are fractionation trajectories, e.g. all terrestrial compositions plot on the indicated line. The various fields cannot all be connected by fractionation trajectories and thus cannot have been derived (by fractionation) from a common source, i.e. the oxygen isotopes were not homogenously distributed throughout the solar system. The dashed line (slope approximately unity) is a mixing line populated by various minerals in Ca–Al-rich (anhydrous) inclusions in C2 and C3 meteorites, and indicates that the primordial inhomogeneity is preserved on a microscopic scale. Reproduced from Podosek (1978).

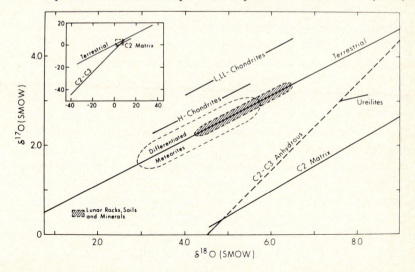

within an individual inclusion. All other elements are not only uniform within the inclusions but also isotopically normal, with but one presently known exception: Ti compositions are anomalous and variable from one inclusion to another and are also interpreted as true isotopic anomalies (cf. Niederer *et al.*, 1980).

A small subset of refractory inclusions is chemically and mineralogically normal but isotopically bizarre. While most inclusions display essentially the same O anomaly pattern and can be accounted for by only two distinct nucleosynthetic components, in a few cases one must postulate additional O components or severe fractionation of one of those found in ordinary inclusions. The latter interpretation is usually preferred because these same inclusions exhibit strikingly large apparent fractionations in elements for which only negligible natural fractionation had previously been observed (Si, Ca, Mg). Elements other than O are uniform within an inclusion but among elements so far examined isotopic anomalies seem by far the rule rather than the exception: the list includes Mg, Si, Ca, Ti, Ba, and Nd and is expected to lengthen. This small group of inclusions is usually designated FUN samples (Fractionation and Unknown Nuclear effects). The anomalies and fractionations differ from one sample to another, and no satisfying explanation has been given for the pattern or origin of either the anomalies, the fractionations, or their joint occurrence.

There are other samples, besides the refractory inclusions, in which anomalies exist on such a microscopic scale. Other than volatiles (see below), however, the only presently known case is again O, specifically O in unequilibrated ordinary chondrites (Gooding *et al.*, 1980).

In addition to the microscopic-scale anomalies described above, there exist anomalies on a much larger scale, and again the prominent element is O. Except as above, all specimens of the earth, moon, and various classes of meteorites have O compositions which are uniform (except for fractionation) within the class but anomalously different from one class to another (Fig. 5.6) typically on the scale of a few per mil (Clayton *et al.*, 1976). This indicates that the terrestrial planet parent bodies bulk compositions characteristically incorporate different proportions of the O components found in the refractory inclusions, and that the solar system was isotopically heterogeneous on a planetary scale as well as a microscopic scale.

For a variety of reasons it is rather difficult to assess and characterize possible isotopic heterogeneity among the volatiles – H, C, N, and the noble gases. In general, there are isotopic variations in these elements, in some cases quite large, on both the microscopic scale and the planetary scale. For H, C, and N the problem is distinguishing possible isotopic heterogeneity from mass-dependent isotopic fractionation. These three are, for the terrestrial planets, trace elements of unusually complex chemistry, and there are many possibilities for substantial fractionation; since they are volatiles, fractionation may also arise in escape from a planetary body. Each of these elements also has only two isotopes, so it is not possible to apply the test of linear mass dependence to distinguish between fractionation and specific nuclear effects. To varying degrees similar remarks apply to the noble gases but the complexity is greatly compounded by the possibility of nuclear effects within the solar system. It is, unfortunately, not possible to make a comparison between isotopic effects in the volatiles and those in the refractory inclusions, since these inclusions are quite poor in volatiles and any that are observed in the inclusions are generally considered to arise in secondary alteration or contamination.

Isotopic structures in noble gases will be discussed in the remainder of this chapter and in Chapter 6, but it is appropriate to make some further points here. It is noteworthy that the noble gases have a considerably longer and more extensive history of anomalous behavior than do the other elements. Xe was the first known case of a planetary scale inhomogeneity, i.e. a meteoritic isotopic composition significantly different from terrestrial composition in a fashion not immediately attributable to normal processes operating within the solar system (Reynolds, 1960; Krummenacher *et al.*, 1962). Xe was also the first known case of a microscopic-scale inhomogeneity, as exhibited by isotopic variations in stepwise heating of a carbonaceous chondrite (Reynolds & Turner, 1964). Isotopic structure complications quickly proliferated, but were not recognized as isotopic anomalies in the sense used here. This was quite justifiable; potential mass-dependent fractionations are common for the noble gases and, much more important, their scarcity makes them extremely susceptible to isotopic modification by nuclear processes acting within the solar system (Section 1.2). Explanations for isotopic variations in the noble gases were accordingly sought in those terms, sometimes successfully and sometimes not. Even when isotopic inhomogeneity was explicitly invoked to account for noble gas observations (e.g. Black, 1972b), little support for this idea was forthcoming. It was not until after observation of O data which were unambiguously incompatible with primordial homogeneity (Clayton *et al.*, 1973) that inhomogeneities in the noble gases could also be accepted widely. Even now it is clear that a number of noble gas isotopic effects reflect primordial inhomogeneity but it is still very difficult to resolve these from fractionation and nuclear effects arising within the solar system. At present, noble gases are generally acknowledged to have the most widespread and extensive isotopic anomalies known; it is worth noting that there is no apparent reason for this, specifically that the traditional explanation for scarcity as the root cause of isotopic complication does not apply in any immediately obvious way. Indeed, it would be expected that the tendency of noble gases to partition into the gas phase would make them among the most readily homogenized elements. That this is not the case is presumably closely linked to the trapping mechanisms by which noble gases are incorporated in planetary materials, and this is a subject of active current research but for which no useful generalizations are yet available.

5.5   Light gas isotopery

The light gases He, Ne, and Ar have sufficient features in common that it is convenient to discuss them as a group. A number of specific compositions referred to in the text are summarized in Table 5.4; in discussions of Ne compositions it is also convenient to refer to the three-isotope diagram in Fig. 5.7. The general approach in this section will be to describe first the major observed differences between *in situ*, solar, and planetary components and then to examine in more detail the structures observed within the trapped components.

In meteorites the only *in situ* Ne component is spallation. Its composition is usually only modestly variable, within about 10% (but also see Smith *et al.*, 1977), according to target chemistry and shielding conditions, and for most purposes its production rate can be considered well known. Qualitatively, the principal compositional feature of spallation Ne is a near equality of all three isotopes; this contrasts strongly with the composition of any known trapped component, especially with the scarcity of ^{21}Ne. Spallation Ne is frequently an important component in extraterrestrial samples, and it is particularly noteworthy that ^{21}Ne is the noble gas isotope which is usually the most sensitive to addition of a spallation component. It is thus easy to identify spallation ^{21}Ne and use it as the basis for an exposure age or an index for estimating the spallation contribution to some other isotope. Conversely, it is generally difficult to resolve trapped ^{21}Ne, since some other independent measure of exposure is required to make a correction for spallation ^{21}Ne. Rarely, short exposure ages can be determined independently by undersaturation of spallation radionuclides such as ^{26}Al (0.73 Ma). Another technique is gas measurement in two or more parts of the same rock and assumption of similar exposure for these parts, e.g. light and dark lithologies in a breccia, or bulk and selective dissolution residues; in such cases spallation-dominated ^{21}Ne in a gas-poor part serves as the basis for correction of spallation ^{21}Ne and thus estimation of trapped ^{21}Ne in a part richer in gases.

If there were only a single primordial Ne composition in the solar system, all Ne analyses would define a single linear correlation in Fig. 5.7, reflecting mixture of a trapped component with spallation Ne. Even the earliest data for Ne in meteorites showed this not to be the case: instead, data populated a finite field, roughly the triangular region outline in Fig. 5.7a. Thus, trapped Ne in meteorites

Table 5.4. *Isotopic compositional data for light noble gas components*

Component	$^3\mathrm{He}/^4\mathrm{He}$ ($\times 10^{-4}$)	$^{20}\mathrm{Ne}/^{22}\mathrm{Ne}$	$^{21}\mathrm{Ne}/^{22}\mathrm{Ne}$	$^{36}\mathrm{Ar}/^{38}\mathrm{Ar}$	References
'Solar' trapped					
Solar wind[a]	4.0 ±0.4	13.6 ±0.3	0.032 ±0.004	–	Geiss et al. (1972)
Lunar soil[a,b]	3.7 ±0.1	12.9 ±0.1	0.0313 ±0.0004	5.33 ±0.03	Eberhardt et al. (1972)
Component B[a]	3.9 ±0.3	12.52 ±0.18	0.0335 ±0.0015	5.37 ±0.12	Black (1972a)
Component C	4.1 ±0.8	10.6 ±0.3	0.042 ±0.003	4.1 ±0.8	Black (1972a)
Component D	1.5 ±1.0	14.5 ±1.0	–	6 ±1	Black (1972a)
'Planetary' trapped					
Component A[a,e]	1.43[c] ±0.20	8.2 ±0.4	0.024 ±0.003	5.31[c] ±0.05	Black & Pepin (1969)
Neon-E	–	<0.01	<0.0001	–	Meier et al. (1980)
Earth	>0.5	9.80 ±0.08	≤0.0290 ±0.0002	5.320 ±0.013	Table 2.1, Sections 6.4–6.6
In situ					
Spallation[d]	2000	0.85	0.92	0.67	–

a Corresponding elemental composition given in Table 5.3.
b Surface-correlated composition for 12001 ilmenite.
c He value is that deemed 'planetary' by Reynolds et al. (1978). Ar value is for total Ar in Murray WIIA-1, on which He evaluation based; experimental uncertainty is relatively large, but not larger than interpretational uncertainty in assignment of a planetary Ar composition.
d Representative values; for a discussion of range, see Smith et al. (1977).
e Alaerts et al. (1980) propose the existence of components A1 ($^{20}\mathrm{Ne}/^{22}\mathrm{Ne} = 8.70 \pm 0.11$, $^{21}\mathrm{Ne}/^{22}\mathrm{Ne} = 0.024 \pm 0.001$), and A2 ($^{20}\mathrm{Ne}/^{22}\mathrm{Ne} = 8.54 \pm 0.08$, $^{21}\mathrm{Ne}/^{22}\mathrm{Ne} = 0.035 \pm 0.001$); see text for discussion.

was found to be variable, spanning approximately $8 \leqslant {}^{20}Ne/{}^{22}Ne \leqslant 13$. Initially, the data were interpreted in terms of an 'unfractionated' primordial ${}^{20}Ne/{}^{22}Ne \approx 13$ and a heavier trapped Ne in meteorites derived from it by isotopic fractionation, e.g. in diffusion or escape from a gravitational field. Difficulties encountered by this hypothesis were the very great degree of loss needed to produce so large an isotopic effect and the absence of corresponding effects in other elements. Pepin (1967) argued that stepwise heating analyses exhibited trends corresponding to two preferred compositions or distinct components, designated A and B to avoid genetic implications, and thus that the meteorite data in general were best interpreted in terms of mixing. Accumulating evidence has strengthened the evidence and the distinct component mixing hypothesis is generally accepted: fractionation may be significant in specific instances but cannot be invoked to account for the major trapped compositional variations observed.

Neon B is characteristically found in samples with solar elemental abundance ratios, and it is now generally recognized as solar Ne. Its specific composition can be determined by intersection of various trend lines and also by applying relatively small spallation corrections to the gas-rich dark fractions of breccias on the basis of spallation Ne gas-poor light fractions.

Neon A composition is characteristic of carbonaceous chondrites which display elemental abundances in the planetary pattern, and 'Neon A' is now synonomous with 'planetary Ne'. In contrast to Neon B, its specific composition cannot be determined explicitly. The meteorite data determine a mixing line AS (Fig. 5.7) but in the absence of any independent estimate of spallation ${}^{21}Ne$ the end-member composition A is constrained only to lie to the left of observed data. The usual value for Neon A is calculated with the assumption that terrestrial atmospheric Ne is also a mixture of Neon A and Neon B, so that Neon A is the intersection of the meteorite trend line AS with the line connecting air and Neon B. Other interpretations are possible (see below and Section 6.5).

Resolution of trapped Ne into A and B components is now a common means of identifying modest additions of solar to planetary gas, since this isotopic criterion is more sensitive and definitive than elemental ratios. It should be noted, however, that this calculation reflects an assumption, not a rigorous demonstration. As discussed below, both solar and planetary components appear complex on closer examination, so such a formal resolution is best regarded as only approximate and suggestive rather than conclusive.

The model for distinct A and B composition for Ne could be valid without specifying the origin of these components, but it would also be incomplete. Given that Neon B apparently represents solar or primordial Ne, a separate account must be made for Neon A. A fractionation origin is still possible but encounters the same problems as the continuum fractionation viewpoint. The

Fig. 5.7. Display of prominent Ne compositions (data from Table 5.4). Fig. 5.7B illustrates tielines to spallation and Neon E compositions (shown on Fig. 5.7A); tickmarks indicate effect of adding 1% and 2% (of ^{22}Ne) spallation to Neon B and Neon A. Dashed line is a fractionation trajectory.

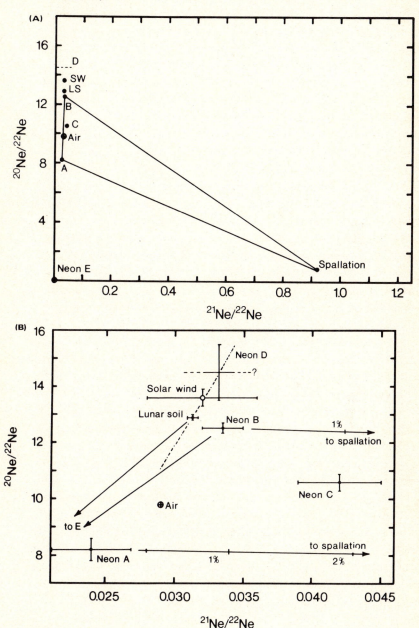

matter is far from resolved, but the present consensus is that Neon A is an example of isotopic heterogeneity (see below) in the sense of Section 5.4.

For He there are two prominent *in situ* components: α decay of U and Th and spallation, characterized by very low and relatively very high $^3He/^4He$ ratios, respectively. Together they often dominate any trapped He, typically the radiogenic component accounting for most of the ^4He and spallation most of the ^3He. The production of each component can be predicted, ^4He from U-Th abundance and ^3He from spallation ^{21}Ne, but He has a well-known tendency to be lost by diffusion, so accurate prediction of *in situ* He abundances is difficult. Assessment of trapped He is thus restricted to relatively few cases of very high trapped gas contents.

Nevertheless, there are enough cases in which solar gas contents are sufficiently high that trapped He dominates *in situ* He. By analogy with Ne, solar He is usually designated Helium B. Identification of planetary trapped He is more difficult, since the $^4He/^{20}Ne$ ratio is similar in both solar and planetary gas, so that even when trapped ^3He is identified the distinction between solar and planetary He must be made on the basis of Ne isotopic composition. Even so, Anders *et al.* (1970) observed covariation of trapped He and Ne compositions in carbonaceous chondrites and identified a planetary He composition distinct from that of solar He (also cf. Black, 1970). More recently, Reynolds *et al.* (1978) clearly identified planetary He concentrated in a selective dissolution residue. Again by analogy with Ne, planetary He is usually designated Helium A.

Planetary and solar He differ substantially in composition (Table 5.4). As with Ne, an explanation might be sought in fractionation or isotopic heterogeneity, but a satisfactory explanation invoking neither is available in terms of nuclear reactions in the sun rather than in gas-poor planetary material, so that He is an interesting and possibly unique case in which primordial composition is better preserved in planetary material than in the sun. ^3He in the sun is enriched by deuterium consumption, $^2D(p, \gamma)^3He$, and then exported to planetary material in the solar wind. If all the deuterium in the solar wind source region has now been converted to ^3He, the difference between Helium A and Helium B corresponds to an initial ratio $D/^4He = 2.5 \times 10^{-4}$; for $He/H = 0.1$, this indicates $D/H = 2.5 \times 10^{-5}$, comparable to other estimates (Reeves, 1974). D/H in terrestrial materials is significantly higher ($\sim 1.6 \times 10^{-4}$), presumably because of fractionation. The difference between Helium A and Helium B is significant on a very large scale, since both D and ^3He are believed produced in big-bang rather than stellar nucleosynthesis and their abundances are important parameters in cosmogonic theory (cf. Reeves, 1974).

The only *in situ* component for ^{36}Ar and ^{38}Ar is spallation, which usually makes a proportionally smaller contribution to Ar than to He and Ne (cf. Figs. 5.1 and 5.2) so that trapped Ar can be studied in a wider range of samples. Mazor *et al.* (1970) and Black (1971) observed small variations in trapped ^{38}Ar/^{36}Ar which appeared to correlate with trapped ^{20}Ne/^{22}Ne, but the variations are not easy to interpret in terms of mixing solar and planetary gas and their reality can be challenged (Smith *et al.*, 1977). It is usually considered that solar and planetary gas have the same ^{38}Ar/^{36}Ar ratio. If any difference exists, it is small and not reliably correlated with other parameters.

Radiogenic ^{40}Ar is ubiquitous and effectively precludes observation of primordial ^{40}Ar in any available samples of solar system materials. ^{40}Ar is apparently produced only in very low abundance in stellar nucleosynthesis; Cameron (1973) estimates a primordial ratio ^{40}Ar/^{36}Ar $= 2 \times 10^{-4}$, and values as low as ^{40}Ar/^{36}Ar $= 1.4 \times 10^{-3}$ are observed (Begemann *et al.*, 1976). More generally, radiogenic ^{40}Ar overwhelms any primordial ^{40}Ar. Even surface-correlated noble gases in lunar soils and breccias, mostly trapped solar wind, contain significant ^{40}Ar (^{40}Ar/^{36}Ar of order unity, but variable) which originates as radiogenic ^{40}Ar in the lunar interior, is released into the transient lunar atmosphere and is evidently ionized and electromagnetically implanted in lunar soils.

On closer examination it is evident that there is compositional variation within solar Ne. Even two rather direct measurements of solar composition, accumulation in lunar soil and collection on foil during the Apollo missions, do not coincide, and both are different from the formal composition of Neon B (Table 5.4 and Fig. 5.7). It may be that Neon B is slightly undercorrected for spallation, but in any case such variations are small and quite plausibly attributable to fractionation, either in differential retention effects in solar wind trapping or in the solar wind itself because of mass-dependent acceleration. Variations in solar ^{3}He/^{4}He are also observed; origin by fractionation is plausible for He as well, of course, and Eberhardt *et al.*, (1972) further suggest a possible secular trend in ^{3}He/^{4}He which may reflect progressive deuterium burning in the solar wind source region.

It is not clear to what extent variations in solar compositions reflect a continuum or a mixture of discrete components. Black & Pepin (1969) and Black (1972a) argued that compositional variations in solar Ne should be interpreted in terms of a distinct component, Neon C; this composition is notably rich in ^{21}Ne and cannot be produced by fractionation of other solar compositions. Black (1972a) suggested that Neon C was the direct accumulation of solar flare Ne and further suggested that Neon B might be a mixture of Neon C and solar wind Ne. These suggestions remain speculative and even the existence of such a component is questioned; Smith *et al.*, (1978), for example, suggest an alternative interpretation in terms of solar flare *spallation* Ne, i.e. an *in situ* nuclear component rather than a trapped component. Black (1970, 1972a) also found evidence for very light (high $^{20}Ne/^{22}Ne$) solar Ne which can be interpreted in terms of two additional components, a rare (unnamed) component suggested to be an 'atmospheric' reimplanted Ne on a meteorite parent body, analogous to reimplanted lunar ^{40}Ar, and Neon D, which he suggested represents a very early solar wind, lighter than modern solar wind because the latter is affected by progressive fractionation (Rayleigh distillation) in the solar wind source. These identifications have been questioned, however, and it remains unclear whether identification of distinct components is warranted and indeed whether any solar Ne lighter than modern solar wind need be invoked at all (cf. Hohenberg *et al.*, 1970; Smith *et al.*, 1978).

Black (1970; 1971; 1972a) also reports identification of C and D components for He and Ar corresponding to the Ne components. Such correlations are difficult to make since different gases must be compared in terms of bulk analyses rather than stepwise heating and the evidence for the existence of such components is often considered marginal. Reported uncertainties for these compositions are in any case large; the principal compositional feature of note is that Helium D, if it exists at all, is quite low in ^3He, presumably reflecting solar composition before modification by deuterium burning.

Planetary Ne also contains isotopic structures, and these have attracted substantial interest. Even the meteorites whose bulk compositions defined the correlation line AS in Fig. 5.7, and thus indicated the existence of Neon A, exhibit stepwise heating compositions not confined to AS. Excursions above AS could be attributed to solar components; excursions below AS require another component, designated Neon E, whose principal feature is high ^{22}Ne (Black & Pepin, 1969; Black, 1972b).

As more data were acquired the limits on Neon E composition were continually moved closer to the origin of Fig. 5.7, and with the isolation of carbonaceous chondrite phases rich in Neon E (Eberhardt, 1974; 1978) it has become clear that Neon E is, for all practical purposes, pure ^{22}Ne (Table 5.4). Addition of Neon E would not significantly change the abundance of any isotope other than ^{22}Ne. Furthermore, the ^{22}Ne is sufficiently monisotopic to eliminate from contention any nuclear process for its production other than decay of ^{22}Na (half-life 2.6 yr): the implied scenario is that before decay ^{22}Na is efficiently separated from any co-produced Ne by chemical means, e.g. condensation into dust while Ne remains in gas. The ^{22}Na could be produced within the solar system (cf. Audouze, *et al.*, 1976), but not easily, and the clear consensus of opinion is that the ^{22}Ne is an extrasolar component never homogenized in the solar system, i.e. an isotopic anomaly. As noted earlier, Black, (1972b) suggested this interpretation but it did not gain support until evidence for anomalies in elements other than noble gases was available. (In principle, the ^{22}Ne may still be *in situ*, in which case it should not be termed a trapped component, but semantic purity is here secondary to historical development and Neon E is universally termed trapped.)

Black (1972b) further suggested that Neon A was itself a composite, a mixture of Neon E and a 'normal' solar system composition. Black specifically suggested Neon D as the normal component, and in this view the light planetary gases are in fact mostly solar: an early solar wind irradiation plus Neon E. Actual solar wind implantation rather than trapping by some other process is not essential to this view of isotopic systematics, of course, and it would be very difficult to confirm experimentally.

The considerations described above raise interesting questions about the nature of planetary Ne. It may be that Neon A is little more than a mixture of normal Ne and Neon E in more or less constant bulk proportions (but unmixed and still separable on a microscopic scale), and of little relevance for planetary bodies other than those in which it is principally characterized, i.e. carbonaceous chondrites. On the other hand, compositions similar to nominal Neon A appear in other classes of meteorites and in the selective dissolution residues in which planetary gases are concentrated, and it may be that Neon A is indeed a widespread and well-defined component, perhaps a distinct nucleosynthetic component, or perhaps ultimately a mixture incorporating Neon E, but one sampled extensively by diverse planetary bodies. Alaerts *et al.* (1980) propose the existence of *two* well-defined planetary Ne components: a Neon A2, of unspecified origin and needed for only one meteorite (Murchison), differing from A1 chiefly in ^{21}Ne content (Table 5.4).

In view of the presumed origin of Neon E as the decay product of ^{22}Na there is no reason to expect any other anomalies correlated with it, nor is there any experimental evidence to suggest any such association. Similarly, if Neon A is itself a distinct nucleosynthetic component (rather than a mixture), and thus a separate anomaly, it bears no presently evident relationship to other anomalies. More generally, there are no apparent complexities in He comparable to those in Ne. It should be noted, however, that in a two-isotope element with two known trapped components as different as Helium A and Helium B and in which trapped gas is only rarely observed because of *in situ* component interferences, it is easily possible that anomalies are present but simply not noticed. Similarly, we have already noted that trapped Ar (i.e. ^{36}Ar and ^{38}Ar) in solar and planetary gas are indistinguishable; in nearly all cases, variations in ^{38}Ar/^{36}Ar other than those arising in spallation are nonexistent or small and plausibly attributed to fractionation. There apparently *is* an isotopic anomaly structure in planetary Ar, however (cf. Frick, 1977), but it seems more closely related to those in the heavier gases (next section) than to the lighter gases; in bulk samples anomalous Ar is evidently quite minor quantitatively, and appears not to account for any observable variation in planetary Ar composition.

5.6   Heavy gas isotopery

The heavier gases Kr and Xe differ from the lighter in a variety of ways, notably in terms of numbers of isotopes, numbers of important components, and degree of complexity involved in attempting to resolve them. This is particularly true for Xe, and 'xenology' (the term was introduced by Reynolds, 1963) has attracted a large measure of attention and interest in cosmochemistry. The study of Kr has many similar if more subtle aspects, and with some latitude much of the study of Kr can be treated as a branch of xenology.

Our concern in this section does not lie with known *in situ* nuclear components, but it should be appreciated that there are several which are often significant. Spallation contributions are generally proportionally smaller than for the lighter gases, but even so often require qualifications in discussions involving the low abundance light isotopes of both Kr and Xe, which are particularly sensitive to addition of spallation components. Also, cosmic-ray exposures of relevant samples sometimes occur beneath sufficient shielding that secondary neutrons become important, and significant contributions of ^{80}Kr, ^{82}Kr, ^{128}Xe, and ^{131}Xe from capture reactions on ^{79}Br, ^{81}Br, ^{127}I, and ^{130}Te/^{130}Ba, respectively, can greatly complicate the problem of correcting for a 'spallation' component. Radiogenic components are also important. In particular, ^{129}Xe from ^{129}I is nearly ubiquitous in meteorites and is usually of such potentially large and variable magnitude that many discussions of general isotopic structure simply ignore ^{129}Xe. Both Xe and Kr lie near the actinide fission yield peaks and the heavier unshielded isotopes, those for which there is no stable heavier isobar (cf. Fig. 3.1 d,e), terminate the decay chains. Significant effects from spontaneous fission of ^{244}Pu are common in meteorites with low gas contents, and in some cases spontaneous fission of ^{238}U and/or neutron-induced fission of ^{235}U are important as well. Consideration of more general isotopic structures, such as that below, must be based on data in which these effects can be either corrected for or shown to be negligible, and often doubt remains that this has been done properly. As noted earlier (Section 5.4), the large number of known nuclear components helps foster an attitude that an unexplained isotopic effect may be due to still another nuclear component yet unknown.

From the earliest observations it was clear that meteoritic Xe had significantly different composition from terrestrial Xe (Reynolds, 1960). This generalization was strengthened by observation of essentially identical compositions in several carbonaceous chondrites (Krummenacher *et al.*, 1962; Eugster *et al.*, 1967). This composition was designated by the acronym AVCC (AVerage Carbonaceous Chondrite) and, since the meteorites involved had a clear planetary elemental abundance signature, the term is equivalent to 'planetary Xe'. A specific representative composition is that for the carbonaceous chondrite Murray (Table 5.5). As data accumulated and became more precise it became evident that AVCC composition was characteristic of planetary Xe in all classes of primitive meteorites (as well as some whose 'primitiveness' was questionable), not just carbonaceous chondrites. It also became evident, however, that this generalization was valid only in the sense of the distinction between meteoritic and atmospheric Xe, and that different meteorites contained trapped planetary Xe of distinctly different bulk compositions (cf. Table 5.5).

It was quickly apparent (cf. Fig. 5.8) that to first order AVCC appeared related to atmospheric Xe by a severe fractionation, more than 3% per mass unit, except that, relative to appropriately fractionated air, AVCC had excess abundances at the heavy, unshielded isotopes (Krummenacher *et al.*, 1962). It was thus reasonable to suppose that there was in fact a genetic relationship, specifically that meteorites contained a 'primitive' Xe component, related to air Xe by fractionation, on which was superposed a second component. There was also a clear suggestion that this second component, consisting of the unshielded isotopes, was in fact a fission component; this second component was commonly designated CCF (Carbonaceous Chondrite Fission). It should be noted that the variations in total planetary Xe compositions now evident can, to first order at least, be interpreted as somewhat greater or lesser CCF abundances relative to primitive Xe.

The first stepwise heating analyses (Reynolds, 1963; Reynolds & Turner, 1964) showed that planetary Xe compositions were variable among the release fractions, and so demonstrated that AVCC was not a single component well before this could be appreciated on the basis of variable total planetary Xe compositions (Fig. 5.9). Isotopic ratio variations were correlated and interpretable as reflecting mixture of just two components, one of them plausibly a fission component, so such observations strongly supported the CCF hypothesis described above. While the observed correlations could not be used for an unambiguous determination of the primitive composition, they could determine CCF composition if it were assumed to be an actual fission component: if so, the abundance of shielded ^{130}Xe must be nil and CCF composition determined by extrapolation of correlations to vanishing ^{130}Xe abundance. Pepin's (1968) report that CCF composition so calculated also had vanishing abundances of the other major shielded isotope, ^{128}Xe, further supported this hypothesis.

Fig. 5.8. Isotopic compositions (IT designates 'intermediate temperature' extraction) of Xe in various meteorites (Novo Urei is a ureilite, all others are carbonaceous chondrites), displayed as per mil variations of observed isotope ratios (normalized to ^{130}Xe) from the corresponding ratios in air. The lines are compositions which can be obtained by fractionating air Xe. The data suggest that to first order air and meteoritic Xe are related by fractionation: this model adequately explains the relative abundances of ^{124}Xe, ^{126}Xe, ^{128}Xe, and ^{130}Xe, possibly ^{131}Xe and ^{132}Xe as well, but definitely not ^{129}Xe, ^{134}Xe, and ^{136}Xe (see text). Reproduced from Pepin & Phinney (1983).

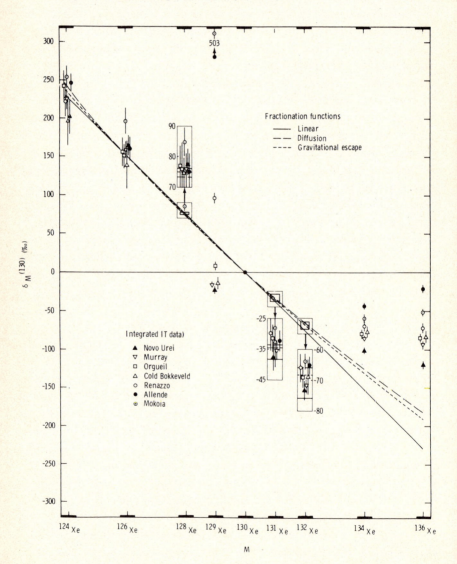

Usefully precise solar Xe compositions (Table 5.5), based on solar-gas-rich meteorites (Marti, 1969) or surface-correlated gas in lunar mare soils (Eberhardt *et al.*, 1970; 1972; Podosek *et al.*, 1971), were not available until after the features of xenology described above were well established. From Fig. 5.10 it is evident that planetary Xe is essentially identical to solar Xe in the relative abundances of the light, shielded isotopes but contains relatively greater abundances of the heavy, unshielded isotopes. This observation also strongly supports the CCF hypothesis, and also clearly suggests that solar Xe is the fission-free primitive component in planetary Xe.

This is a relatively tidy picture, but, to be sure, there are difficulties as well. An obvious flaw is that no plausible explanation has yet been advanced for the origin of the steep fractionation in air Xe. Equally striking is the absence of a suitable parent for CCF: no known fissioning nuclide produces Xe of the right composition or could plausibly have been present in sufficient abundance in the early solar system to account for the relatively large CCF concentrations observed, nor would it be expected that any actinide parent would have sufficient

Fig. 5.9. Xe compositional variations observed in stepwise heating of the Renazzo meteorite (Reynolds & Turner, 1964). Point A is atmospheric composition, point P is a hypothesized trapped (primordial) composition. The lines labeled eq. (7) are fractionation trajectories, whence the shaded area is the compositional field allowed for admixture of fractionated air and fractionated AVCC Xe; the effect of adding known fission components (^{244}Pu, ^{238}U) is also indicated. None of these effects can account for the observed variations. Reproduced from Funk *et al.* (1967).

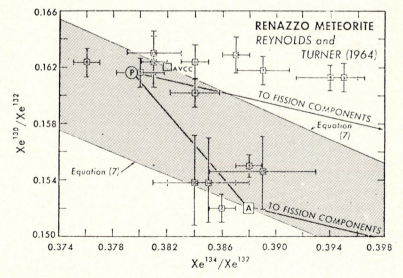

Table 5.5. Selected compositional data for xenon

Composition	^{124}Xe	^{126}Xe	^{128}Xe	^{129}Xe	^{130}Xe	^{131}Xe	^{132}Xe	^{134}Xe	^{136}Xe
Solar									
SUCOR[a]	2.89 ±0.04	2.63 ±0.06	50.9 ±0.3	637.1 ±1.8	≡100	499.0 ±2.2	606.6 ±2.1	225.2 ±0.9	182.7 ±0.6
Pesyanoe[b]	2.97 ±0.13	3.21 ±0.13	50.3 ±1.1	630.9 ±9.4	≡100	495.2 ±6.7	606.1 ±7.3	221.2 ±3.2	178.8 ±2.8
Planetary									
Murray[c]	2.86 ±0.92	2.56 ±0.03	50.6 ±0.2	644.0 ±2.0	≡100	505.5 ±1.5	616.0 ±2.0	234.6 ±0.8	197.0 ±0.9
Allende[d]	2.91 ±0.02	2.57 ±0.02	51.5 ±0.2	1115.5 ±2.8	≡100	507.8 ±0.9	620.7 ±0.8	244.1 ±0.4	209.4 ±0.3
Kenna[e]	2.89 ±0.03	2.54 ±0.02	50.8 ±0.2	635.8 ±1.5	≡100	502.7 ±1.0	613.6 ±1.1	231.3 .±0.6	191.6 ±0.6
Terrestrial									
Air[f]	2.337 ±0.007	2.180 ±0.011	47.15 ±0.05	649.6 ±0.6	≡100	521.3 ±0.6	660.7 ±0.5	256.3 ±0.4	217.6 ±0.2
Hypothetical									
U-xenon[g]	2.95	2.54	50.9	628.7	≡100	499.6	604.8	212.9	166.3

[a] Surface-correlated Xe in lunar soils; Podosek *et al.* (1971).
[b] 1000 °C fraction; Marti (1969).
[c] Podosek *et al.* (1971).
[d] Drozd *et al.* (1977).
[e] Wilkening & Marti (1976).
[f] Table 2.2
[g] Pepin & Phinney (1983).

geochemical coherence with Xe to account for the narrow observed range of CCF proportions in planetary Xe. There are equally substantial if less obvious difficulties as well. In the simple model described, there are three independent means of determining CCF composition (assuming it to be devoid of ^{130}Xe): extrapolation of the isotopic correlations found within planetary Xe (cf. Fig. 5.9), subtraction of solar Xe from planetary Xe (cf. Fig. 5.10), and subtraction of fractionated air Xe from planetary Xe (cf. Fig. 5.8). As it turns out, no two of these calculations agree. Put another way, neither solar Xe nor fractionated air

Fig. 5.10. Isotopic compositions of planetary, solar, and terrestrial Xe, displayed as per mil variations of observed isotope ratios (normalized to ^{130}Xe) in air and the carbonaceous chondrite Murray from the corresponding ratios in SUCOR, a solar Xe composition calculated to be surface-correlated Xe in a lunar mare soil. The dashed line, illustrating linear fractionation, is primarily for reference (not a 'best fit') and suggests that air and solar Xe *cannot* be related solely by fractionation. Reproduced from Podosek (1978).

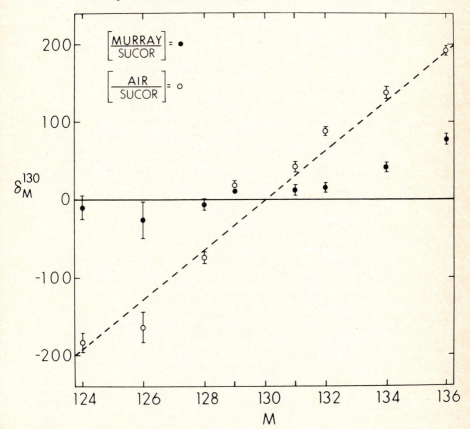

Xe can be considered one of the end members in meteoritic Xe (i.e. they do not fall on the correlations), nor can air and solar Xe be related simply by fractionation.

These difficulties are not necessarily fatal to the basic idea, however. The shielded isotope match between air Xe and fractionated planetary/solar Xe is sufficiently good that origin by fractionation is generally accepted even in the lack of a creditable mechanism. Two main approaches have been taken to the CCF parent problem. One is to postulate that the parent is a superheavy element with the requisite abundance, nuclear properties, and chemical behavior (e.g. Anders & Heymann, 1969). The second is to retain the basic idea of a heavy-isotope-rich nuclear component in planetary Xe but abandon the postulated origin by fission, which has the advantage of relaxing constraints on its composition but the disadvantage of leaving its origin unspecified; in such cases the name is usually also changed to avoid the fission implication, e.g. to Xenon X (Sabu & Manuel, 1980a) or Xenon H (Pepin & Phinney, 1983). A third approach, which attracted little support, asserted that the second component in planetary Xe was not an independent nuclear component at all but was an extremely fractionated (by unknown mechanism) version of primitive Xe (e.g. Hennecke & Manuel, 1971; Kuroda *et al.*, 1974). The problem of the isotopic mismatch among primitive planetary Xe, solar Xe, and air Xe can be resolved by postulating that neither of the latter two are entirely primordial but incorporate known nuclear components such as ^{244}Pu and/or ^{238}U fission. For the earth, with a comparatively low planetary noble gas abundance, this is not only plausible but quite reasonable, and is indeed believed to be the case. For solar Xe composition determined from lunar mare soils, it is also plausible to suppose incorporation of lunar fission (or other) Xe in surface-correlated gas just as for radiogenic ^{40}Ar; such an effect is clearly present in lunar highland samples (e.g. Behrmann *et al.*, 1973; Bernatowicz *et al.*, 1978). It is unclear, however, whether or to what extent nominal solar Xe compositions actually are contaminated by such nuclear components.

It should be noted that all these developments in xenology can be made within the context of primordial homogeneity, as indeed they were, at least in principle. The prevailing view was that there was one principal (primordial) Xe component, albeit an elusive one, one nuclear component (CCF or equivalent) whose origin was unclear but easily imaginable to have been within the solar system, and several potentially important nuclear components clearly known to be produced within the solar system, and that all observed compositions could be accounted for by fractionation and superposition of these components. The frustrating lack of a clear resolution of which components contributed to which compositions could be (and was) blamed on the inherent difficulties of working with data of finite precision, the possibility of air contamination in meteorites, and especially on the embarrassing wealth of available components. Subsequent developments have shown that there are still more components, that isotopic uniformity is untenable, and that this basic model is oversimplified, but it may yet be the case that this model is essentially correct at least in terms of accounting for total Xe compositions of both meteorites and the larger terrestrial planets.

Reynolds & Turner (1964) noted and commented on the circumstance that, in their stepwise heating data for Renazzo, variations in ^{124}Xe correlated positively with ^{136}Xe. Such covariation is inconsistent with the CCF hypothesis because ^{124}Xe (as well as ^{126}Xe and ^{128}Xe) should be absent from a fission component and thus should correlate negatively with ^{136}Xe. This observation was evidently forgotten or ignored subsequently, however, presumably because ^{124}Xe and ^{126}Xe are typically measured with relatively poor precision and elevated abundances in these isotopes are suggestive of spallation. An important advance in xenology came when Manuel *et al.* (1972) called attention to similar but much more pronounced correlation in stepwise heating data from Allende (Fig. 5.11). They noted that the presence of light isotopes was incompatible with a fission origin and so named the second component X (the complementary, 'primitive' component analog was later named Y) and considered it to be a distinct nucleosynthetic component, i.e. an isotopic anomaly. Pepin & Phinney (1983) have stressed the observations of different isotopic correlations in different meteorites, particularly that in the C3V carbonaceous chondrites Allende and Mokoia the second component is not only more prominent (relative to the primitive component) but also contains enhanced abundances of the light isotopes (Fig. 5.12). In their view the data are best explained in terms of two extra components, an H component consisting of the heavy, unshielded isotopes (and thus the successor of CCF) and an L component consisting of the lighter isotopes, specifically those (^{124}Xe, ^{126}Xe, ^{128}Xe, ^{131}Xe) which terminate β^+ decay chains (cf. Fig. 3.1e). It is thus both plausible and consistent to postulate that ^{130}Xe, the only isotope with stable isobars of both higher and lower atomic number (Fig. 3.1e), is absent in both H and L components and so compositions can still be computed by extrapolating correlations to vanishing ^{130}Xe (Fig. 5.12). In this model the H and L components are separable and thus not always combined into a single anomalous component X.

In contrast to the case for the 'second' component, the actual composition of the 'first' component – the long-sought primitive Xe – on the other end of the correlations cannot be determined on the basis of a plausible nuclear regularity. An appropriate constraint can be applied, however, by postulating the conviction noted above: that there exists a primordial Xe component such that meteoritic compositions can be generated by adding variable amounts of a single additional component and atmospheric composition can be generated by adding a plausible

nuclear component to a fractionated version of this same primordial component.
Pepin & Phinney (1983) have reported a multidimensional correlation in which
this can be done consistently, and which thus identifies a primordial component
designated U-xenon (Table 5.5). Meteoritic compositions are generated by addition
of their H component; air Xe can be generated from fractionated U-xenon by addition of a fission component, and the fission composition can be inferred (rather
than assumed) to be similar to that of ^{244}Pu (Fig. 6.2). This is a very important

Fig. 5.11. Isotopic compositions observed in stepwise heating of
carbonaceous chondrites. The linear correlation requires the existence
of a Xe component (upper right) which cannot be simply fission,
spallation, or a fractionated version of common (solar, AVCC, air) Xe.
Reproduced from Manuel *et al.* (1972).

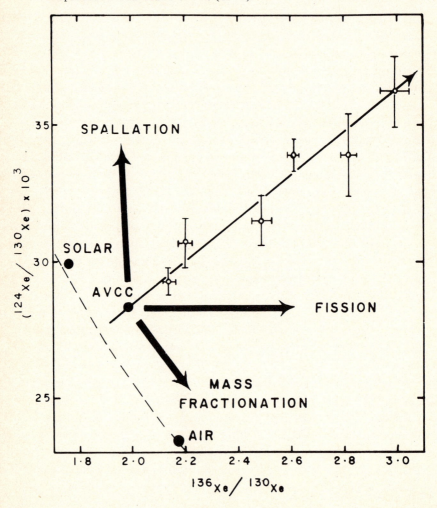

Fig. 5.12. Xenon component compositions calculated by extrapolation of observed isotopic correlations (in meteorites) to vanishing abundance of ^{130}Xe. Many meteorites yield the same composition, defining an H-xenon component (dashed line) containing only the heavy isotopes unshielded by more neutron-rich isobars: ^{136}Xe, ^{134}Xe, ^{132}Xe, and ^{131}Xe. Others (Allende, Mokoia) yield a different composition (solid line) interpreted as a superposition of the same H-xenon and another component, L-xenon, containing only the light isotopes unshielded by more neutron-deficient isobars: ^{124}Xe, ^{126}Xe, ^{128}Xe, and ^{131}Xe. Cf. Figs. 5.8 and 5.11. Reproduced from Pepin & Phinney (1983).

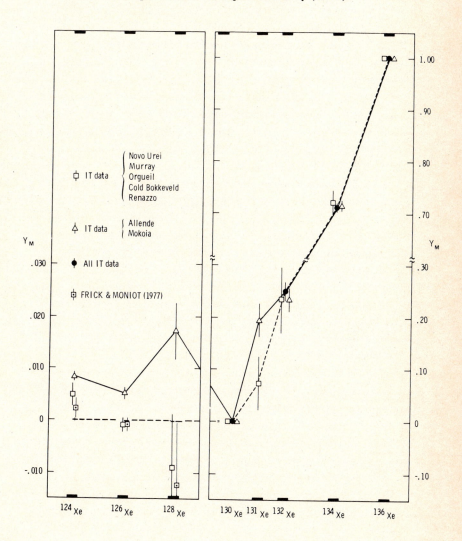

M

development, but some qualifications are necessary as well, especially since it is difficult to assess how closely this complicated procedure actually constrains U-Xe composition. For one thing, not all meteoritic compositions can be accounted for so simply (see below), and in particular not even U-Xe can be considered a pure component: some meteoritic data, in their generalized model, require subtraction of a light-isotope component from U-Xe. Moreover, U-Xe is notably deficient in the heavy isotopes and in their model solar Xe must include a substantial admixture of H-Xe (about 10% of the ^{136}Xe). Thus, H-Xe could not be a trace component rendered prominent in meteorites only by their general depletion of volatiles but must be a major component of Xe in the sun (but, by hypothesis, *not* in the earth).

It is not yet clear whether the Pepin & Phinney model is adequate resolution of these long-standing problems in xenology. Even if it is, it evidently reflects bulk averaging of Xe components in macroscopic samples, since more recent experiments indicate that the component structure of Xe is more complicated still. Selective dissolution experiments, particularly those involving various oxidizing treatments, often yield residues in which isotopic effects are very much greater than those found in stepwise heating analyses of bulk meteorites (e.g. Lewis *et al.*, 1975; Frick & Moniot, 1977; Matsuda *et al.*, 1980). Characteristically, the effect is pronounced enhancement of both heavy and light isotopes (Fig. 5.13). It appears that the effect is an enrichment of the H and L components relative to 'primitive' Xe, but important questions about component organization are raised by the observation that correlations among the selective dissolution residues are not the same correlations that are observed in stepwise heating analyses of bulk meteorites (e.g. Frick & Moniot, 1977; Pepin & Phinney, 1983). Furthermore, it appears that observations cannot be accounted for in terms of single H and L components (plus a 'primitive' component), but rather that each is variable or composite; it may be, however, that such complexities are ultimately accountable in terms of already familiar components such as spallation in L-Xe and ^{244}Pu fission in H-Xe (cf. Pepin & Phinney, 1983).

In addition to H- and L-Xe components, whose manifestations seem quite general, there appear to be still other components in at least some samples. In particular, selective dissolution residues from the Murchison meteorite contain enhanced abundances of the middle Xe isotopes, notably ^{128}Xe and ^{130}Xe (Srinivasan & Anders, 1978; Alaerts *et al.*, 1980); these observations are interpreted in terms of an s-Xe component, so named because it can be associated with stellar nucleosynthesis by the s-process (slow neutron addition). There may also be a component greatly enriched (perhaps monisotopic?) in ^{124}Xe (Lewis *et al.*, 1979; Pepin & Phinney, 1983), but its identification is only tentative.

Fig. 5.13. Xenon compositions observed in (oxidizing) etch residues in the Allende meteorite. Isotopic ratios, normalized to ^{130}Xe, are shown as multiples of the corresponding ratios in solar Xe. These data illustrate extreme enrichment of both the heavy-isotope-enriched and the light-isotope-enriched components (cf. Figs. 5.11 and 5.12). Reproduced from Lewis & Anders (1981).

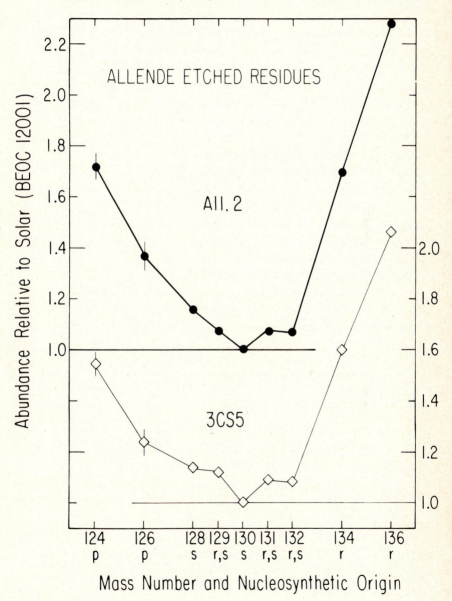

Considerable research has been directed to identifying the host phases of the various Xe components seen in meteorites, but as yet no generally accepted picture has emerged. There is also substantial uncertainty about the ultimate origin of these components. s-Xe is presumably a distinct nucleosynthetic component (s-process), i.e. an anomalous component. Similarly, 'monisotopic' ^{124}Xe, if it exists, is presumably an anomalous component, but one for which there is no obvious nucleosynthetic process. Most researchers consider H- and L-Xe to be anomalous components as well; while some (cf. Sabu & Manuel, 1980a) consider them a single component (Xe-X), most consider them separable components with separate nucleosynthetic origins, roughly speaking the p-process (proton addition) for L-Xe and the r-process (rapid neutron addition) for H-Xe. As reviewed by Anders (1981), however, it can also be argued that neither H nor L components are anomalous, but rather that L-Xe is extremely fractionated 'primitive' Xe and that H-Xe is the fission product of a volatile superheavy element (and thus should retain its original designation, CCF).

As noted earlier, isotopic effects in Kr frequently parallel those in Xe and so are usually thought to be associated with them. The Kr effects tend to be more subtle, however, and thus more difficult to resolve. Accordingly, it is often considered that Kr isotopery will be understood when and to the extent that the corresponding xenology is understood.

As with Xe, meteorites with planetary elemental abundance patterns had essentially the same Kr isotopic compositions. Since the most accurate measurements could be made on carbonaceous chondrites, which typically have the most gas, an AVCC Kr composition (Table 5.6) was used to represent it; again as with

Table 5.6. *Selected isotopic composition data for krypton*

Composition	^{78}Kr	^{80}Kr	^{82}Kr	^{83}Kr	^{84}Kr	^{86}Kr
Solar BEOC-12[a]	0.593 ±0.005	3.89 ±0.02	20.05 ±0.08	20.09 ±0.07	≡100	30.50 ±0.07
Planetary AVCC[b]	0.597 ±0.005	3.92 ±0.03	20.15 ±0.08	20.17 ±0.08	≡100	30.98 ±0.08
Terrestrial Air[c]	0.609 ±0.002	3.960 ±0.002	20.22 ±0.01	20.14 ±0.02	≡100	30.52 ±0.03

[a] Surface-correlated composition in lunar mare soil (Eberhardt *et al.*, 1972).
[b] Eugster *et al.* (1967).
[c] See Table 2.2.

Xe, however, this composition is not restricted to carbonaceous chondrites, but characterizes planetary Kr in other classes of meteorites as well.

Planetary (AVCC) Kr is distinct from air Kr. Within error limits, however, the difference between these compositions can be understood as a relatively slight fractionation, about 0.5% amu, *without* any superposed specific isotopic effects (cf. Fig. 5.14). It should be noted that the Kr fractionation is not only much smaller than the corresponding inferred Xe fractionation, it is in the opposite sense. This circumstance (along with the absence of any apparent fractionation between terrestrial and planetary Ar) has contributed to the difficulty in construction of a model which would account for the differences between terrestrial and planetary noble gases in general.

Various determinations of solar Kr composition differ somewhat, but the differences are evidently accountable in terms of fractionation (possibly instru-

Fig. 5.14. Compositions of planetary and terrestrial Kr, normalized to ^{84}Kr, expressed as per mil variations from solar Kr (cf. Fig. 5.10). All data from Table 5.6.

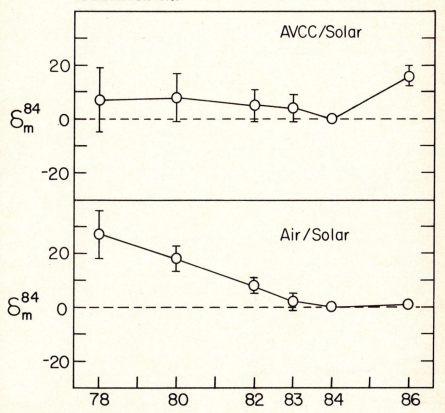

mental?). Nevertheless, it is clear that planetary Kr is indistinguishable from solar Kr (or a slightly fractionated version of it) except that it contains a 1-2% excess of ^{86}Kr (Fig. 5.14). The situation is quite analogous to that for Xe (Fig. 5.10). The model which accounts for meteoritic Xe in terms of a primitive Xe plus superposed CCF Xe is naturally extended to interpret meteoritic Kr in terms of a primitive (solar) Kr plus a superposed CCF Kr, thus accounting for the ^{86}Kr excess (fission ^{84}Kr and ^{83}Kr would be expected to be much smaller than fission ^{86}Kr in the absolute sense and thus cause only undetectably small isotopic enrichments).

Kr isotopic composition variations in stepwise heating of meteorites are also observable, but are much smaller than the corresponding effects in Xe and thus more difficult to characterize. However, the same selective dissolution residues that show large isotopic effects also show significant isotopic effects

Fig. 5.15. Isotopic ratios (ordinate) in noble gases released in stepwise heating of selective dissolution residues of the Orgueil and Murray meteorites, displayed as function of cumulative fractional release of denominator isotope (abscissa). The shaded areas indicate, for He and Ne in Orgueil, the maximum effect (product of isotopic ratio variation and quantity of gas) of *in situ* spallation contributions (negligible for the heavier gases). Reproduced from Frick & Moniot (1977).

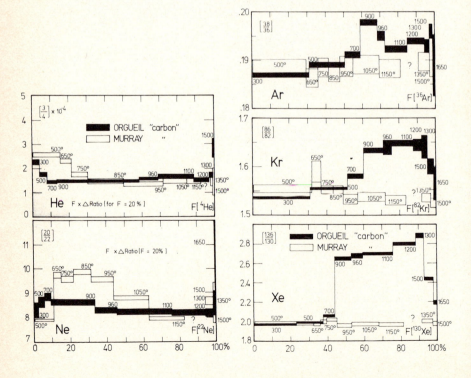

in Kr, and it is now amply clear that planetary Kr, like Xe, is not a single uniform component, but rather displays a complex internal structure. The Kr isotopic variations are well-correlated with those in Xe (Fig. 5.15). The extreme effects in Kr, as in Xe, are characterized by enrichment of heavy isotopes; unlike the case for Xe, however, these are accompanied by depletion of the light isotopes rather than enrichment (Fig. 5.16). To first order, at least, planetary Kr structure is compatible with a two-component mix: a 'primitive' component (solar?) plus one other, enriched in heavy and depleted in light isotopes; whether this simplicity is valid in detail, or whether other components are present, is not yet resolved.

Interpretations for Kr parallel those for Xe. If H-Xe and L-Xe, the components responsible for most of the effects, are anomalous nucleosynthetic components, they are presumably accompanied by an anomalous Kr component; it should then be noted that the anomalous Kr which correlates with anomalous Xe is rich in heavy isotopes, as is H-Xe, but is not accompanied by a light isotope enrichment analogous to L-Xe. If H-Xe and L-Xe are not anomalous but are (superheavy) fission and strongly fractionated solar components, the same interpretation can be applied to Kr (Fig. 5.16); in this case, however, it should be noted that the

Fig. 5.16. Isotopic compositions of an oxidizing dissolution residue of the Allende meteorite (solid squares) and of solar gas (open circles), Xe data normalized to ^{130}Xe and Kr data to ^{82}Kr, expressed as per mil deviations from the corresponding ratios in atmospheric Xe and Kr. Reproduced from Frick (1977).

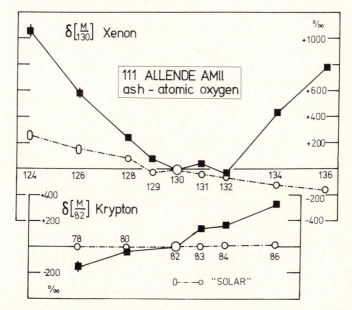

Kr and Xe fractionations are in opposite sense. Alaerts *et al.* (1980) identify an anomalous s-Kr accompanying s-Xe, but it is more difficult to characterize quantitatively.

Correlations have been sought – but not found – between the Xe and Kr isotopic variations described above and the He and Ne isotopic variations described earlier. While there are thus no apparent He and Ne anomalies which correspond with the Xe and Kr anomalies (?), it should be recalled that it would be rather difficult to detect such structures in these systems (Section 5.5). There is, however, an apparently anomalous Ar component, of high $^{38}Ar/^{36}Ar$, which does correlate with the Xe and Kr structures (Fig. 5.15). In this sense Ar appears more like the heavier gases Kr and Xe than like the lighter gases He and Ne. This feature strengthens the same association already suggested by coherence in *elemental* abundance ratios in planetary gases (Section 5.2).

The great effort and interest which noble gas isotopic structures have attracted should not be allowed to obscure the generalizations which emerged in the pre-anomaly period of cosmochemical study. Planetary noble gas compositions in bulk meteorites, the data base presumably most relevant for comparison with noble gases in the earth and the other major terrestrial planets, are only narrowly variable in comparison with the compositions observed in stepwise heating and selective dissolution residues. This can be attributed to the presence of anomalous components in quantities small in comparison with 'normal' gases and/or in relatively constant proportions to the 'normal' gases in macroscopic samples.

A rather different perspective is advocated by Sabu & Manuel (1980a, and earlier references cited therein). These authors note a correlation, in the selective dissolution residues, between heavy gas isotopic structures and elemental abundances, particularly of the light gases He and Ne (Fig. 5.17). They propose the thesis that the observations reflect mixing of two components: a Y component consisting of isotopically 'normal' (solar) heavy gases (Ar, Kr, and Xe) in planetary proportions but essentially devoid of lighter gases, and a single X component consisting of anomalous Ar, Kr, and Xe and essentially all of what would ordinarily be considered planetary He and Ne. This simplified model is clearly not valid in detail but equally clearly emerges from the gross trends evident in Fig. 5.17. What is not yet clear is whether this generalization has genetic significance or is a trivial consequence of a possible two-carrier (for noble gases) situation in a few meteorites; it is also unclear whether this generalization applies to differences between meteorites and meteorite classes. A suggested (but only suggested) corollary of this model is an absence of anomalous isotopic components within Ne (as noted in Section 5.5, there is no need for anomalous He) and indeed Sabu & Manuel (1980b) argue that this is the case. The Sabu & Manuel X + Y model is also linked to a rather extreme astrophysical scenario. Both the Ne arguments and the astrophysical scenario are generally regarded as unacceptable by other observers. There is, however, no need to link the Ne isotopery and astrophysics with the elemental/isotopic correlations (Fig. 5.17), and these correlations merit continuing attention and investigation.

In closing this section, we consider the special case of ^{129}Xe. In view of the ubiquity (even in lunar surface-correlated Xe) of a potentially large monisotopic component (from decay of ^{129}I), it is difficult to establish firm values for the abundance of ^{129}Xe in any other component. One approach is simply to take the lowest observed abundance as an upper limit to nonradiogenic abundance; this is, for example, the approach taken by Pepin & Phinney (1983) in estimating the ^{129}Xe content of U-Xe (Table 5.5). Alternately, I-Xe dating systematics can be used to infer (by extrapolation) a trapped ^{129}Xe abundance (cf. Fig. 4.9). Meteoritic trapped ^{129}Xe abundances so calculated are sometimes very low (cf. Podosek, 1970a; Drozd & Podosek, 1976). Such low abundances have never actually been observed, however, so their existence remains only conjectural. Accordingly, the best estimate for primordial ^{129}Xe abundance is probably the

Fig. 5.17. Illustration of trends between light noble gas (He and Ne) elemental abundances and Xe isotopic compositions in selective dissolution residues of Allende (filled symbols) and other (open symbols) meteorites. Low values of ^{134}Xe/^{136}Xe correspond to enrichment of H(CCF)–Xe, high values to more typical planetary compositions. Reproduced from Sabu & Manuel (1980a).

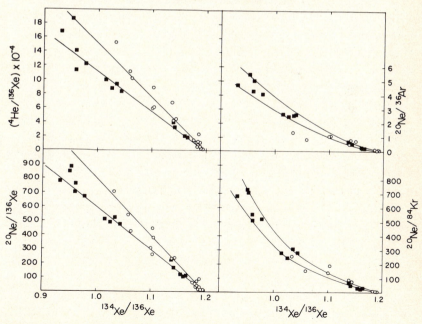

U-Xe value in Table 5.5 (the lowest value observed in stepwise heating of the ureilite Novo Urei).

5.7   Origin of planetary gas

Ever since planetary gas was recognized as a distinct pattern of elemental abundances (Section 5.2) there has been substantial interest in providing a theory of its origin, not only as an end in itself but for what constraints a viable theory would place on conditions and processes in the early solar nebula. A very wide variety of models has been offered, invoking a number of fundamentally different processes. Nevertheless, no consensus has emerged: either the right solution has not yet been proposed, or it has not yet been recognized. No model for just the noble gases alone has been widely accepted, and the situation is further complicated for more general models which attempt to treat other volatiles along with the noble gases, particularly when the comparison between meteorites and planetary atmospheres is considered as well (cf. Anders & Owen, 1977; Bogard & Gibson, 1978; Owen *et al.*, 1983).

While it is usually conceived that there is 'a' planetary gas component, i.e. a single entity, this is not necessarily the case. There are certainly variations among noble gas compositions designated planetary, both from one meteorite class to another and among members of a given class. In general, however, the planetary signature is sufficiently clear that the variations are usually ascribed to somewhat different parameter values, e.g. temperature or pressure, governing the outcome of a universal (but unknown) process common to all the meteorites. Perceptions of commonality are subjective, however, and the possibility that different processes produce accidentally similar (but not identical) results in different meteorites cannot be excluded. Thus, for example, Göbel *et al.* (1978) have identified diamond as a major carrier of trapped noble gases in ureilites; the gases in ureilite diamond have a composition which is clearly identifiable as 'planetary' in spite of differences from the planetary gas composition characteristic of carbonaceous chondrites (Fig. 5.18). The planetary gas carrier in carbonaceous chondrites, the subject of substantial research interest, is presently not well characterized but it is certainly not diamond. Ureilite diamond is evidently shocked graphite, but whether the ultimate origin of trapped gases in both classes of meteorites can be attributed to the same basic process remains an open question. In a different vein, it has often been noted that in the range of planetary gas compositions the heavier gases Xe, Kr, and Ar are relatively coherent with each other, as are the lighter gases Ne and He, and that the biggest compositional variations are those of the proportion of light to heavy gases, i.e. in the Ne/Ar ratio. Similarly, the selective chemical attack experiments noted in Section 5.2 suggest that the light and heavy gases often reside primarily in different carrier phases. It may thus be argued that the ultimate origin of trapped Ne and He is different from that of Xe, Kr, and Ar; this too remains an open question.

The usual approach to the problem is that of accounting for the origin of planetary gas within the solar system, ultimately from the gases ambient in the solar nebula. The requirements are to produce both the composition and concentration levels observed in meteorites. This approach ignores the isotopic evidence for distinct nucleosynthetic components of extrasolar origin (Sections 5.5 and 5.6). At least in a limited sense this remains a legitimate procedure since these anomalous components still seem a minor component in comparison with normal gases. These anomalous components were either locally concentrated in the nebula gas, in which case the problem of how they were trapped is the same as that for all the gases, or they were retained in presolar solid carriers and never were in the gas phase in the early solar system. In the latter case, the gases were trapped in some extrasolar environment such as interstellar space, red-giant atmospheres, supernova shock fronts, etc. In either case the problem of accounting for normal planetary gases remains within the solar system.

In some early models planetary gas was essentially a secondary feature in that it was produced within meteorite parent bodies of substantial size as a residue of the depletion of original gases which were much more abundant and more nearly solar in composition. Suess (1949) suggested that planetary gas could have been generated as the residue of gravitational escape from the parent body. Gravitational escape is very sensitively dependent on atomic mass, however, and unreasonably high temperatures and/or low gravitational fields must be postulated to avoid elemental fractionations much more severe than those actually observed (Suess, 1962); it also predicts isotopic fractionation effects contrary to observations. The model of gravitational escape has thus largely been abandoned. Similarly, it might be supposed that planetary gas is the residue of diffusive loss during elevated temperatures in the parent body metamorphism. Similar objections can be raised. Such loss might also be expected to show sharp cutoff temperatures, below which loss is minimal and above which loss is substantial, with different chemical species showing different cutoff temperatures. The expected pattern would thus be close to all-or-none depletion, and furthermore with markedly different patterns in response to the wide range of degrees of metamorphism exhibited by various meteorites. This is again contrary to observation. As a rule, valid in general if not in detail, volatiles in primitive meteorites – volatiles in general, not just noble gases – exhibit relative abundances much more nearly constant than absolute abundances. The consensus view is that meteoritic volatile patterns reflect a two-component mixture: a volatile-bearing component (the matrix) which is the same for all members of a given class and relatively slightly different from one class to another, and an additional component (chondrules, also metal) which is essentially volatile free; varying degrees of dilution by volatile-free material thus produce variations in overall abundance but little variation in composition of volatiles (cf. Larimer & Anders, 1967). In this view volatile abundance patterns, including the planetary noble gas pattern, are not secondary – products of modification within the parent bodies – but are instead primary – produced by the interactions of small solid particles with the ambient solar nebula, before those particles accrete into planetary bodies big enough to preclude further interactions of their interiors with the nebula.

Hypotheses for the origin of planetary gases thus usually are constrained by general models for solar nebula evolution. The solid particles destined to become meteorites (the volatile-bearing matrix) evidently equilibrated with nebula gas down to temperatures which vary somewhat according to exact model calculations and meteorite class but which are generally quoted in the range 300 K to 500 K; these temperature estimates are based primarily on volatile (but not noble gas) abundances. The models generally specify total gas pressures within an order of magnitude of 10^{-4} atm. For the cosmic abundances listed in Table 5.1, the corresponding noble gas partial pressures are 1.2×10^{-5} atm of ^{4}He, 1.7×10^{-8} atm of ^{20}Ne, 5×10^{-10} atm of ^{36}Ar, 1.5×10^{-13} atm of ^{84}Kr, and 1.2×10^{-15} atm of ^{130}Xe. These pressures are quite low, and require extremely efficient trapping of noble gases. The gas concentrations in Ivuna (Table 5.2), for example, correspond to effective distribution coefficients ranging from about $8 \, \mathrm{cm^3 \, STP \, g^{-1} \, atm^{-1}}$ for ^{4}He to $6 \times 10^5 \, \mathrm{cm^3 \, STP \, g^{-1} \, atm^{-1}}$ for ^{130}Xe. Distribution coefficients even higher, by some three orders of magnitude, are seemingly required by the observations (Section 5.2) that at least in carbonaceous chondrites most of the minerals are nearly devoid of noble gases and that a very small fraction of their mass (of the order of 0.1%) carries most of the gas inventory.

An immediately obvious possibility is simple solubility equilibrium between solids and ambient gas. A solubility model also has the advantage of a natural explanation for the actual trapping: it is easily imagined that solubility equilibrium could be established at high temperature and then quenched – i.e. the dissolved gases trapped – as temperatures fell. Since the first experimental test (Kirsten, 1968), however, the solubility model is one which had to be advocated in spite of, rather than because of, empirical data. Kirsten's results for enstatite melt, and subsequent results for similar systems (Table 4.3), indicate elemental fractionation in the wrong direction and distribution coefficients far too low, by factors of about 10^{-5} for He and 10^{-11} for Xe. Moreover the enthalpies of solution are positive (solution is endothermic), so that the elemental fractionations would be worse and the distribution coefficients still lower (assuming that the gases are less soluble in solids than melts) at the relevant nebular temperatures.

While data are still sparse, it seems quite clear that equilibrium solubility in the major anhydrous silicate minerals of meteorites is totally inadequate to produce planetary noble gases. An alternative possible solvent family is hydrated minerals such as the phyllosilicates common in carbonaceous chondrites. Here the only relevant data are for serpentine (Zaikowski & Schaeffer, 1979). The serpentine distribution coefficients are indeed higher and the elemental fractionation in the right direction (Table 4.3), but if serpentine is representative the quantitative evaluation is still negative. Elemental fractionation across the range He-Xe is only about an order of magnitude in serpentine (cf. Fig. 9.9) but about five orders of magnitude in planetary gas (Figs. 5.2 and 5.4). Absolute distribution coefficients for serpentine are still far too low as well, by a factor of about 2×10^{-9} for Xe for example. From such considerations it seems very difficult to defend the solubility hypothesis, and its advocacy must be based on the *ad hoc* postulate that an appropriate carrier phase, identity presently unknown, has the appropriate distribution coefficients, i.e. relative values to produce the planetary pattern and absolute values of unparalleled magnitude.

As described in Section 4.5, there have been a number of experimental studies, mostly prompted by interest in the origin of planetary gas, in which various kinds of condensed materials synthesized in an atmosphere containing noble gases are found to have trapped some of these gases. The gases are presumably distributed in the interior of the materials, but it would be misleading to describe them as dissolved since it is questionable, in fact rather unlikely, that thermodynamic equilibrium is involved. Usually the dependence on ambient partial pressure is not investigated, so effective distribution coefficients must be computed under the assumption that trapped gases are proportional to ambient pressure. Such experiments are interesting in that they may provide or suggest analogs for complex synthesis processes in the solar nebula. In some cases, notably the 'soot' and 'kerogen' data of Frick *et al.* (1979), the relative distribution coefficients are in fact quite suggestive of the planetary pattern (cf. Fig. 8.6); Frick *et al.* also call attention to the particular relevance these forms of carbonaceous materials may have to the suggested carbonaceous carrier of meteoritic gases. While such analogs are thus attractive by these criteria, the inferred distribution coefficients are still far too low in absolute terms; the Xe distribution coefficient for 'soot', for example, is $1.7 \, cm^3 \, STP$ $g^{-1} \, atm^{-1}$, very high in comparison with other trapping coefficients but still several orders of magnitude too low to account for trapping planetary gases from the solar nebula.

Trapped gases in terrestrial sediments provide a similar analog. Relative to a presumably atmospheric source reservoir, trapped gases in some sedimentary rocks exhibit an elemental fractionation pattern suggestive of planetary gases (cf. Figs. 8.2–8.4). Again, however, the analogy fails in absolute terms; even the highest Xe contents known in sedimentary rocks (Table 8.1 and Fig. 8.7) are less than 10^{-7} cm^{3}STP/g (of ^{130}Xe), at air pressure corresponding to distribution coefficients less than about 25 cm^{3}STP g^{-1} atm^{-1}.

Another obvious candidate mechanism for acquisition of nebular gases is physical adsorption. Strictly speaking, any adsorption model is incomplete in that adsorption can be invoked only to provide an initial concentration of gas and establish the elemental fractionation pattern; some other mechanism must be postulated to account for the trapping. Possible trapping mechanisms include shock implantation and grain growth by phase change, chemical reaction, or condensation from vapor. In any case, there is certainly empirical evidence that mere contact between solid and gas can cause 'trapping', presumably mediated by adsorption, whether or not any plausible trapping mechanism is known (e.g., Niemeyer & Leich, 1976; Yang *et al.* 1982; Yang & Anders, 1982a,b; also cf. Section 4.2). Some hypothesized mechanisms even suggest multiple adsorptions, e.g. trapping of an adsorbed layer by grain growth, adsorption on the new surface followed by more trapping, etc., so that an effective distribution coefficient for trapping might even be larger than the corresponding distribution coefficient for simple adsorption.

Important features of the adsorption model are that the distribution coefficients can be made arbitrarily high by appropriate stipulation of temperature, and that adsorption easily produces the steep elemental fractionations characteristic of planetary gas. Fanale & Cannon (1972) showed that adsorption on meteoritic material at about 100 K would produce both absolute and relative distribution coefficients appropriate for planetary gas (cf. Table 4.2), or rather at least the heavy gases. An illustration of compositional trends appropriate for various temperatures is given in Fig. 5.18.

The qualitative features of adsorption make it an attractive model for the origin of the planetary gas pattern. In detail, however, there are qualifications and limitations. For known sorbents, attainment of the requisite distribution coefficient for, say, Xe typically requires sufficiently low temperature (less than 100 K) that the corresponding elemental fractionation is more severe than even the planetary pattern (Fig. 5.18), especially for Ne and presumably even more so for He. The fractionation problem can be lessened by postulating higher temperature but only at the expense of postulating very efficient multiple adsorption or very high nebular pressure. If adsorption is indeed involved in producing the heavy planetary gases, it seems rather likely that some other mechanism must be invoked for the lighter gases Ne and He. The low temperatures required constitute another problem, not only because it seems rather unlikely that temperatures ever fell that low in the asteroid belt (where meteorites, including carbonaceous chondrites, are generally believed to have originated) but because such low temperatures would produce chemistries dominated by major element ices. Comets seem more appropriate, and it is indeed sometimes suggested that carbonaceous chondrites are cometary residue, but this view greatly exacerbates the problem of how the gases came to be trapped, especially for meteorites other than carbonaceous chondrites.

A quite different model for the origin of planetary gases, usually cited as 'ambipolar diffusion', has been proposed by Jokipii (1964) (also see Arrhenius & Alfvén, 1971, and Hostetler, 1981). The premise is that if gas of cosmic composition is partially ionized, strong chemical fractionation can be effected by loss of neutral species from a region in which the ions are held by a magnetic field. The ion composition depends on the mechanism and extent of ionization and on the details of electronic configuration, but for moderate ionization of a coherent chemical family such as the noble gases the dominant factor governing elemental fractionation between the ions and their original reservoir is $e^{\delta \epsilon / kT}$, where T is the relevant temperature and $\delta \epsilon$ is the difference in (first) ionization potential. Since ionization potentials decrease monotonically from He to Xe (Table 4.1), the ion population is fractionated in favor of the heavy gases. A temperature of about 10^4 K produces a quite respectable match to planetary composition; the effect of slight variation in temperature is illustrated in Fig. 5.18. Jokopii (1964) argues that the relevant astrophysical parameters are plausible: the minimum magnetic field required is about 5×10^{-5} gauss, variable over a characteristic distance scale of 1 AU, and ionization is assumed to be by electron impact (T is then the electron temperature) with solar radiation providing the necessary energy source.

The ambipolar diffusion model is also incomplete in that it provides an explanation for compositions but not for concentrations. We may imagine an extensive region of the solar system containing solids and a residue of initial gas whose composition is controlled by ionization potential; in this model the noble gases are already in planetary proportions. This is an important advantage of the model: since the reservoir already has the right composition, it is easier to understand that various materials could sample this same composition at a wide variety of concentrations. Still, some trapping mechanism must be invoked to get the gases into the solids, presumably without much further elemental fractionation. Since noble gas partial pressures will be lower than in the original gas, this mechanism must be quite efficient indeed.

A noteworthy feature of all the major models discussed above – solution, adsorption, trapping during synthesis, ambipolar diffusion – is that they are chemical models seeking to explain elemental fractionations. None of them predicts significant isotopic differences between initial (solar) and final (plane-

Fig. 5.18. Illustration of noble gas elemental fractionation patterns. The ordinate is depletion of cosmic abundance, as in Figs. 5.1–5.5, except for normalization to unity at Xe (equivalent to a scale change); the abscissa differs from Figs. 5.1–5.5 in the location of each gas. The upper figure (a) is for adsorption at the indicated temperatures on 'highly graphitized carbon black' (based on data of Sams *et al.*, 1960). The lower figure is for ion abundance ratios; 'plasma conditions' designates the case where ionization is by electron impact, for which the relevant temperature is the electron temperature. Reproduced from Göbel *et al.* (1978).

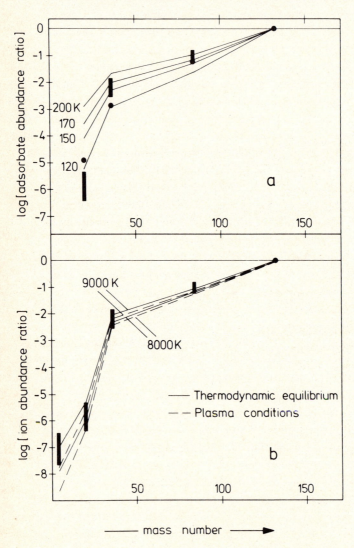

tary) compositions, possibly except for isotopic fractionation effects such as occasionally appear by ill-understood mechanisms (cf. Fig. 4.5). Presumably specific isotope effects are to be ascribed to nucleosynthetic heterogeneities (Sections 5.4–5.6) and/or *in situ* nuclear components in the host solids.

We have already noted (Section 5.3) that while the noble gases in the atmospheres of the major terrestrial planets Venus, Earth, and Mars are present in rather different absolute amounts, their compositions have notable planetary affinity. It could be imagined that the solids which accreted into the major planets contained planetary noble gases just like the solids which accreted into meteorite parent bodies, and that subsequent planetary outgassing (cf. Section 12.2) deposited these gases in the atmospheres. If so, the problem of the ultimate origin of the noble gases in planetary atmospheres is the same as that of meteoritic gases. There are, however, differences between meteoritic gases and those in the major planets, both isotopic and elemental. The possibility that these differences might be primary rather than secondary (established within the planets) invites consideration of how major planet gases might differ fundamentally from meteoritic gases. If such differences exist they presumably can be attributed to position within the solar system or planetary size or both. There are also fewer constraints on trapping mechanisms. For the earth, interior and atmospheric gases are isotopically identical (Section 9.3) except for *in situ* nuclear components; we cannot make the comparable statement for elemental composition, however (Sections 9.2 and 11.8). Thus, if the earth formed by heterogeneous accretion, we cannot say that the gases presently in the interior were originally trapped by the same mechanism and in the same abundances and proportions as those now in the atmosphere. For Venus and Mars, of course, we have no information at all about interior noble gases.

As one example, Pollack & Black (1979) propose that Venus, Earth, and Mars have progressively lower total inventories of noble gases because of solar system position. They infer that whatever trapping mechanism was involved was proportional to ambient partial pressures, and that these three planets trapped gases in the same compositions but at greatly different concentrations because pressures (at the relevant time and temperature) were lower farther from the sun. Without further details it is not clear what prediction would be made for the asteroid belt, but this view at least suggests that noble gas trapping for Venus, Earth, and Mars was basically different from what it was for meteorites.

A rather different model is suggested by Sill & Wilkening (1978). They point out that if equilibrium is maintained to low temperatures (<80 K) the first clathrate (Section 4.6) to form in solar nebula composition will be methane in water, and that this methane clathrate would dissolve substantial amounts of the heavy noble gases and could do so in approximately planetary proportions. Noble gas concentration in such low temperature clathrate would be so great that the earth's atmospheric inventory of heavy noble gases (Xe, Kr, and Ar) could be provided by infall of clathrate from the outer solar system which amounted to less than 1 ppm of the earth's mass. The same infall would not account for other terrestrial volatiles, including Ne, which would have to originate in other processes, presumably those arising in equilibration at the more conventional terrestrial accretion temperature around 500 K. The Sill & Wilkening model illustrates not only the potential decoupling of light from heavy noble gases, but also the potential decoupling of meteorites and larger planets: it does not provide an explanation for trapped gases in meteorites but, as a model for planetary atmospheres, it has no need to do so. This model does not, however, readily explain the trends observed among Venus, Earth, and Mars.

Another rather different model has been proposed by Ozima & Nakazawa (1980). As discovered independently by Safronov (1969), Hayashi (1972), and Goldreich & Ward (1973), condensed grains in the solar nebula should sediment toward the median plane and, when a critical density (about $10^{-7}\,\mathrm{g/cm^3}$) is exceeded, spontaneously aggregate into bodies ('planetesimals') of substantial size, about $10^{18}\,\mathrm{g}$ (radius about 5 km). The planetesimals would subsequently grow more slowly by further accretion and aggregation, ultimately leading to both meteorite parent bodies and the larger terrestrial planets. Ozima & Nakazawa point out that since lithostatic pressures in planetesimals would be rather low (central pressures of about 1 bar for 10 km radius, 100 bar for 100 km, etc.) the planetesimals would be expected to be rather porous and that the pores would be filled with nebular gas, assuming that nebular gas had not yet been dissipated. Ultimately the growth of the planetesimals would seal the pores and presumably some of the gas would be captured. The abundance of this natural supply of nebular volatiles would depend on porosity and nebular temperature and pressure; for plausible values the pore space gas would correspond to a depletion factor of about 10^{-11} to 10^{-12} relative to cosmic abundance. This factor would be the same for all chemical species, and clearly could supply only a trivial amount of observed amounts of most volatiles; it is, however, of the same order of magnitude as the Ne depletion in the earth (cf. Fig. 5.5).

Ozima & Nakazawa point out that gases in the pores could be concentrated by the planetesimal's own gravitational field; in particular, if diffusional equilibrium could be maintained so that each gas species made its own equilibrium adjustment to the gravitational field, substantial elemental fractionation could result. As illustrated in Fig. 5.19, a planetesimal radius of several hundred kilometers would enrich Xe by a few orders of magnitude but have little effect on He and Ne. A key question for this model is thus whether pores remain accessible and gravitational equilibration remains possible until a radius of several hundred kilometers is achieved. Ozima & Nakazawa (1980) estimated a gravitational equilibrium time of about $R^2/(4 \times 10^9)$, in years for R in centimeters, and a characteristic timescale for planetesimal growth is about 10^4 yr (Nakagawa, 1978), so it appears possible that the relevant conditions may be met for planetesimals up to one to two hundred kilometers, but the uncertainties in the important parameters are high enough to preclude any definitive assessment. An additional important question is whether, in fact, accretion did precede dispersal of the nebula.

The model of Ozima & Nakazawa (1980) is not a theory for the origin of planetary gas in meteorites, both because meteorite parent bodies are not believed to have grown to the right size and because the compositional trends in planetesimal gas do not match those in planetary gas (Fig. 5.19). As seen from the comparisons in Fig. 5.19, however, it is possible to imagine a mixture of planetary and planetesimal gas such that planetary gas made the major contribution to Ar and Kr and planetesimal gas made the major contribution to He,

Fig. 5.19. Average noble gas abundances which would be present in the pore spaces of a planetesimal (open triangles) of the indicated radius if thermodynamic equilibrium were maintained with a nebula of solar composition at the surface of the planetesimal. The ordinate is depletion factor relative to cosmic abundance, i.e. the same as in Figs. 5.1–5.5. The calculations were made for mass ratio of solid phase to gas phase of 0.003 43, planetesimal density 2 g/cm^3, porosity 33%, and gas temperature 225 K. Abundance patterns for the earth's atmosphere (bold solid line) and typical planetary abundance (dashed line) are shown for comparison. Reproduced from Ozima & Nakazawa (1980).

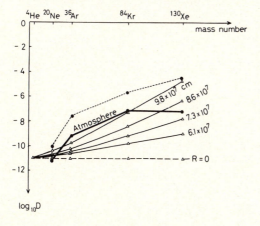

Ne, and Xe. Ozima & Nakazawa suggest such a mixture as the origin of the earth's noble gases. An important feature of this hypothesis is its prediction for isotopic effects. Along with the elemental fractionation in planetesimal gas there will be concomitant isotopic fractionation favoring heavier isotopes, and moreover, the fractionation is strongest for the heaviest element. Ozima & Nakazawa calculate that a planetesimal radius of about 500 km would produce isotopic fractionation characteristic of the difference between solar and terrestrial Xe (Section 5.6). The corresponding Ne fractionation would be minimal, so in this model terrestrial Ne composition would reflect a mixture of planetesimal Ne (unfractionated, thus of solar composition) and planetary Ne (Neon A), which is consistent with observation. This model also predicts Ar and Kr fractionation, and even though planetesimal Ar and Kr would be diluted with planetary Ar and Kr, the fractionation, particularly for Kr, should be observable, and it is not. The model also predicts a terrestrial Xe/Kr ratio at least as high or higher than planetary (cf. Section 5.3).

6   Component structures

6.1   Introduction

The noble gases found in air have been described in Chapter 2; a survey of the noble gases found in other classes of terrestrial materials is conducted in Chapters 7-10. This chapter is a preliminary overview of the general character of terrestrial noble gases; its focus will be the attempt to account for observed noble gas data in terms of identifiable sources, i.e. components. The basic approach will be to attempt resolution into components seen to exist in the solar system at large (Chapter 5) and components which are modifications of the earth's original inventory of noble gases and attributable to the earth's own evolutionary processes.

In practice, this chapter will concentrate on isotopic comparisons. Considerations of total elemental abundances on a large scale have already been introduced in Section 5.2 and will be pursued in succeeding chapters. In detail, there are many processes which affect elemental abundances, and they are generally difficult to understand quantitatively. In contrast, there are fewer processes capable of generating significant isotopic effects, and on the whole they are simpler and easier to treat quantitatively. Concentration on isotopic structures thus affords some optimism for quantitative understanding.

There are two ways to approach the problem of isotopic comparisons, each with some advantages in specific cases. One way is to imagine various possible isotopic effects, and then inquire whether they are likely to be quantitatively significant. This is the approach taken in Sections 6.2 and 6.3, and the earlier discussion of isotopic fractionation (Section 4.9) is part of the same process. Alternatively, we may compare isotopic compositions in various reservoirs and, where there are differences, seek plausible explanations for those differences. This is the basis orientation of Sections 6.4-6.9, as it is in most of the discussion in Chapter 5.

6.2   The radiogenic component

It is well known that radiogenic ^{4}He and ^{40}Ar are the primary ingredients in many terrestrial noble gases. There are also other radiogenic gases of interest in geochemistry. The discussion below is summarized in Table 6.1 (also cf. Table 3.1).

U presently produces ^{4}He at the rate of 1.21×10^{-7} cm^3 STP g(U)$^{-1}$ yr^{-1}, 4.0% of it from ^{235}U, the balance from ^{238}U. The present ^{4}He production from Th is 2.87×10^{-8} cm^3 STP g(Th)$^{-1}$ yr^{-1}. Although Th and U are certainly not always found in the same proportions, the Th/U ratio in most common rocks is sufficiently uniform that it is often convenient to think of them as a single source of ^{4}He. For Th/U = 3.3 (by weight) the combined present production of ^{4}He is 2.15×10^{-7} cm^3 STP g(U)$^{-1}$ yr^{-1}, 44% of it from Th. In the discussion below, the ^{4}He from 'U' will include the Th contribution in this proportion.

^{238}U has a spontaneous fission branch which produces ^{136}Xe at the present rate 5.08×10^{-16} cm^3 STP g(U)$^{-1}$ yr^{-1}; other heavy isotopes of Xe and Kr are also produced (Table 3.2). For gas produced by U we have, for current production, ^{136}Xe/^{4}He $= 2.3 \times 10^{-9}$.

The α particles emitted in the U and Th decay series are sufficiently energetic to induce nuclear reactions in the lighter elements. Other reactions can be induced by neutrons which are also produced in the decay chains (proportionately very few neutrons are attributable to the fissions; they are mostly due to (α, n) reactions). Hundreds of α or n reactions could be itemized. In general, however, the final products of these reactions are not very abundant and are unobservable except when they are noble gases (Section 1.2). This still leaves a large field. All the halogens can produce noble gases by (n, γ) reactions, for example, and there are many other specific reactions which produce noble gases. Even so, the effects are often visible only in special cases where nonradiogenic gas abundances are more than usually low and either U and/or Th or some special target element is present in abnormally high abundance. Some examples are included in Table 3.3. Two special cases of less restricted interest are described below.

The various species generated in α- or n-induced reactions can be denoted, with some liberty, as radiogenic, since they are ultimately attributable to radioactive decay. With further liberty, they may be considered a 'component' since they are produced together, in proportion to U abundance. In this sense, the radiogenic component associated with U decay includes not only ^{4}He but also the Xe and Kr isotopes produced in fission and all the α and n products. For the latter it is rather difficult to be very precise about the composition of this radiogenic component, and not very rewarding either, since the composition (i.e. the ratio of some specific product to ^{4}He) obviously depends on the abundance of the relevant target. For the α reactions, the short range of the α dictates a sensitive dependence on the microscale distribution of the target. For the n reactions, the geometry dependence is less but the total chemistry, not just

Table 6.1. Radiogenic terrestrial noble gases

Isotope	Principal source	Current production	Air abundance	Remarks
^{4}He	α decay	2.15×10^{-7} cm^3 STP g (U)$^{-1}$ yr^{-1}	100%	Includes Th for Th/U = 3.3; cf. Table 3.1, note (b), and Section 12.5
^{3}He	^{6}Li (n, α) ^{3}H	^{3}He/^{4}He $\approx 10^{-7}$	2%	Cf. Section 12.5
^{21}Ne	^{18}O (α, n) ^{21}Ne	^{21}Ne/^{4}He $\approx 4 \times 10^{-8}$	$\leqslant$4%	Atmospheric fraction (if degassed) for 18 ppb U; cf. Section 6.3
^{136}Xe	^{238}U fission	^{136}Xe/^{4}He = 2.3×10^{-9}	$\leqslant$1%	Plus other Xe and Kr isotopes (cf. Table 3.2)
^{40}Ar	^{40}K decay	3.89×10^{-12} cm^3 STP g (K)$^{-1}$ yr^{-1}	100%	^{4}He/^{40}Ar = 5.5 for K/U = 10^4; cf. Section 11.5
^{129}Xe	^{129}I decay	Nil	7%	Initial ^{129}I abundance $\geqslant 6.41 \times 10^{-15}$ g g^{-1}
^{136}Xe	^{244}Pu fission	Nil	4%	Initial ^{244}Pu abundance $\geqslant 3.32 \times 10^{-11}$ g g^{-1}
^{86}Kr	Fission ?	?	1%	Identification questionable; no plausible source

the target-to-U ratio, is important, since many elements are involved in producing the neutrons, moderating them, and competing as neutron sinks. Furthermore, the products will be different elements and generally located in different minerals, and so will be relatively easily separable from ^{4}He and from each other. This remark also applies, of course, to the fission Xe and Kr.

One of the special cases of interest is ^{3}He, i.e. radiogenic He is not pure ^{4}He. Early observations (Aldrich & Nier, 1948) of continental rocks and gas wells indicated dominance of radiogenic ^{4}He but also some ^{3}He; the ^{3}He/^{4}He ratio was usually lower than the air ratio, but variable and sometimes higher than the air ratio. Morrison & Pine (1955) developed the idea of scarce-isotope production by α- and n-induced reactions described above, and found a quite satisfactory agreement between the observations and theoretical calculations on this basis, including experimental demonstration of the neutron production. The reaction principally responsible for production of ^{3}He is ^{6}Li$(n, \alpha)^3$H. The Morrison & Pine (1955) value for radiogenic He in 'normal' (continental) rock is ^{3}He/^{4}He $= 10^{-7}$, which is the figure generally cited (Table 6.1). More recent calculations for production of radiogenic ^{3}He have been made by Gerling *et al.* (1971), and many observations are reviewed by Tolstikhin (1978). Radiogenic ^{3}He/^{4}He ratios are quite variable with rock type, ranging up to 10^{-5} (in the Li mineral spodumene) and down to 10^{-9}, with a 'typical' value seemingly somewhat closer to 10^{-8} than to 10^{-7}.

The second special case is production of ^{21}Ne. Wetherill (1954) observed large excesses of ^{21}Ne (and other isotopes) in U- and Th-rich minerals. He attributed the excess primarily to the ^{18}O$(\alpha, n)^{21}$Ne reaction, and was able to show that this reaction could account for the quantities he observed. Since O is abundant and ubiquitous, this reaction should produce ^{21}Ne at a fairly uniform rate, not sensitively dependent on chance association of target with U and Th. Heymann *et al.* (1976) have recently repeated the production rate calculation; the figures they cite correspond to a production ratio ^{21}Ne/^{4}He $= 4.2 \times 10^{-8}$, somewhat higher than Wetherill's rate. There will also be radiogenic production of the other Ne isotopes, ^{20}Ne from ^{17}O(α, n) and ^{22}Ne from ^{19}F(α, n) and ^{25}Mg(n, α), but these will generally be produced at lower absolute rates than ^{21}Ne, and since ^{21}Ne has the lowest abundance in normal Ne it is by far the most conspicuous radiogenic isotope of Ne. Rison (1980a) presents calculations for the production rates of all the Ne isotopes; his figure for ^{21}Ne is about 2.1×10^{-14} cm^3 STP g(U)$^{-1}$ yr^{-1}, corresponding to ^{21}Ne/^{4}He $= 1.0 \times 10^{-7}$.

The other major radionuclide which produces a noble gas is ^{40}K. In contrast to the case for U and Th the decay of ^{40}K is relatively simple and sedate and produces only ^{40}Ar (and ^{40}Ca). The present production rate of ^{40}Ar is 3.89×10^{-12} cm^3 STP g(K)$^{-1}$ yr^{-1}.

Albeit to a lesser extent, K, like Th, tends to be geochemically coherent with

U, at least in crustal rocks, typically occurring in proportion $K/U \approx 10^4$ by weight (Wasserburg *et al.*, 1964). To the extent that this is true, radiogenic gas will be produced currently with $^4He/^{40}Ar = 5.5$. Because of the shorter lifetime of ^{40}K, this production ratio was lower in the past; integrated over 4.55×10^9 yr the $^4He/^{40}Ar$ ratio is 2.1. For chondritic proportions, $K/U \approx 8 \times 10^4$, current production is $^4He/^{40}Ar = 0.69$. A number of often cited as characteristic of oceanic rocks is $^4He/^{40}Ar \approx 10$ (but see Section 11.5).

It is of great interest to consider how the radiogenic production described above, integrated over the age of the earth, relates to global inventories of the noble gases. Since construction of a truly global inventory is not possible, comparison is usually made with the atmospheric inventory. Such considerations are obviously closely tied to the global inventories of U, Th, and K. This is itself a matter of keen geochemical and geophysical interest, partly on general principles but more so because these elements are the major sources of current heat production in the earth. More detailed consideration is taken up in Chapter 12.

For any reasonable model of U and Th abundances, the integrated production of ^4He exceeds the present atmospheric inventory by a factor of several thousand. It has thus been concluded that He is not retained in the atmosphere (Section 12.7). The corresponding radiogenic ^3He is apparently responsible for only a small part of the atmospheric budget (Section 12.5).

The K content of the earth is a matter of great interest (and contention). As first noted by Turekian (1959), the atmospheric ^{40}Ar inventory sets a lower limit to total K in the earth. For the currently accepted geochronological parameters (Steiger & Jäger, 1977), this lower limit is 77 ppm, and corresponds to complete degassing of radiogenic ^{40}Ar (and so presumably also all the other noble gases) from the earth. At the other extreme, the highest K content currently entertained is the chondritic value, about 900 ppm; this would correspond to only 9% degassing of radiogenic ^{40}Ar.

Heymann *et al.* (1976) estimate that, for 18 ppb U, radiogenic ^{21}Ne is about 4% of the atmospheric inventory. The figure would be more or less in proportion as assumed U is greater or smaller, and is probably accurate within less than a factor of two (e.g. with Rison's (1980a) production rate, radiogenic ^{21}Ne could be 10% of the atmospheric inventory). Whether this fraction of actual atmospheric ^{21}Ne is radiogenic depends on whether the radiogenic ^{21}Ne has entered the atmosphere. For the same figure, 18 ppb U, production of fission ^{136}Xe is 3.6×10^{14} cm^3 STP, about 1.1% of the atmospheric inventory.

The considerations above have started by identifying possible sources and estimating their magnitudes; as described in subsequent sections of this chapter, there are no problems in finding room in the observations to accommodate these sources. Working in the other direction, the observed Xe composition of air suggests excesses of ^{129}Xe and of ^{136}Xe (Section 6.7). These are interpreted as being radiogenic, and are attributed to decay of ^{129}I and spontaneous fission of ^{244}Pu, respectively. Both of these are now extinct, but are plausibly considered to have been present at the time of separation of solar gas from solids destined to become earth (Chapter 13). The initial abundances needed to produce the atmospheric effects are noted in Table 6.1.

While the requisite amount of ^{244}Pu is plausible, and the effects at other Xe isotopes also suggest ^{244}Pu (Section 6.7), it is interesting to compare quantities of fission ^{136}Xe with other candidates for its production. ^{238}U, as noted above, can produce 1.1% of atmospheric ^{136}Xe for 18 ppb U; this U concentration could not realistically be more than doubled, so an upper limit to ^{238}U contribution to ^{136}Xe is about 2%. To produce, say, 1% of atmospheric ^{136}Xe by neutron irradiation of ^{235}U requires a fluence of 1.3×10^{17} neutrons/cm^2 today (1.4×10^{15} neutrons/cm^2 4.55×10^9 yr ago). Aside from the difficulties of producing the neutrons, such fluence would have other noticeable effects, e.g. consumption of ^{157}Gd and substantial production of other noble gas species, and it can be concluded that ^{235}U is not a significant source of atmospheric Xe.

As noted in Section 6.8 there may be an approximately 1% excess of ^{86}Kr in air. The most likely way to produce ^{86}Kr is fission, but there are quantitative difficulties. Allowing a 5% fission contribution to atmospheric ^{136}Xe, a 1% contribution to ^{86}Kr requires $^{86}Kr/^{136}Xe = 5.1$ in the source, too high for fission. Even allowing for 25 times more fission-bearing Xe hidden in the interior or in sediments (Section 5.3), with no hidden Kr, the ratio is reduced only to $^{86}Kr/^{136}Kr = 0.20$. ^{244}Pu could presumably be invoked to produce that much fission, but its ^{86}Kr yield is low, $^{86}Kr/^{136}Xe = 0.02$. ^{238}U is closer ($^{86}Kr/^{136}Xe = 0.12$) and ^{235}U adequate ($^{86}Kr/^{136}Xe = 0.32$), but by the considerations in the previous paragraph, the quantities involved are prohibitive. A very modest (and essentially undetectable) enhancement of ^{86}Kr in air is reasonable, but if there really is a 1% excess of ^{86}Kr in air, the absence of any plausible source constitutes a substantial problem.

It is interesting to consider the rate of accumulation of radiogenic gas in comparison with trapped gas levels. A rock containing 1% K produces *in situ* ^{40}Ar at about 4×10^{-14} cm^3 STP g^{-1} yr^{-1}; for a trapped ^{40}Ar concentration of 10^{-6} cm^3 STP/g or less (cf. Sections 9.4 and 11.3), *in situ* ^{40}Ar accumulation reaches this level in 25 Ma or less. A U concentration of 1 ppm produces ^4He at the rate 2×10^{-13} cm^3 STP g^{-1} yr^{-1}; in 25 Ma this accumulates to 5×10^{-6} cm^3 STP/g, near the upper end of the range of trapped ^4He concentrations (cf. Fig. 9.2). Thus, *in situ* radiogenic ^{40}Ar and ^4He are generally dominant (more abundant than trapped ^{40}Ar and ^4He) in any igneous rock older than about 10^8 yr, or even substantially less if trapped concentrations are low, and trapped radiogenic ^{40}Ar and ^4He (Section 9.4) are typically readily observable only in young rocks of age less (preferably much less) than about 10^7 yr.

6.3 The spallation component?

Cosmic rays (mostly 'galactic' but also 'solar') are sufficiently energetic that they can cause a variety of nuclear processes, generating a set of final products loosely termed 'spallation' products (Chapter 3). In general, the effects are sufficiently small that there is no observable depletion of the target species or enhancement of the product species, except when the normal background of the product species is nil (short-lived radionuclides) or very low (noble gases).

The characteristic attenuation depth for cosmic-ray spallation production is of the order of 10^2 g/cm^2. Meteorites and returned lunar samples have been exposed to cosmic rays at shallow shielding, and spallation noble gas components are typically prominent in these samples. Air (10^3 g/cm^2) is optically thick to cosmic rays, however; noble gas spallation production at the surface of the earth is observable in some samples of unusual chemistry (Takagi *et al.*, 1974; Srinivasan, 1976) but on a global scale spallation at the surface of the earth is quantitatively unimportant and the only significant production occurs in air. The best known result is ^{14}C. The scarceness of suitable heavy targets in air precludes significant production of any noble gas species other than ^{3}He. (There will be some production of the lighter Ar isotopes from ^{40}Ar and ^{4}He is actually produced at several times the rate of ^{3}He, but these are negligible contributions to the relevant inventories.) Spallation ^{3}He (about one third of it originally made as ^{3}H) is a small but nontrivial part of the atmospheric budget of ^{3}He (Section 12.5).

The spallation gases imported in meteorites falling at the present rate are negligible (Heymann *et al.*, 1976), and any significant production at the surface of an airless early earth unlikely (Bernatowicz & Podosek, 1978). There is a more substantial possibility, however, that solids destined to become the earth might have acquired perceptibly enhanced noble gas inventories by exposure to cosmic rays before accretion. To account for, say, 10% of present atmospheric ^{21}Ne inventory by spallation, preaccretionary earth materials would need an exposure to cosmic rays (at present intensities and assuming no gas shielding) for about 10^8 yr if distributed in size like present asteroids (Heymann *et al.*, 1976) or for about 10^3 yr as small ($\leqslant 1$ m) bodies (Bernatowicz & Podosek, 1978). These times are plausible, but theories of early solar system evolution are not sufficiently well developed to predict whether they are likely. In practice, the logic works in the reverse direction: Heymann *et al.* (1976) were the first to suggest that spallation in preaccretionary earth materials might be important, as suggested by a possible excess of atmospheric ^{21}Ne relative to likely primordial Ne (Section 6.5), and if the spallation contribution can be resolved independently it will provide a nice constraint on the preaccretionary history of the earth.

Bernatowicz & Podosek (1978) noted that if terrestrial ^{21}Ne contains a significant spallation contribution, so in all likelihood does terrestrial ^{3}He. For spallation on chondritic targets, ^{3}He/^{21}Ne ≈ 7.5. The fractional ^{3}He contribution corresponding to a given spallation contribution to ^{21}Ne depends on the He/Ne ratio in the target, and we have essentially no observational constraints for preaccretionary earth, nor can a spallation contribution to ^{3}He be resolved from a ^{3}He/^{4}He ratio dominated by radiogenic ^{4}He. Falling back on analogy, if we suppose preaccretionary earth to have the planetary primordial composition in Table 5.3, then ^{3}He/^{21}Ne $= 9$. A 10% spallation contribution to ^{21}Ne thus corresponds to an 8% contribution to ^{3}He. The ^{3}He contribution would be lower or higher in proportion as primordial gas is richer or poorer in ^{3}He. The noble gas isotopes most sensitive to spallation are ^{21}Ne and ^{3}He, so the effect would be smaller for the other gases: for 10% spallation ^{21}Ne, we would have 8% for ^{3}He, 0.3% for ^{22}Ne, negligible (<0.1%) for all other species (Bernatowicz & Podosek, 1978).

The possible presence of a significant spallation contribution to the terrestrial inventory of noble gases must be considered speculative. There may be no need for a ^{21}Ne contribution beyond the primordial one (Section 6.5) and even if there is a need there is at least one alternative candidate to fill it (Section 6.2). The converse proposition can be supported more definitively, and deserves explicit statement: the spallation contribution to terrestrial noble gases is no more than of the order 10^{-1} for ^{21}Ne and (probably) ^{3}He, of the order 10^{-3} or less for other species, and the cosmic-ray exposure of preaccretionary earth materials is no more than about 10^3 yr as small bodies or 10^8 yr as asteroid-sized bodies.

6.4   Argon

Ar is the easiest terrestrial noble gas to describe, in the sense of involving the fewest complications in accounting for its origin. In brief, essentially all terrestrial ^{36}Ar and ^{38}Ar is primordial and all terrestrial ^{40}Ar is radiogenic.

In extraterrestrial materials, there are variations in ^{38}Ar/^{36}Ar which must be ascribed to primordial inhomogeneity (Section 5.5), but they are small and apparently uncommon. In whole-rock samples, ^{38}Ar/^{36}Ar in planetary Ar varies only narrowly if at all, and is indistinguishable from ^{38}Ar/^{36}Ar in solar Ar. The terrestrial ^{38}Ar/^{36}Ar ratio is the same as in planetary/solar Ar (Table 5.4). Nor is there any reason to expect any significant nuclear component in terrestrial ^{36}Ar or ^{38}Ar, either radiogenic (Section 6.2) or spallation (Section 6.3). We can conclude that no special case need be made for terrestrial ^{36}Ar and ^{38}Ar: it is the same as that found elsewhere in the solar system. This conclusion is perhaps trite, but it should be noted that it is not so easy to make the same statement for other noble gases.

Terrestrial primordial Ar is evidently principally planetary rather than solar. Since planetary and solar Ar are not appreciably different isotopically, this evaluation rests mostly on elemental abundance comparisons (Section 5.3).

There are, to be sure, real if small variations in $^{38}Ar/^{36}Ar$ in the earth. Some of these are attributable to fractionation (see below). Radiogenic ^{36}Ar and ^{38}Ar have also been observed (Fleming & Thode, 1953; Wetherill, 1954) in rocks rich in U and/or Th; the specific reactions involved are not firmly known, but reasonable estimates of production rates indicate that radiogenic ^{36}Ar and ^{38}Ar will be negligible in normal rocks and in the earth as a whole. Whether or not air accounts for most of the total primordial Ar inventory, there is little doubt that $^{38}Ar/^{36}Ar$ in the earth as a whole is the same as the air value.

In contrast, the primordial abundance of ^{40}Ar is essentially nil (Section 5.5) and the usual dominance of ^{40}Ar in the earth and in other samples of the terrestrial planets is due to decay of ^{40}K in an environment very strongly depleted in Ar relative to K.

Since ^{40}Ar and ^{36}Ar in air have separate origins, the atmospheric $^{40}Ar/^{36}Ar$ ratio reflects both the original composition of the earth and its subsequent evolution, and it is accordingly a key parameter in many considerations of that origin and evolution (cf. Chapters 12 and 13). It is worth explicit note that $^{40}Ar/^{36}Ar$ in modern air should be unique: the present air value will not be found in the atmospheres of other planets, in meteorites, nor in the interior of the earth, nor even in terrestrial air of an earlier epoch.

Particularly in view of the very wide range of values a $^{40}Ar/^{36}Ar$ ratio might assume in a variety of environments, the uniqueness of air Ar is useful in removing ambiguities about the possible occurrence of air contamination and isotopic fractionation. Specifically, if some sample is observed to contain trapped Ar with the composition of modern air (cf. Dalrymple, 1969), there is a very strong presumptive case that the source of the Ar is indeed modern air. By the same logic, if a sample contains Ar with composition consistent with that of fractionated modern air Ar (cf. Krummenacher, 1970; Kaneoka, 1980), it may be concluded that it actually is fractionated modern air Ar. That samples can be contaminated with air or isotopically fractionated air is not especially surprising. There are, however, cases for other gases where no comparable (or comparably sensitive) criterion can be used to disinguish between fractionation/contamination and sampling of a reservoir of air composition but nevertheless distinct from air. It is thus useful to note that Ar provides an unequivocal demonstration that air fractionation/contamination can and does occur, whether or not a quantitative explanation of the effect is available.

6.5   Neon

As for the other noble gases, discussions of the origin of terrestrial Ne are usually framed in terms of the components identified in extraterrestrial

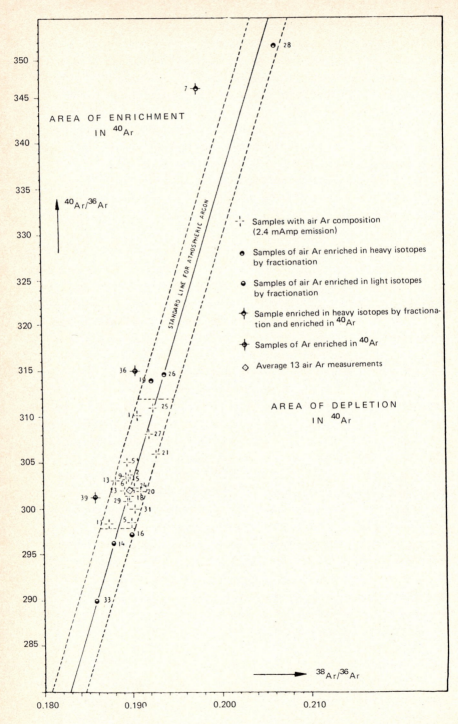

materials. The wide range of possibilities (Section 5.5) leads to a correspondingly large degree of ambiguity in accounting for the origin of terrestrial Ne. Moreover, for Ne more than for the other gases, it is possible to question whether air Ne adequately represents the initial or present total terrestrial Ne inventory.

The isotopic composition of air Ne does not match that of any preferred composition identified in extraterrestrial samples (Fig. 5.7). We might thus pose at least three different hypotheses for the origin of terrestrial Ne: (i) fractionated solar Ne, (ii) a mixture of Neon A and solar Ne, (iii) a mixture of Neon E and solar Ne.

The first hypothesis - that air Ne is fractionated solar Ne - amounts to ignoring the analogy between terrestrial gases and the planetary gases observed in meteorites and postulating that the origin of noble gases (or at least Ne) is fundamentally different in the earth than in meteorites. It also requires an additional *ad hoc* postulate to produce the fractionation. Alternatively, it might be supposed that both terrestrial and meteoritic Ne are fractionated solar Ne, and that they differ primarily in that terrestrial Ne is less fractionated (Fig. 5.7). This modification does not relieve the need for an additional postulate to explain the fractionation, however, and in either case the similarity of elemental abundance ratios in air and in meteoritic planetary gases (Section 5.3) must be dismissed as coincidence. Fractionation does not seem to be an acceptable explanation for the major affects observed in meteorites (Section 5.5); in the absence of independent evidence favoring this hypothesis we, along with most other workers in the field, do not favor the view that terrestrial Ne is fractionated solar Ne. (As noted in Section 6.7, there is strong evidence for severe fractionation in terrestrial Xe, but no adequate explanation for the origin of the fractionation. It might thus be argued that terrestrial Ne could be fractionated as well, by the same (unknown) mechanism. The absence of corresponding effects (Sections 6.4 and 6.8) in the gases of intermediate atomic weight - Ar and Kr - strongly suggests decoupling, however, i.e. that separate interpretations must be advanced for Ne and Xe.)

The second and third hypotheses noted above are variants of the customarily accepted proposition that terrestrial noble gases are to be interpreted in terms of component structures observed in extraterrestrial materials. They differ primarily in the interpretation of meteoritic planetary Ne, i.e. Neon A (Section 5.5). If Neon A is itself composite, a mixture of Neon E with planetary Ne of solar composition

Fig. 6.1. Ar compositions in recent volcanics. The solid line is the locus of compositions obtainable by fractionation of air composition. Data are *not* corrected for instrumental discrimination. Analyses of air Ar lie in a narrow field (dashed lines): the observation of sample analyses outside this field but on the fractionation trend indicates natural isotopic fractionation of Ar. Reproduced from Krummenacher (1970).

whose prevalence is a reflection of mixing in constant proportions, then terrestrial Ne is presumably a mixture of those same two components, differing from meteoritic Ne in that it contains a somewhat smaller proportion of Neon E. Since Neon E is not known to be associated with any other noble gas components, this interpretation does not lead to a prediction of any other differences between terrestrial and meteoritic planetary gases. If, on the other hand, Neon A is a distinct component which characterizes planetary Ne in general, then terrestrial Ne presumably should be viewed as (planetary) Neon A plus a superposed solar Ne component which must be invoked to account for its elevated $^{20}Ne/^{22}Ne$ ratio (Fig. 5.7). The solar component would presumably be added to dispersed dust or planetesimals before they accreted to form the earth. The inferred solar component would be relatively modest, and not affect elemental abundances very much, i.e. would not elevate the Ne/Ar ratio out of the planetary range and into the solar range (Sections 5.2 and 5.3). The solar component would make only a negligible contribution to the heavier gases Ar, Kr, and Xe, but would probably make a significant contribution to the original terrestrial inventory of He (cf. Section 6.6).

The discussion above is based entirely on the attempt to account for the $^{20}Ne/^{22}Ne$ ratio in air. These various interpretations make different predictions for the abundance of primordial ^{21}Ne (cf. Fig. 5.7). If primordial terrestrial Ne is a fractionated version of solar Ne or a mixture of solar Ne with a Neon A whose ^{21}Ne content is actually given by the nominal Neon A compositions such as listed in Table 5.4, then the present ^{21}Ne abundance in air is all or very nearly all primordial. It is possible, however, that primordial ^{21}Ne is significantly less than the present air abundance. This would be the case if terrestrial Ne were a mixture of Neon E and solar Ne, for example. Similarly, the original determination of Neon A composition (Pepin, 1967; Black & Pepin, 1969) employed the assumption that Neon A was on the extension of the supposed mixing line between solar and terrestrial Ne (Fig. 5.7), and thus also the assumption that all terrestrial ^{21}Ne is indeed primordial. If this assumption is relaxed, the difficulty in identifying small amounts of spallation ^{21}Ne in meteorites (Section 5.5) suggests the possibility that primordial planetary Ne composition actually lies farther to the left along the AS mixing line in Fig. 5.7 than is indicated by the nominal Neon A composition. This would indeed be the case, for example, if Neon A were actually a mixture of Neon E and solar Ne (Fig. 5.7). As pointed out by Heymann *et al.* (1976), in such cases primordial Ne might be significantly poorer in ^{21}Ne than air Ne, and an atmospheric excess of ^{21}Ne, perhaps up to approximately 10%, would have to be accounted for by other means. This possibility is only conjectural but it is certainly plausible, and at least two sources for excess ^{21}Ne, radiogenic and spallation components (see below), can be invoked.

There are two prominent ways in which Ne composition can be modified in the terrestrial environment: addition of radiogenic Ne (Section 6.2) and isotopic fractionation (cf. Section 11.6). It should be noted that while Ne has three isotopes and can thus be checked for the linearity which is the characteristic signature of fractionation (Section 4.9), such a check is often difficult or ambiguous in practice. Because of the low abundance and thus poor measurement precision of ^{21}Ne, and the possible presence of a radiogenic component, it is usually difficult to use the ^{21}Ne abundance to verify or refute the suggestion that a variation in ^{20}Ne/^{22}Ne arises by fractionation. It is thus relevant to emphasize that Ar observations definitely document the occurrence of isotopic fractionation among terrestrial reservoirs (Section 6.4).

Radiogenic Ne (Section 6.2) includes all three isotopes, but because of the scarcity of ^{21}Ne in nonradiogenic Ne a radiogenic component is most easily recognized by enhanced ^{21}Ne abundance. In most circumstances the presence of radiogenic Ne predicted on the basis of radiogenic ^{4}He (Table 6.1) is too low to be detectable. In some cases of very high U/Th and/or very low nonradiogenic Ne, however, the isotopic effects of added radiogenic Ne can be quite large (cf. Wetherill, 1954; Emerson *et al.*, 1966).

For the earth as a whole, radiogenic ^{21}Ne is apparently not negligible. Since the target is ^{18}O, the production of radiogenic ^{21}Ne is not as subject to geochemical caprice as is that of the other minor radiogenic gases (Section 6.2 and Table 3.3). Wetherill (1954) noted that total production of radiogenic ^{21}Ne in the age of the earth should amount to 1–2% of the air inventory, but this was largely ignored until relatively recently. Heymann *et al.* (1976) recalculated the production rate and adopted a higher U/Th abundance, and concluded that total radiogenic ^{21}Ne should be about 4% of the air inventory. Whether this fraction of air ^{21}Ne is actually radiogenic depends, of course, on whether or not radiogenic gas produced in the interior has been efficiently degassed into the atmosphere.

Heymann *et al.* (1976) have also called attention to the possibility (discussed above) that primordial Ne, or even primordial Ne plus radiogenic Ne, may not account for all the ^{21}Ne in air. They suggest that in such a case the excess ^{21}Ne could be accounted for by a spallation component. As above, this possibility is only conjectural, but it is also quantitatively plausible (Section 6.3).

Quite definite variations in Ne compositions are sometimes observed in igneous rocks, mostly in the sense of compositions somewhat lighter than air. The simplest explanation is fractionation, and Kaneoka (1980) has pointed out that the magnitudes involved are plausible: almost all data have ^{20}Ne/^{22}Ne $\geqslant$ 10.3, the value for single-stage diffusion with $m^{1/2}$ dependence (cf. Section 4.6). The corresponding ^{21}Ne/^{22}Ne ratios are generally consistent with fractionation, but often with added radiogenic Ne (Section 11.6 and Fig. 11.9) so, as noted earlier, ^{21}Ne measurements cannot usually be relied upon to demonstrate fractionation.

The question of Ne fractionation is important because alternative interpretations are possible. Thus, it is sometimes asserted, e.g. by Craig & Lupton (1976), that light Ne represents solar Ne rather than fractionated air Ne. Craig & Lupton consider that primordial Ne has solar composition (^{20}Ne/^{22}Ne ≈ 13) and that air Ne is heavier because of isotopic fractionation attendant on extensive gravitational escape of atmospheric Ne early in earth history. They therefore interpret Ne lighter than air (up to ^{20}Ne/^{22}Ne = 10.3 in their Kilauea gas) as indicating admixture of juvenile solar Ne with contaminant air Ne. (They use the term 'primordial' in the sense in which we use 'solar'; this seems inappropriate since it carries a connotation that air Ne is not primordial.) Craig & Lupton do not discuss the possibility of producing light Ne by fractionation nor the possibility that primordial terrestrial Ne might have a composition other than solar. As follows from our earlier discussion, we disagree with the Craig & Lupton (1976) interpretation. It is certainly possible but seems unnecessarily complicated, since there is no independent evidence for the required global fractionation and the present elemental abundance of Ne in air would again have to be considered only coincidentally in the planetary range (Section 5.3). There is, on the other hand, ample evidence for local fractionation in individual samples and a clear expectation, based on the meteorite observations, that primordial Ne in the earth *should* be heavier than solar Ne. All in all, we find no compelling evidence that either the present or any past total terrestrial Ne inventory has a ^{20}Ne/^{22}Ne ratio significantly different from that of modern air.

6.6   Helium

In sources of continental affinity, He is evidently at least predominantly radiogenic. In rocks, observed ^{4}He can be accounted for by *in situ* production in U–Th decay; in fact, as is well known, ^{4}He is often deficient relative to predicted accumulation in ages defined by other chronometers, i.e. ^{4}He is prone to diffusive loss. The ^{4}He is accompanied by ^{3}He in ratios generally less than the atmospheric ratio but variable, and on the whole consistent with radiogenic production also (Section 6.2); a review is given by Tolstikhin (1978). He is also sometimes prominent in natural gas wells and emanations (Chapter 10); here also the He is apparently radiogenic, and while the geological circumstances leading to gas well accumulation may be unusual, special circumstances for producing the He are evidently unnecessary, production in local rocks being sufficient (cf. Morrison & Pine, 1955). Altogether, any other He which may be present in continental materials is masked by a dominant radiogenic component.

In rocks more closely associated with the mantle, i.e. more directly and more recently derived from the mantle, 'excess He' is common (Section 9.4). This simply means that they contain more He than can be produced by *in situ* radio-active decay in their ages (emplacement ages), and the effect is understandably more prominent in younger rocks. The excess He is inherited from whatever the appropriate mantle reservoir is (Chapter 11). This is the same phenomenon as 'excess ^{40}Ar' and is likewise particularly associated with submarine volcanics.

In the 'mantle He' component, ^{4}He is accompanied by ^{3}He in ratios that are variable in different samples but typically around ^{3}He/^{4}He $\approx 1.2 \times 10^{-5}$ (cf. Chapters 9 and 11). This is an order of magnitude higher than the air ratio and so not attributable to air contamination. It is also far too high to characterize radiogenic He (Section 6.2), so the mantle component He is, by default, generally considered to be primordial. Specifically, Gerling *et al.* (1971) and Tolstikhin *et al.* (1974) have given detailed consideration to the possible production of such elevated ^{3}He by nuclear processes within the solid earth; they find no plausible mechanisms and so support the assertion that it is primordial. Actually, it is not particularly startling that the solid earth should contain primordial He, at least in the sense that the quantities involved are not large in comparison with other noble gases (in air) long considered primordial. The observation of primordial He in the mantle is nevertheless quite important: it shows that the earth cannot be completely degassed of its original volatile inventory; another way of saying the same thing is that it is the only case in which a primordial (nonradiogenic) noble gas observed in a rock can with complete assurance be judged juvenile rather than simply representing atmospheric contamination.

It is not exactly easy to predict what the primordial He composition of the earth should be. The planetary elemental abundance pattern (Section 5.3) of the other noble gases (except for Xe) in air suggests that primordial He in the earth should also be planetary, and thus have Helium-A composition (Table 5.4), ^{3}He/^{4}He $= 1.4 \times 10^{-4}$. Air Ne composition suggests the possibility of solar wind irradiation of earth materials in the dispersed state before accretion (Section 6.5); if so, and if this irradiation occurred after deuterium-burning (Section 5.5), an initial composition richer in ^{3}He, up to Helium-B composition, ^{3}He/^{4}He $= 3.9 \times 10^{-4}$, would be possible (cf. Section 9.8), but if before deuterium-burning the composition should still be Helium A. Also as suggested by air Ne, there may be a nontrivial spallation component (Section 6.3) which would elevate the initial ^{3}He/^{4}He ratio somewhat.

In geochemical literature discussing primordial He in the mantle, there is sometimes a connotation (or an assertion) that both the ^{3}He and ^{4}He are primordial. This is almost certainly incorrect. The most likely primordial He composition, Helium A, has ^{3}He/^{4}He $= 1.4 \times 10^{-4}$, and the uncertainties range to compositions richer in ^{3}He. Any composition leaner in ^{3}He is most plausibly interpreted as a mixture of radiogenic He and primordial He. There being no evidence for any primordial He leaner in ^{3}He than Helium A anywhere else in the solar system, nor any dearth of U–Th to supply radiogenic He to the mantle, there is no basis for invoking a ^{3}He-poor primordial component for the earth. For a typical mantle ^{3}He/^{4}He $= 1.2 \times 10^{-5}$, resolution into Helium A and radiogenic He indicates that the ^{3}He is indeed essentially all primordial but that the ^{4}He is 90% radiogenic.

Also, in most literature discussions of mantle He, and in this section, 'primordial' implicitly means 'present when the earth formed'. More generally in cosmochemical literature, and elsewhere in this book, primordial has the more restricted meaning 'present when the solar system formed'. If terrestrial He contains a solar wind component or a spallation component enhanced in ^{3}He by virtue of nuclear reaction in the solar system, as mentioned above, then the earth's original inventory of ^{3}He is primordial only in the first sense, but not the second. For most of the ramifications that have come to be associated with primordial ^{3}He, the distinction, if indeed there is a difference, is irrelevant; it is not totally academic, however, since it affects estimates of what other noble gases might be expected to be associated with ^{3}He.

At plate margins, fumaroles and hot springs often contain He enriched in ^{3}He relative to air, indicating a He contribution from the mantle (Chapter 10). Surface waters in general contain He in solubility equilibrium or nearly so with air, but groundwater sometimes contains prominent excess radiogenic He extracted from its host rocks, and seawater in general contains varying amounts of excess 'mantle He', with its characteristic high (relative to air) ^{3}He/^{4}He ratio (Chapter 7). Primordial ^{3}He was, in fact, first detected and identified as such by virtue of excess ^{3}He in seawater (Clarke *et al.*, 1969). The seawater observations are important for a number of reasons, not the least of which is that excess He in seawater provides the only available basis for quantitative estimation of present juvenile flux of any atmospheric species, radiogenic or primordial, noble gas or not.

He entering the atmosphere does not accumulate, but escapes with a characteristic time of 10^6 yr; a discussion of the atmospheric He budget is presented in Section 12.5. Probably excepting prephotosynthetic O_2, He is the only atmospheric species whose total inventory is significantly influenced by escape. The sources and escape mechanisms for ^{3}He and ^{4}He are not closely coupled, and for ^{3}He the major source may not even be geologic. Atmospheric He is important in geochemistry in its role as a source for other samples (e.g. water and sediments), but its abundance and composition are essentially an accident of no great significance. Neither in abundance nor in composition does atmospheric He bear any particularly close relationship to its geological sources, and both abundance and composition are probably significantly variable on a geologically short timescale.

6.7   Xenon

The attempt to account for terrestrial isotopic composition in terms of
isotopic compositions observed elsewhere in the solar system runs into far greater
complexities for Xe than for any other noble gas (indeed, for any other element).
There are three major and relatively well-characterized kinds of Xe in the solar
system: terrestrial (air) Xe, planetary Xe (trapped Xe in primitive meteorites),
and solar Xe (solar wind implanted in meteorites and lunar samples). As described in more detail in Section 5.6, all three are quite distinct in their isotopic
compositions (Table 5.5) and relationships between them are far from obvious.

Even casual inspection of the data suggests that the largest effect which distin-
guishes terrestrial from extraterrestrial Xe is isotopic fractionation (Figs. 5.8 and
6.2). In detail, fractionation cannot account for all the differences and specific
isotope effects must be invoked as well, but the suggestion for fractionation is
so strong that it is generally accepted even without a comprehensive theory for
the origin of terrestrial Xe. More specifically, the relative abundances of four
isotopes – ^{124}Xe, ^{126}Xe, ^{128}Xe, and ^{130}Xe – are essentially identical in solar and
(most) meteoritic planetary Xe. For these same four isotopes, fractionation *is*
adequate to account for relative abundances in air Xe, without the need for
specific isotope effects, with reasonable allowance for experimental uncertainties
and uncertainty in the exact functional form of the fractionation (Fig. 5.8). It
is noteworthy that these four isotopes are also those which are shielded by more
neutron-rich isobars (Fig. 3.1f). This nuclear feature suggests the basic premise
which has become a major feature of xenology: there is a dominant 'primitive'
Xe component common to both planetary and solar Xe and, in fractionated
form, to terrestrial Xe as well, and the differences among these observed compositions are generated by superposed nuclear components constituted of the heavy
isotopes which can be produced from neutron-rich progenitors.

Although some early investigators considered terrestrial Xe to be primitive
and thus meteoritic Xe to be fractionated, continually growing appreciation of
the general primitive nature of meteorites and recognition of the nature of solar
Xe indicate that this provincial view is untenable. It is clear that terrestrial Xe
is the special case which must be explained in terms of what is known about Xe
in the rest of the solar system.

The inferred fractionation is severe, about 3.7%/amu (if linear). Such a fractionation does not emerge in any natural way as a prediction of general models
for planetary formation, but *ad hoc* assumptions to produce the fractionation
can be inserted readily. Since its light isotopes are depleted, terrestrial Xe can be
imagined as the fractionated residue from diffusive loss (small grains in space)
or from loss controlled by a gravitational field (exospheric loss from a planetesimal

or even an almost fully accreted earth). Xe isotopic fractionation also emerges from the gravitational stratification model of Ozima & Nakazawa (1980) (see Section 5.7).

A significant problem with any model which produces fractionation as a result of loss, aside from justifying the astrophysical scenario involved, is that any loss process severe enough to so deplete the light Xe isotopes will surely result in substantial loss of Xe and effectively total loss of lighter species, specifically the lighter noble gases. This difficulty can be circumvented by further postulating that terrestrial gases are a mixture of the fractionated Xe plus gases added after the fractionation, presumably normal planetary gases are judged from the relative proportions of Ne, Ar, and Kr (Section 5.3). This approach generates other problems: the original fractionation must have been even more severe because present Xe is diluted by added normal Xe, and the total abundance of Xe should then be greater than the planetary abundance (relative to the lighter gases). The abundance of Xe in air is in fact low rather than high in this sense (Section 5.3), however. This problem can be averted by still another postulate that Xe in air is selectively depleted relative to the other gases, a plausible premise but one difficult to defend on the basis of actual observation (cf. Section 12.6). Any comprehensive model must also explain a much smaller but opposite sense fractionation in Kr (Section 6.8) and the absence of fractionation in Ar.

The specific isotope differences in the heavy Xe isotopes clearly suggest a fission origin. No known fissioning nuclide could plausibly have been present in sufficient abundance to produce the observed effects, however. Irrespective of the quantities involved, no known fission Xe composition could account for the differences between any two of solar, planetary, and terrestrial Xe. Moreover, no single component of any composition could account for the differences among all three kinds of Xe. Candidate models are thus usually framed in terms of multiple heavy isotope components, and even so it is difficult to construct a model without allowing for at least one heavy isotope component in all three kinds of Xe. This is, in fact, not at all unreasonable: a nontrivial fission Xe component in the earth, from ^{244}Pu and/or ^{238}U, is indeed to be expected (Section 6.2), meteoritic planetary Xe has long been known to display a complex internal structure involving a heavy isotope component (Reynolds, 1963; Reynolds & Turner, 1964), and even surface-correlated Xe in lunar mare soils, the most precise estimates of solar Xe composition available, may contain a nontrivial admixture of lunar fission Xe (Podosek *et al.*, 1971; Drozd *et al.*, 1976). In such a generalization the Xe problem is certainly soluble in principle. For several reasons, however – finite uncertainties in data, the difficulties in characterization of the structures within planetary Xe (Section 5.6), uncertainties in the exact fractionation function, and the large number of degrees of freedom involved with so many components – a satisfactory quantitative resolution of this problem has long remained elusive.

An extensive synthesis and proposed resolution of the terrestrial Xe problem are presented by Pepin & Phinney (1983). They infer a composition for the primitive component, designated U-xenon (Table 5.5), such that superposition of other components identified in meteorites can account for both the meteorite data and solar Xe as well, and such that addition of fission Xe to fractionated U-Xe accounts for atmospheric Xe composition (Fig. 6.2). There are a number of difficulties with this model (cf. Section 5.6) and it is incomplete in that it is strictly formal and does not attempt to explain the reason for the fractionation or the earth's lack of the extra Xe components found in meteorites (and the sun!). It nevertheless has no competitors with comparable detail. In the Pepin & Phinney model the terrestrial fission Xe is inferred (not assumed) to be primarily from ^{244}Pu and to account for 4.7% of ^{136}Xe; both of these inferences are quite reasonable.

Relative to fractionated air Xe, a number of meteorites show apparent deficiencies of ^{129}Xe (cf. Fig. 5.8). This is generally interpreted as indicating that the earth, in common with most meteorites, incorporated ^{129}I which has subsequently decayed to ^{129}Xe and thus made a significant contribution to air Xe (Marti, 1967). Taking the lowest observed meteoritic abundance of ^{129}Xe to set an upper limit to (nonradiogenic) primordial ^{129}Xe, and allowing for the fractionation (cf. Fig. 6.2), the corresponding (lower limit) radiogenic contribution to atmospheric ^{129}Xe is 6.8% (Pepin & Phinney, 1983).

6.8 Krypton

The isotopic composition of terrestrial Kr, as represented by Kr in air, is not exactly the same as that in any known extraterrestrial reservoir, but the relevant relationships seem fairly straightforward. Within errors, planetary (meteorite) Kr and solar Kr are identical in the relative abundances of all isotopes except ^{86}Kr (Section 5.6); for the same isotopes, terrestrial Kr seems to be a slightly fractionated (about 0.5%/amu) version of this general solar system reservoir (Fig. 5.14). As with Xe, there is no quantitatively justifiable explanation for the origin of the fractionation, but that it is indeed fractionation, rather than some more complex interpretation, seems reasonable nevertheless. It should be noted that the inferred Kr fractionation is considerably less severe than the Xe fractionation and is in the opposite direction.

As described in Section 5.6, planetary Kr differs from solar Kr in that it contains about 1–2% more ^{86}Kr (Fig. 5.14). The presumptive interpretation is that planetary Kr is solar Kr plus an additional component, one relatively rich in ^{86}Kr but possibly also containing ^{84}Kr and ^{83}Kr in abundances small enough

Fig. 6.2. Xe isotopic compositions displayed as per mil deviations from a reference composition, with isotope ratios normalized to ^{130}Xe (cf. Fig. 5.8). The reference composition is U-xenon (Table 5.5 and Section 5.6) fractionated to match the relative abundances of the shielded isotopes in air. Reproduced from Pepin & Phinney (1983).

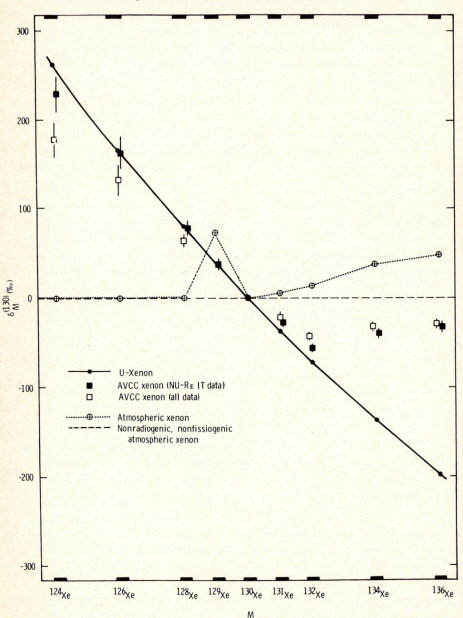

U-Xenon

AVCC xenon (NU-R$_E$ IT data)

AVCC xenon (all data)

Atmospheric xenon

Nonradiogenic, nonfissiogenic atmospheric xenon

that the isotopic comparison is not significantly affected. The added Kr might be an anomalous Kr component or might be a CCF Kr component, alternatives analogous to those for Xe. If it is hypothesized that terrestrial Kr is fractionated planetary Kr, the compositional match is satisfactory and no additional assumptions concerning specific isotope components are required. It should be noted, however, that this view assumes that the heavy Kr component present in meteorites is also present in the earth; this proposition is reasonable enough in itself, but contrasts with the case of Xe, for which it seems that the prominent heavy isotope component present in meteorites is absent in the earth (Sections 5.6 and 6.7). Alternatively, it may be hypothesized that the earth lacks this heavy isotope Kr component, i.e. that terrestrial Kr is predominantly fractionated solar Kr. In this case, however, terrestrial Kr apparently contains excess ^{86}Kr (relative to fractionated solar Kr), coincidentally also about a 1–2% enrichment (Fig. 5.14), which must be accounted for in terms of some other source. In terms of known nuclear components the only candidate is fission, which is qualitatively plausible but quantitatively inadequate (Section 6.2).

Only two known nuclear components can plausibly be invoked to modify terrestrial Kr compositions: fission and neutron capture by Br (cf. Chapter 3). It is highly unlikely that these components are significant in the total Kr inventory, and even local enhancements in specific samples are quite rare. With such rare exceptions, all measurements of Kr in terrestrial samples yield either air composition or slightly fractionated air composition. We may thus conclude that, possibly excepting fractionation effects, 'terrestrial' Kr composition is that of air Kr.

6.9   Excess ^{129}Xe (and ^{136}Xe?)

As was described in Section 6.7, it appears that atmospheric Xe contains ^{129}Xe in abundance about 7% greater than the presumed primordial Xe originally trapped in earth materials. This 7% excess ^{129}Xe is generally presumed to result from decay of ^{129}I, a now extinct (half-life 17 Ma) primordial radionuclide. Since ^{129}I was present only in low abundance (^{129}I/^{127}I ≈ 10^{-4}) early in the solar system's history, its decay would make no appreciable difference in Xe composition until after the general chemical fractionation which characterizes the terrestrial planets (Sections 5.1 and 12.1), i.e. the depletion of volatiles (including depletion of Xe relative to I). The implication of this excess ^{129}Xe in the earth's atmosphere is then that for earth materials this fractionation took place relatively early, before the extinction of ^{129}I. This excess ^{129}Xe is often spoken of as a remarkable phenomenon. It should not be so considered. The phenomenon of early condensation has been long and well established on the basis of meteorite studies which

show that ^{129}Xe excesses are the rule rather than the exception and which constrain the promptness of condensation much more severely on the basis of other
short-lived radionuclides (notably ^{26}Al) than they do on the basis of ^{129}I. If
anything, the phenomenon of excess ^{129}Xe in air is remarkable not because it
exists at all but because the excess is not greater than it actually is (see Section
12.7).

There is another phenomenon also loosely described as 'excess ^{129}Xe'. There
are a number of observations (see Chapters 9 and 10) of terrestrial Xe in which
^{129}Xe is present *in excess relative to air.* Such excesses are also generally ascribed
to decay of ^{129}I. It is important to appreciate the difference between this kind of
excess ^{129}Xe and the effect described in the previous paragraph. Overabundance
of ^{129}Xe in any sample, relative to air, is indeed a remarkable phenomenon, one
which has not received its due attention outside the community of noble gas
specialists. It implies that there exist reservoirs within the earth that have not
been mixed (with whatever other parts of the earth were degassed to form the
atmosphere) since the time when ^{129}I was still appreciably extant, which could
not have been more than a few times 10^7 yr after the formation of the earth.
Furthermore, these reservoirs have not only maintained their isotopic integrity
for more than 99% of the earth's history, they can be sampled today!

Most of the laboratories which report terrestrial noble gas observations
also routinely analyse meteorites. In particular, this is true for the laboratories
which have reported ^{129}Xe excesses relative to air. It may be that one reason
why this effect has received relatively little attention has been an unarticulated
suspicion that this effect is an artifact, a contamination of terrestrial noble gas
samples with meteoritic noble gases in which large ^{129}Xe excesses are common-
place. This ambiguity has recently been removed. Smith & Reynolds (1981)
have confirmed the existence of excess ^{129}Xe (10% higher in natural CO_2 well
gas than in air) by analysis in an experimental system of which no part had
previously been used for meteorite analysis; the effect is clearly not explainable
as meteoritic contamination.

It was also shown in Section 6.7 that atmospheric Xe also contains more of
the heavy isotopes (≈ 5% excess at ^{136}Xe) than its probable primordial composition. With only somewhat less certainty, the excess heavy isotopes are ascribed
to spontaneous fission of ^{244}Pu, another short-lived (82 Ma) radionuclide present
in the early solar system. The quantities involved are quite reasonable (cf.
Sections 6.2 and 12.7). The situation is quite analogous to the case for ^{129}Xe
and ^{129}I. Accordingly, we would expect also to find some samples which contain
^{244}Pu-fission excesses beyond the amount present in air; if anything, ^{244}Pu-
fission excesses should be larger and more widespread than the ^{129}Xe excesses
because of the significantly longer half-life of ^{244}Pu. Curiously, this is not so: no

terrestrial samples have been observed to have ^{244}Pu-fission Xe excesses relative to air, not those with the ^{129}Xe excesses nor any others. Perhaps those reservoirs which have been isotopically isolated since very early in the earth's history were strongly depleted in Pu relative to I. Even this would not really solve the problem, however, since isolation of a reservoir with very low Pu/I before full decay of ^{129}I would also precede full decay of longer-lived ^{244}Pu, so the Xe in this reservoir should be *depleted* in heavy isotopes relative to air, and there are no such observations. The lack of large ^{244}Pu-fission excesses poses problems for formal degassing models which are considered in Section 13.8.

As recently suggested by Staudacher & Allègre (1982), there may actually be a loose correlation between excesses, relative to air composition, of ^{129}Xe and ^{136}Xe (see Section 11.7). Staudacher & Allègre (1982) estimate a radiogenic excess ratio ^{136}Xe/^{129}Xe ≈ 0.2 and *assume* that the parent is ^{244}Pu. This value is essentially identical to the radiogenic contributions to air (0.23); even if this view is correct, the ratio is still surprisingly low (Section 13.8). Also, it does not readily fit the most extreme observed enrichments of ^{129}Xe, in well gas (Section 10.3), which indeed have accompanying enrichments of ^{136}Xe, but from ^{238}U not ^{244}Pu. However, it would also be possible to argue that the parent is ^{238}U. This is somewhat less difficult to accommodate in degassing models (Section 13.8) but then accentuates the problem of lack of ^{244}Pu excesses. Even if the meteorite analogy is abandoned to the extent that no assumptions are made about the composition of initial nonradiogenic terrestrial Xe, the presence of ^{244}Pu in the early solar system suggests that there *should* be ^{244}Pu-fission Xe excesses.

6.10   Anomalies?

This chapter has been devoted to the attempt to identify observed occurrences of terrestrial noble gases in terms of primordial components – those known, principally on the basis of meteorite studies, to be the stuff of which the solar system in general is made – plus specific nuclear components which were generated within the solar system, and plausible physical and chemical processes which act to modify their distributions. There are some difficulties or potential difficulties, for instance the nature of the earth's original Xe content and the nature of the process which produced such a large apparent isotopic fractionation (Section 6.7), or the absence of a suitable source for excess ^{86}Kr in air (Sections 6.2 and 6.8) if, indeed, there is such an excess. Nevertheless, on the whole, the enterprise is successful. The important questions and problems are those of accounting for the distribution of components, not of what the components are. Put another way, the questions that are raised are of the 'when, where, and how' variety; these are pursued in Chapters 7–13.

By and large, there seems not too much problem with the 'what', and there seems little need to invoke a *Deus ex machina* to account for any fundamental difficulties about where the earth's noble gases come from.

In this section we describe two potential anomalies which would belie this generalization. In both cases it is widely considered that the problems are interpretational or experimental, not real anomalies. Nevertheless, the questions raised are sufficiently profound that they merit consideration here.

The first case is that of Ar in iron meteorites. Hennecke & Manuel (1977) observed that stepwise heating Ar compositions in iron meteorites (metal phase) varied in accord with two-component mixing: one component is plausibly cosmic-ray spallation, the other is richer in ^{40}Ar and is consistent with air composition. Such observations, of both other gases and other extraterrestrial samples, are not uncommon in samples with very low trapped gas contents. The usual interpretation is that the air-like component (when compositionally distinguishable) is exactly what it appears to be: air. Hennecke & Manuel, however, argue that the iron meteorites contain true extraterrestrial trapped Ar of air composition.

The customary interpretation (Section 6.4) is that primordial ^{40}Ar is nil and that air composition, ^{40}Ar/^{36}Ar $= 295$, reflects the earth's low total Ar/K ratio and its history of degassing. It asks too much of coincidence to suppose that some other planetary body has evolved to the same ^{40}Ar/^{36}Ar ratio. Hennecke & Manuel accordingly assert that $^{40/36}$Ar $= 295$ is a *primordial* composition common to earth and iron meteorites. This view is at odds with much of the 'established' precepts of geochemistry. For the earth, the implication is that there has been no significant evolution of Ar composition by addition of radiogenic ^{40}Ar. For the solar system as a whole the implications are even more profound: such a high ^{40}Ar/^{36}Ar ratio cannot, in any nucleosynthetic theory, be produced in stellar nucleosynthesis. The only known way to produce such Ar is the conventional way: by ^{40}K decay in a planetary environment of very low Ar/K. In the absence of any compelling reasons to assign an extraterrestrial origin to air-like Ar in iron meteorites, support of so exotic a hypothesis seems unjustified.

Air-like Ar in iron meteorites might be atmospheric contamination in or on the sample or might simply be a higher than expected blank in the extraction system. Fisher (1981) explicitly suggests the latter, specifically fluxing of the crucible upon melting of the iron. In an experiment designed to minimize this potential artifact he also finds air-like Ar in an iron meteorite; its release pattern does not support arguments for extraterrestrial origin, however, and he concludes that in his data as well as those of Hennecke & Manuel the air-like Ar is indeed air.

The second case is that of josephinite, a very unusual and rare, perhaps unique, rock whose characteristic mineralogy is an intimate intergrowth of metallic Fe-Ni alloys with andradite garnet. So far it has been found only as placers in streams cutting through a partly serpentinized alpine peridotite in Josephine County, Oregon (USA). Two models for its origin have been advocated: (i) a byproduct of the serpentinization, (ii) a sample of the earth's core, later qualified to at least an exceptionally deep mantle origin. The debate has been conducted mostly in terms of geochemical and petrologic arguments. Josephinite merits particular attention here because of the special nature of noble gas data which have been used in the arguments. Experimental data have been reported by Downing *et al.* (1977), Bochsler *et al.* (1978), Bernatowicz *et al.* (1979), and Craig *et al.* (1979); as stressed by Bernatowicz *et al.*, the noble gas data have not yet been of particularly great use in resolving the question of the origin of josephinite but have in turn raised questions that transcend even that of whether or not a sample of the core is accessible on the surface.

Downing *et al.* (1977) observed primordial ^{3}He in josephinite, ^{3}He/^{4}He up to 4.0×10^{-5} in total sample, notably high but not without parallel (Section 11.4) and reasonable considering the presumed 'mantle' origin of alpine-type peridotites. They also assert a number of quite unusual effects: (i) anomalous primordial noble gases, notably air-like Ar (as above), (ii) a K-Ar age of 4.6×10^9 yr, (iii) excess ^{129}Xe, in the sense indicated in Section 6.9, furthermore supposed to represent *in situ* decay of ^{129}I. Bernatowicz *et al.* (1979), in contrast, observed noble gases judged 'entirely unremarkable', and further assert that all the unusual effects cited by Downing *et al.* are either misinterpretations or unjustifiable advocacy of 'exciting' interpretations over quite satisfactory but unexciting explanations; they conclude that there are no especially unusual features evident in either their data or those of Downing *et al.*

The situation is quite different for the data of Bochsler *et al.* (1978), who found two exceptional isotopic effects which cannot be understood as 'ordinary'. One effect was *very* high ^{3}He/^{4}He, up to 3.9×10^{-4} in total sample (higher in individual temperature steps), higher than the planetary value and, within the framework of compositions observable elsewhere in the solar system, understandable only as solar He (cf. Section 5.5) or spallation He (cf. Section 6.3). More remarkable still, they observed an unparalleled enrichment of ^{21}Ne, up to ^{21}Ne/^{20}Ne $= 0.038$ in total sample, 13 times the air ratio; excess ^{21}Ne was accompanied by smaller excess of ^{22}Ne (0.1–0.2 times excess ^{21}Ne if ^{20}Ne is air). Bochsler *et al.* considered various normal nuclear processes which might be responsible, including radiogenic Ne (cf. Section 6.2) and spallation Ne (cf. Section 6.3), but found no suitable mechanism. If the effect is real and not attributable to some overlooked nuclear process in the earth, the only alternative is an anomalous nuclear component (Section 5.4). If the Ne is anomalous, the He is likely to be so also. This would be the *only* (known) case of a primordial inhomogeneity in *any* element needed to account for isotopic variations between different samples of the earth. Bochsler *et al.* (1978) noted that no comparable Ne component is known in extraterrestrial materials, whence this anomalous component would be unique to the earth, and Bernatowicz *et al.* (1979) noted that, aside from the origin of josephinite, the existence of any such sample sets very severe constraints on models for the origin of the earth, requiring preservation of isotopic integrity throughout the history of the earth and yet recent accessibility on the surface.

Bernatowicz *et al.* (1979) found no Ne anomalies and Craig *et al.* (1979) found neither Ne nor He anomalies, and both raised the possibility of experimental artifact in the Bochsler *et al.* data. Craig *et al.* specifically suggested scavenging of previous sample gas (notably Ne produced in a nuclear reactor) in RF discharge in H_2 released from the josephinite.

The issues raised by the josephinite data are profound, but the situation is not yet resolved. Pending definitive experimental clarification, it seems best to consider noble gases in josephinite to be unremarkable, the previously spectacular results ascribed to misinterpretation by Downing *et al.* (1977) and probable experimental artifact by Bochsler *et al.* (1978).

7 Water

7.1 Introduction

As a good first-order generalization, the noble gases found in natural waters are acquired from air and are present in concentrations approximately consistent with air equilibration. Solubility data (Tables 7.1–7.4 and Fig. 7.1) are thus of central importance in evaluating noble gas observations in water. A recent comprehensive review and data evaluation for the general phenomenon of gas solution in water is given by Wilhelm *et al.* (1977).

On the whole, noble gases exhibit about the same order of magnitude of solubility in water as do other gases which do not react chemically with the water. Ar, in particular, is approximately as soluble as the major atmospheric gases: its solubility (pure water at 0 °C) is 2.26 times that of N_2 and 1.09 times that of O_2. As a group, however, the noble gases exhibit a fairly wide spread in solubilities, with the characteristic features of strongly increasing solubility and temperature dependence of solubility with increasing atomic weight. This signature, combined with the useful feature that (with exceptions discussed later) they are conservative - no sources or sinks in organisms or other material in seawater and unlikely to participate in complex chemical reactions - makes the noble gases useful in a variety of geochemical studies.

A noteworthy feature of such studies is that they frequently make the most stringent demands encountered in noble gas geochemistry for high-precision absolute elemental abundances. Particularly in marine studies the effects of interest are manifested in elemental abundance variations measured in per cent. For Ar, accuracy at the required level can be achieved in gas chromatography; for mass spectrometric abundance determinations at the per cent level, isotope dilution is usually necessary. The technical problems are also different from those in most other noble gas spectrometry: ample gas amounts are available but special care must be taken to insure complete extraction of gases from the water sample and then to separate the water vapor from the noble gases. Care must also be exercised in sampling, since a small air contamination can produce large distortions (cf. Table 7.5).

To highlight differences between observed and expected gas concentrations, a delta value (usually in per cent) representation is often used:

$$\Delta \equiv (c/c^* - 1) \times 100\% \qquad (7.1)$$

where c is observed concentration and c^* is the expected or equilibrium concentration. Generally, c^* is calculated for water at a specified temperature in equilibrium with vapor-saturated air at 760 torr total pressure, in which case Δ is the 'wet-saturation anomaly'. For deep ocean water the relevant temperature is the potential rather than the *in situ* temperature. Representative values of c^* are included in Table 7.6.

Failure of a given water sample to reflect equilibrium with the air with which it was last in contact can be ascribed to a variety of causes. An obvious and important case is non-conservation, e.g. addition of nonatmospheric gases (Sections 7.5 and 7.6). Equally obvious is simple lack of equilibration for kinetic reasons, e.g. rapid isolation of glacial melt, which would be difficult to treat quantitatively. This evidently is not too severe a problem, however, since equilibration times are apparently fairly short, e.g. a day or so for the top few meters of seawater (cf. Broecker, 1974); in any case we are unaware of any saturation anomalies ascribed to kinetic equilibration failure. There are, however, a number of additional effects which are likely to be significant and can be described quantitatively. These are discussed below and illustrated in Table 7.5.

(i) *Pressure effects.* Variations spanning a few per cent or so can be generated by total barometric pressures different from 760 torr and by relative humidities different from 100%. A given pressure variation will produce the same Δ for all gases, independent of temperature.

(ii) *Temperature effects.* Anomalies will arise if the wrong temperature is used to calculate c^*. This could be due to evaporative effects at the air–water interface or to a change in water temperature after isolation from air (other than that from adiabatic compression). The apparent anomaly will be greater for the heavier gases and greater at lower temperature, reflecting the temperature dependence of c^*.

(iii) *Bubbles.* Bubbles will produce supersaturation because of the hydrostatic pressure excess. At one extreme, if only a small fraction of the air in a bubble disolves before it escapes to the surface, this will produce the same effect as an increase in pressure, item (i) above. The other extreme, when the bubble completely dissolves, is denoted *air injection*. The resultant Δ will be inversely proportional to c^* and so be greater for the lighter gases and smaller at lower temperatures, although Δ for He and Ne will not be very sensitive to temperature.

(iv) *Mixing.* Because the solubility–temperature curves are concave from above (Fig. 7.1), mixing of waters equilibrated at different temperatures will produce apparent supersaturation relative to the temperature of the mix. The effect will be greatest for the heavier gases, but will be minimal for He and Ne, and in the normally encountered range is greater at lower temperatures.

Because of the variety of effects which can and apparently do contribute to saturation anomalies, it is not possible to determine the origin of a given anomaly from a single datum. Resolution becomes feasible if there are data for as many gases (in a single sample) as there are effects, however. If all the effects are relatively small, the net anomaly Δ is approximately the sum of the Δs from the individual effects. For He, Ne, and Ar in particular, for which the most extensive data are available, plausibly likely mixing anomalies are small (cf. Table 7.5). If it is assumed that only pressure and temperature variations and air injection, items (i) (ii), and (iii) above, contribute to the apparent saturation anomalies of a given water sample, then for each gas the observed Δ is a homogeneous linear combination of ΔP, ΔT, and Δa, representing pressure, temperature, and air injection. The coefficients depend only on the solubility (and temperature) and are different for each gas (Table 7.5). Thus, knowledge of ΔHe, ΔNe, and ΔAr permits inversion and determination of ΔP, ΔT, and Δa. Formal application of this approach is illustrated by Craig & Weiss (1971) (cf. Fig. 7.2).

Observed saturation anomalies can thus be resolved into contributions from the effects described above as long as the data are sufficiently precise. In practice, the effects cannot be resolved as clearly as might be hoped, and, as noted by Craig & Weiss (1971), the resolution is quite sensitive to experimental uncertainties.

7.2   Solubility data

The early geochemical literature concerned with noble gases in water relied heavily on the solubility data of Morrison & Johnstone (1954) for pure water and Konig (1963) for seawater. Subsequent and more accurate experimental data differ from these early standards by a few to several per cent.

Weiss (1970a) reviewed and evaluated the literature data then available for N_2, O_2, and Ar and fitted them to a smooth functional dependence. His results are reported in terms of the Bunsen coefficient β, defined as the quantity of gas (in cm^3 STP) dissolved in unit volume (1 cm^3) under unit partial pressure (1 atm). Pure water data were fitted to an integrated van't Hoff equation:

$$\ln \beta = a_1 + a_2/T + a_3 \ln T \tag{7.2}$$

Because of the 'salting out' effect, gases are some 25% less soluble in seawater than in fresh water (Table 7.4). The effect is often rendered in terms of the Setchenow relation

$$\ln \beta = b_1 + b_2 S \tag{7.3}$$

where b_1 gives $\ln \beta$ in pure water and b_1 and b_2 are independent of salinity S.

Combining these relations and scaling the temperature, Weiss (1970a) fitted the data to

$$\ln \beta = A_1 + A_2(100/T) + A_3 \ln(T/100)$$
$$+ S[B_1 + B_2(T/100) + B_3(T/100)^2] \qquad (7.4)$$

Parameters for this equation for N_2, O_2, and Ar solubility are shown in Table 7.1, scaled for salinity S in per mil ($\%_{00}$).

Fig. 7.1. Solubility of gases in pure water, from Table 7.1 (N_2, Ar) and Table 7.3 (He, Ne, O_2, Kr, Xe). Ordinate of main figure is Henry's Law solubility. Ordinate scales at right show concentrations (in cm^3 STP/g) for equilibrium with 760 torr *dry air* with composition given in Table 2.1.

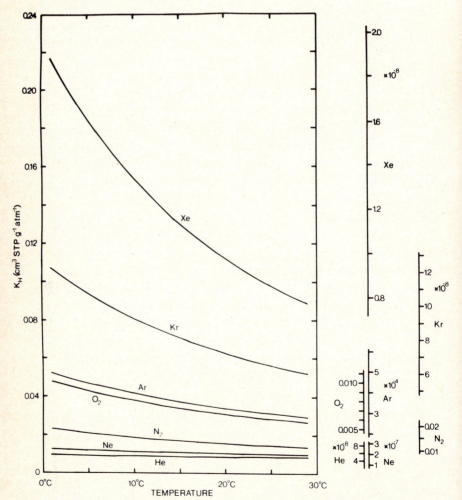

Subsequently, Weiss (1971a,b) presented experimental data for He, Ne, and Ar solubilities, concluding that modification of the Ar solubility parameters was unnecessary and fitting his He and Ne data to eq. (7.4) with results also shown in Table 7.1.

Of more immediate practical utility for geochemical application are moist-air solubilities, the c^* in eq. (7.1). Weiss (1970a, 1971a) parameterized c^* by

$$\ln c^* = A_1 + A_2(100/T) + A_3 \ln (T/100) + A_4(T/100)$$
$$+ S[B_1 + B_2(T/100) + B_3(T/100)^2] \tag{7.5}$$

with results given in Table 7.2, scaled for S in per mil ($^0/_{00}$) and c^* in cm^3 STP/kg under 1 atm total pressure of water-saturated air. The data in Table 7.2 are the usual basis for evaluation of wet-saturation anomalies. Detailed tables of β and c^* at various T and S are given by Weiss (1970a, 1971a) and Kester (1975) also gives tables of c^* calculated from Table 7.2 data.

More recently, Benson & Krause (1976) have published high-precision solubility data for O_2, He, Ne, Kr, and Xe in pure water. Their data were fitted in terms of the Henry's Law constant k in

$$f = kx \tag{7.6}$$

where f is the gas fugacity above a solution in which the mole fraction of the gas is x. They fitted their data to the functional form

$$\ln k = A_0 + A_1/T + A_2/T^2 \tag{7.7}$$

with the coefficients given in Table 7.3. The Benson & Krause O_2 data agree within about 0.2% with the experimental data which were smoothed by Weiss (1970a) to give the Table 7.1 parameters, but their He and Ne solubilities are both 1–2% higher than those of Weiss (1971a).

The basic solubility parameters β (Table 7.1) and k (Table 7.3) are both readily usable for calculation of equilibrium concentrations at arbitrary partial pressures. In principle, correction for volume change by solution and nonideal gas behavior must be made, and it should be noted that the solubility experiments are performed at partial pressure near 1 atm, but for geochemical purposes the corrections are negligible. It is unfortunate that different experimental determinations of the same physical quantity still frequently differ by more than their stated uncertainties. Still, the available solubility data appear sufficiently good, no worse than a per cent or so uncertainty, and often considerably better, for all the noble gases (and N_2 and O_2) in pure water and all but Kr and Xe in salt water.

Kester (1975) tabulates salt water moist-air saturation values for Kr and Xe from the data of Wood & Caputi (1966), assuming the salinity dependence expressed in eq. (7.4). For Kr and Xe normalizations in this chapter we will use

Table 7.1. Solubility of gases in water and in seawater[a]

Gas	A_1	A_2	A_3	B_1	B_2	B_3
He	−34.6261	43.0285	14.1391	−0.042 340	0.022 624	−0.003 312 0
Ne	−39.1971	51.8013	15.7699	−0.124 695	0.078 374	−0.012 797 2
N$_2$	−59.6274	85.7661	24.3696	−0.051 580	0.026 329	−0.003 725 7
O$_2$	−58.3877	85.8079	23.8439	−0.034 892	0.015 568	−0.001 938 7
Ar	−55.6578	82.0262	22.5929	−0.036 267	0.016 241	−0.002 011 4

[a] Values for use in eq. (7.4) for T in Kelvin, salinity S in per mil, Bunsen coefficient β in cm^3 solution under partial pressure of 1 atm; from Weiss (1970a, 1971a).

Table 7.2. Moist-air saturation concentration of gases in water and seawater[a]

Gas	A_1	A_2	A_3	A_4	B_1	B_2	B_3
He	−167.2178	216.3442	139.2032	−22.6202	−0.044 781	0.023 541	−0.003 426 6
Ne	−170.6018	225.1946	140.8863	−22.6290	−0.127 113	0.079 277	−0.012 909 5
N$_2$	−177.0212	254.6078	146.3611	−22.0933	−0.054 052	0.027 266	−0.003 843 0
O$_2$	−177.7888	255.5907	146.4813	−22.2040	−0.037 362	0.016 504	−0.002 056 4
Ar	−178.1725	251.8139	145.2337	−22.2046	−0.038 729	0.017 171	−0.002 128 1

[a] Values for use in eq. (7.5) for T in Kelvin, salinity S in per mil, c^* in cm^3 STP in 1 kg of solution under 1 atm total pressure of water-vapor-saturated air; from Weiss (1970a, 1971a).

the pure water solubilities of Benson & Krause (1976) modified by the tabulated salinity dependence in Kester (cf. Table 7.6). Since the Wood & Caputi pure water solubilities agree well with those of Benson & Krause at about 25 °C but are higher by 3–4% at 0 °C, and the salting-out coefficients do not smoothly continue the mass-dependent trend of the lighter gases (Table 7.4), the seawater solubilities of Kr and Xe must evidently be considered uncertain by at least a few per cent.

The only report of experimental determination of noble gas isotopic fractionation upon solution in water is that of Weiss (1970b), who found that ^{3}He is *less* soluble than ^{4}He by (1.2 ± 0.2)% at 0 °C, the fractionation increasing in magnitude by about 0.1% per 10 °C. The effect is in the same sense but much larger than the $\lesssim 0.1$% effects found for O_2 and N_2 by Klots & Benson (1963). On the basis of the reasonably inferred mass dependence we can guess that isotopic fractionation will be quite small, around 0.1%, for Ne and negligible for the heavier gases.

Table 7.3. Solubility of gases in pure water[a]

Gas[b]	A_0	A_1	A_2
He	5.074 6	4 127.8	−627 250
Ne	4.298 8	4 871.1	−793 580
O_2	4.060 5	5 416.7	−102 6100
Kr	3.632 6	5 664.0	−112 2400
Xe	2.091 7	6 693.5	−134 1700

[a] Values for use in eq. (7.7) for T in Kelvin, giving Henry's Law constant k as the ratio of fugacity (in atm) to solute mole fraction x; from Benson & Krause (1976).
[b] Wilhelm *et al.* (1977) tabulate Rn solubility data.

Table 7.4. Relative gas solubilities in seawater and fresh water[a]

Temperature (°C)	He	Ne	N_2	O_2	Ar	Kr	Xe
0	0.809	0.786	0.749	0.766	0.764	0.72	0.73
25	0.838	0.825	0.785	0.798	0.798	0.77	0.78

[a] Tabulated quantity is c^*_{35}/c^*_0, where c^*_0 is wet air saturation solubility in pure water and c^*_{35} is corresponding value for seawater (35$^0/_{00}$ salinity). Data from Weiss (1970a, 1971a) (cf. Table 7.2) except for Kr and Xe, from Wood & Caputi (1966), as tabulated by Kester (1975). Data are for c^* by weight water; values for c^* by volume water will be slightly different.

7.3 Seawater

Materials, such as dissolved gases, without known sources or sinks in the ocean are said to be conservative. To first order, the conservative gases dissolved in seawater, specifically including the noble gases, are in solubility equilibrium with air, surface waters at ambient conditions, and deep waters at the conditions of their formation (when they were isolated from air). Dissolved O_2, because of its role in biological (and thanatological) processes, is nonconservative, and its variations in seawater are well known and widely studied. Dissolved N_2 is a marginal case: sources and sinks are evidently possible in special cases, but in the open ocean N_2 can apparently be considered nearly conservative. In any case, observations of dissolved N_2 are generally quite close to saturation values; saturation anomalies up to a few per cent have been reported, both positive and negative and without clear correlation with other oceanographic variables. Reviews for O_2 and N_2, as well as other nonreactive gases, are readily available, e.g. Broecker (1974) and Kester (1975).

Representative data for noble gases in seawater are given in Tables 7.6 and 7.7. Much of the early data is apparently characterized by analytical difficulties, with interpretations further hampered by lack of adequate solubility data. As the experimental observations, and particularly the solubility data, became reliable at the 1% level, the general features of dissolved noble gas distribution became evident (cf. Craig & Weiss, 1971; Bieri, 1971). As noted, dissolved noble gases (and N_2) are present at approximately the moist-air saturation values, but there are saturation anomalies at the level of a few to several per cent (Table 7.7), well beyond stated experimental uncertainties. The light gases He and Ne, in particular, show consistently positive anomalies. This is suggestive of air injection (Table 7.5), and indeed the consensus is that in surface waters in

Table 7.5. *Illustrative apparent wet-saturation anomalies (%) in seawater for various effects*

Effect		ΔHe	ΔNe	ΔAr	ΔKr	ΔXe
True partial pressures 1% higher than nominal; any temperature		1.0	1.0	1.0	1.0	1.0
Temperature increases 1 °C after	at 0 °C	0.5	0.9	2.6	3.5	4.2
equilibration	at 25 °C	0.2	0.7	1.7	2.1	2.6
Air injection, 1 cm³ (STP, dry)/l	at 0 °C	12.9	10.0	2.4	1.2	0.6
	at 25 °C	13.9	12.1	4.0	2.3	1.3
Mix equal amounts of water at 0 °C and 10 °C		0.2	0.3	1.3	2.0	2.8
Mix equal amounts of water at 20 °C and 30 °C		0.1	0.1	0.7	1.1	1.5

Table 7.6. Selected data for noble gas abundances (cm³ STP/g) in water

Sample	Notes	Temperature (°C)	He (10⁻⁸)	Ne (10⁻⁷)	Ar (10⁻⁴)	Kr (10⁻⁸)	Xe (10⁻⁸)
Fresh water							
Air saturation	a	0	4.90	2.25	4.98	12.57	1.95
Air saturation	a	25	4.41	1.78	2.84	6.22	0.83
Groundwater (Israel)	b	53	997	2.03	3.06	7.61	1.01
Sea water							
Air saturation	a	0	3.97	1.77	3.80	9.05	1.43
Air saturation	a	25	3.70	1.47	2.26	4.79	0.65
Pacific surface water	c	22	3.84	1.55	2.42	5.30	—
Pacific deep water	d	1	4.42	1.86	3.73	9.04	—
Atlantic deep water	e	4	4.22	1.85	3.51	8.43	—
Pacific deep water	f	1	—	1.77	3.06	7.31	1.08
Red Sea brine	g	56	1 400	1.4	—	—	—

a Values for equilibrium with 760 torr total pressure of air saturated with water vapor. He, Ne, Ar data from Table 7.2; Kr, Xe data from Tables 7.3 and 7.4. Sea water salinity taken to be 35°/∞.
b Hamat Gader spring water; Mazor (1972).
c Average of top 7 samples, Piquero station 3 (14° S 102° W), by Bieri & Koide (1972); cf. Table 7.7.
d Average of deep water (>2000 m) samples, three Piquero stations, Bieri & Koide (1972) (their table 3).
e Average of bottom 3 samples, AII-20 Station 944 (16° N 59° W), by Bieri *et al.* (1968); cf. Table 7.7.
f Average of samples M-39 and M-26, Expedition Monsoon, by Mazor *et al.* (1964). NB These data indicate saturation anomalies ≈20% which would probably be considered questionable at present.
g Atlantic II sample 714-G7, Lupton *et al.* (1977a); data tabulated are cm³ STP per gram of H₂O.

Table 7.7. *Apparent wet-saturation anomalies*[a] *in marine waters*

Water (Expedition-Station)	Location	Surface water				Deep water				References
		ΔHe	ΔNe	ΔAr	ΔKr	ΔHe	ΔNe	ΔAr	ΔKr	
Atlantic (AII-20-944)	16° N 59° W	4.8	5.0	2.7	3	5.2	5.5	1.9	7	Bieri *et al.* (1968)
Drake Passage (Zapiola-C-14)	58° S 60° W	4.0	5.0	3.2	10	–	–	–	–	Bieri *et al.* (1968)
Pacific (NOVA-1)	34° N 146° W	3.3	2.0	−0.4	2	6.5	1.6	−5.6	−3	Bieri *et al.* (1968)
Pacific (NOVA-4)	9° N 178° E	2.3	1.6	0.0	4	7.6	3.8	−3.1	0	Bieri *et al.* (1968)
Pacific (Piquero-3)	14° S 102° W	4.3	5.0	3.1	5	11.9	6.3	1.7	5	Bieri & Koide (1972)
Pacific (Carrousel-14)	33° S 73° W	–	4.7	−2.5	–	8.6	3.8	−0.2	–	Craig *et al.* (1967)

[a] Anomalies in per cent, normalized to Table 7.2 (He, Ne, Ar) or to Tables 7.3 and 7.4 (Kr).

general, and in Atlantic deep waters, the anomalies are attributable to air injection by bubbles, with allowance for modest pressure variations and postequilibration temperature changes. Supersaturation due to air injection of the order of $0.5-1.0 \text{ cm}^3/\text{l}$ is common in the studied waters. Air injection, or the other plausible mechanisms for producing anomalies described above (Section 7.1 and Table 7.5), can account for only about two thirds of the average ΔHe in Pacific deep waters, however (Craig & Weiss, 1971; Bieri, 1971). The balance, about 3% of the saturation concentration, is now generally attributed to nonconservative behavior of He, specifically to the addition of juvenile (radiogenic) He from the suboceanic mantle or crust (Section 7.5). This conclusion is probably the most important result to emerge from general noble gas elemental abundance studies of seawater.

Both Craig & Weiss (1971) and Bieri (1971) stress that analysis of the He anomalies must be based on comparison with these of Ne and Ar. Craig & Weiss (1971) made a formal analysis expressing each of ΔHe, ΔNe, and ΔAr as a linear sum of terms in ΔP, ΔT, and Δa (Section 7.1). A convenient graphical presentation (Fig. 7.2) is the relation between $\Delta\text{He} - \Delta\text{Ne}$ and $\Delta\text{Ne} - \Delta\text{Ar}$. Since pressure variations are the same for all the gases, ΔP drops out of these differences, and the graph can be analyzed in terms of a grid of intersecting loci of constant ΔT and constant Δa. (The two families are not orthogonal and their positions and angle of intersection depend on temperature; they cross at around $10\,^\circ\text{C}$, so the relative slopes are different in Figs. 7.2a and 7.2b.) From the parameters in Fig. 7.2 it is evident that inferred ΔTs (and sympathetically varying ΔPs) are quite sensitive to uncertainties of a per cent or so. It is clear, however, that saturation anomalies in surface waters and Atlantic deep waters can be understood in terms of plausibly modest values for air injection and pressure and temperature variations, but that those for Pacific deep water cannot. The high ΔHe values cannot be generated by any plausible combination of these mechanisms and so must be interpreted in terms of juvenile (nonatmospheric) excess He.

As noted, the excess He is needed to account for only about one third of ΔHe in Pacific deep waters. It is disquieting when an effect of interest must be observed as a relatively small superposition on another, unrelated, effect, coincidentally of the same order of magnitude, particularly when the separation of the effects is not so very much larger than experimental uncertainties. The existence of excess He is not in doubt, as can be inferred from the observations described in the next two sections, but evaluation of its magnitude is quite sensitive to uncertainties in the data, both the observations and the solubilities. In both respects, the history of noble gas investigations in water has included many cases where systematic errors, as indicated by discrepancies with other experiments, have exceeded stated confidence limits. Assessment of excess He would change, for example, in response to revision of the solubility data. (The *fresh* water He and Ne solubilities of Benson & Krause (1976) are 1–2% higher than those of Weiss (1971a); if it is speculated that the salt-water solubilities should be increased by, say, 1%, this would reduce the inferred air injection term, *but* not the excess He, since ΔHe and ΔNe would both be reduced: cf. Fig. 7.2). As has been noted in the literature, it must also be presumed that Henry's Law solubility can be accurately (to about 1%) extrapolated across more than five orders of magnitude from laboratory measurements at around 1 atm to the partial pressures in air. Revision of the He abundance in air (Table 2.1) by 1% or so would also change the overall picture.

Aside from identifying air injection as a common feature in most waters, and excess He in Pacific deep waters, noble gas elemental abundances have also been used as tracers for water masses and to study mixing in intermediate waters. Discussions are given by Bieri *et al.* (1966, 1968); Craig *et al.* (1967); Craig & Weiss (1968); and Bieri & Koide (1972). On the whole, the arguments cut

Fig. 7.2(a). Correlation of apparent saturation anomalies in seawater; Δ values given in per cent. Solid lines are loci for constant amounts of air injection, labeled by a in cm^3 STP/kg. A second family for constant postequilibrium temperature changes ΔT can be constructed; only the line for $\Delta T = 0$ is shown, but ΔT intersection values on the $a =$ constant lines are indicated. Data are from Bieri *et al.* (1968). Reproduced from Craig & Weiss (1971).

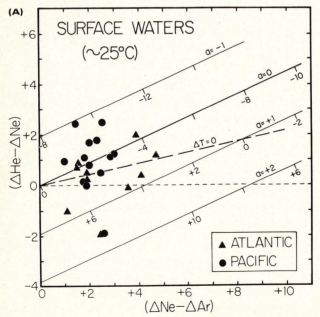

finer and are therefore closer to the limit of experimental uncertainties than
the air injection and excess He features discussed above, and so are less definitive;
in any case, they seem not to have made any large impact on marine science.

To this point in this section, 'noble gases' have meant He, Ne, and Ar, which
are the principal data base. Among the more reliable recent data, Kr data are
scarce (Bieri *et al.*, 1968; Bieri & Koide, 1972), and Xe data essentially nonexistent, presumably because of analytical difficulties. Kr anomalies, normalized
to the saturation concentrations described in Section 7.2, are included in Table
7.7. Good solubility data would be helpful in integrating ΔKr into the
ΔHe-ΔNe-ΔAr analysis, where it might be useful since it is more sensitive to
ΔT and less sensitive to Δa than the lighter gases (Table 7.5).

Xe data would be even more useful in this sense, if there were any. The best
data are probably those of Mazor *et al.* (1964) (Table 7.6). Normalized as for
Kr above, their data yield ΔXe = +7% in surface water and ΔXe = −21% in deep
water. The surface water result is similar to more recent ΔKr (Table 7.7).
Although the Mazor *et al.* data also include approximately 20% saturation
anomalies in Ne and Ar that would be considered spurious today, it is tempting
to speculate that their deep water Xe undersaturation is real, possibly also the

Fig. 7.2(b). Correlation of apparent saturation anomalies in seawater;
cf. Fig. 7.2a. Drake Passage waters are shallow but cold. Data are from
Bieri *et al.* (1968). Reproduced from Craig & Weiss (1971).

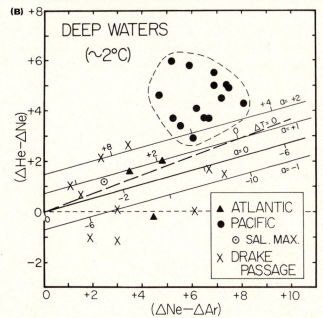

scattered but very low values of Bieri *et al.* (1964), since of all the noble gases
Xe is the best candidate for an adsorptive sink in ocean water (cf. Chapter 4).

As a final note, the noble gases in seawater are derived from air and, like the
water itself, are properly considered part of the terrestrial atmosphere (Chapter
12). Particularly for the more soluble heavy gases, their contribution to the total
inventory is a small but nontrivial, if generally ignored, addition to the air
reservoir (Table 7.8).

7.4   Meteoric water

Noble gases in rainwater are in solubility equilibrium with air and,
although we know of no investigations addressed to the question, there is no
reason to believe any differently for river and lake water. Most work on meteoric
water has thus been on groundwater, which is removed from contact with air,
especially geothermal water (see review by Mazor, 1975). Matters of interest
are the temperature at which the water was last equilibrated with air, whether
it has behaved as a closed system since air equilibration, and indeed whether or
not it is actually meteoric.

Concerning the latter question, the alternative to meteoric water is juvenile
water. Distinction between these two alternatives for the origin of geothermal
waters has been attacked extensively through a variety of geochemical methods,
generally with the result that the water in question is meteoric rather than
juvenile. In the case of noble gases, there is in fact no clear idea of what the
noble gas signature of juvenile water ought to be. If noble gases are observed
in quantities appropriate for meteoric water, or in quantities sensibly related,
this may be taken as presumptive evidence that the origin is indeed meteoric,
although certainly moderate dilution by noble-gas-free juvenile water could not

Table 7.8. *Sea water contribution to terrestrial noble gas inventory*

Gas	Approximate average concentration (cm^3 STP/g)	Total[a] amount (cm^3 STP)	Fraction[b] of air inventory
He	4.0×10^{-8}	5.6×10^{16}	0.3%
Ne	1.7×10^{-7}	2.4×10^{17}	0.3%
Ar	3.5×10^{-4}	4.9×10^{20}	1.3%
Kr	8.5×10^{-8}	1.2×10^{17}	2.7%
Xe	1.1×10^{-8}	1.6×10^{16}	4.5%

[a] For 1.41×10^{24} g of sea water. The fresh-water contribution will be negligible in comparison.
[b] Quantities in air from Table 2.1.

be excluded. In any case, no water has ever been identified or suggested to be anything other than meteoric on the basis of its noble gases.

The general rule is that except for juvenile radiogenic gases (Section 7.5) frequently leached from the host rocks, noble gases are conservative in groundwater. An interesting set of examples (cf. Table 7.6) is the study of thermal waters in the Jordan Rift Valley, Israel, by Mazor (1972). Such waters have observed temperatures up to 63 °C. Relative to their present temperatures they are supersaturated in noble gases; the noble gas contents are, however, quite appropriate for air saturation at surface temperatures, 15–25 °C (Mazor, 1972; also see Benson, 1973). This indicates that noble gas contents have been conserved during their history in groundwater, evidently in excess of 10^4 yr, and furthermore that Pleistocene surface temperatures in this region were much the same as they are today, rather than significantly closer to 0 °C.

In other cases, however, nonconservation is evident. Mazor & Wasserburg (1965), for example, found that gaseous emanations from hot springs in Yellowstone National Park (USA) contained noble gases in relative proportions more or less correct for surface temperature air saturation, indicating, along with other lines of evidence, that the water was meteoric. Mazor & Fournier (1973), however, found that while the same was generally true in most cases for the actual waters, the absolute concentrations in water were typically lower than the air-saturation values, down to 3% in one case. Reviewing this and other cases, Mazor (1975) points out that the absolute abundances are under-saturated not only for surface temperatures but also for observed temperatures, and so could not be attributed to re-equilibration at the elevated temperature. Instead, he suggests, this indicates loss of the noble gases through partitioning into a vapor phase, presumably steam; he also cites at least one case where relative abundances are correct for surface temperature equilibration but the absolute abundances are too high, apparently reflecting recondensation of the vapor after isolation from its source. Mazor (1975) goes on to note that such noble gas saturation anomaly patterns might be useful for diagnosing steam formation in geothermal water systems, a feature of obvious utility in geothermal energy prospecting.

No data for noble gas solubility in ice are available, but solubilities can be expected to be much lower than for liquid water. The only noble gas observations are these of Matsuo & Miyake (1966), who analyzed Ar along with N_2, O_2 and CO_2 in natural ices. They found the major gases and Ar present in roughly atmospheric proportion, evidently contained principally in occluded gas bubbles. The bubbles were present in variable amount, but often of the same order as that expected for exsolution on freezing (about 3% by volume). By the same technique described in Sections 7.1 and 7.3, it is possible to partition observed gases into a solution component and an 'air injection' component, with most of the leverage coming from highly soluble CO_2. Matsuo & Miyake found the solution component to be generally quite small ($<2\%$), so the dominant effect is evidently mechanical occulusion of air bubbles.

Sorption of the heavier noble gases on ice might be interesting (cf. Section 8.6) but there are no relevant data.

In many examples of hot springs, fumaroles, etc., associated with tectonic activity, the water involved is meteoric but the noble gases are, in part, juvenile. Observations are described in Chapter 10.

7.5   Juvenile ^{4}He

The seawater data described in Section 7.3 lead to the conclusion that while marine waters in general show 'explainable' positive saturation anomalies up to several per cent for He and Ne, the He abundances of deep water in the Pacific (but not the Atlantic) were higher still, and could not be explained in terms of conservative behavior after isolation from air. A non-conservative effect, addition of nonatmospheric (i.e. juvenile) He was invoked. On the basis of the best analytical observations and solubility data now available, the excess (juvenile) He contributes typically about 3% to the saturation anomaly ΔHe in Pacific deep water, about one third the total anomaly (the other two thirds due mainly to air injection). Since the element He is mostly the isotope ^{4}He, and mass spectrometers measure ^{4}He anyway, the juvenile He is actually ^{4}He, and was quickly identified as radiogenic ^{4}He. This was not surprising: the effect was predicted before it was observed (Revelle & Suess, 1962), declared in the first experimental observations (Bieri *et al.*, 1964; 1966) even before the more reliable experimental data, solubility data, and resolution techniques could be applied (Craig & Weiss, 1971; Bieri, 1971), and shown to be quite plausible quantitatively (Bieri *et al.*, 1967). The average flux of juvenile ^{4}He into the atmosphere estimated by Bieri *et al.* (1967) was $2-3 \times 10^6$ atom cm^{-2} sec^{-1}. A more recent value, 3×10^5 atoms cm^{-2} sec^{-1} (Craig *et al.* 1975), is described in the next section.

Although the 3% excess He is fairly small and, by itself, arguable, it should be pointed out that there are other cases where supersaturation effects due to addition of juvenile radiogenic ^{4}He are not at all subtle. Hundredfold or more enrichments of He, relative to air saturation, are frequently observed in geothermal groundwaters (Table 7.6), reflecting extraction of radiogenic ^{4}He from host rocks. More recently, Lupton *et al.* (1977a) have observed a striking and, for seawater, unprecedented 380-fold enrichment of He (Table 7.6) in Red Sea brines (discussed further below). In both these cases, it should be noted, the nonradiogenic gases are present at normal saturation levels.

Addition of juvenile radiogenic ^{4}He should be and (presumably) generally is accompanied by juvenile radiogenic ^{40}Ar, although the latter effect is observable only in the more extreme cases. For a nominal ^{4}He/^{40}Ar ≈ 10 in radiogenic gas, a hundredfold enrichment of He in, say, saturated 25 °C seawater requires 3.7×10^{-6} cm^3 STP/g (Table 7.6), while the corresponding 3.7×10^{-7} cm^3 STP/g of ^{40}Ar produces only an 0.16% addition to the dissolved Ar. (The equivalent figure for ^{136}Xe is $\Delta \approx 0.002\%$.) Part of the imbalance is due to the composition of the source gas and part to the low solubility of He, but mostly this results from the scarcity of He in air in comparison with its sources (Section 12.3).

In practice, small enrichments of radiogenic ^{40}Ar can be determined with much greater precision and less ambiguity by examining the ^{40}Ar/^{36}Ar ratio for deviations from the air ratio than by trying to assess small deviations from solubility equilibrium. The same remark applies to the ^{4}He/^{3}He ratio, although, as described below, results proved otherwise.

7.6   Juvenile ^{3}He

Because of its low natural abundance and, in normal practice, severe interferences in mass spectrometric analysis, in the early noble gas analyses of seawater ^{3}He was either ignored or used only as the spike in isotopic dilution. In the past decade, however, ^{3}He data have assumed a position among the most important in terrestrial noble gas geochemistry.

The custom for reporting He isotopic composition is to use a 'ratio anomaly' δ^3He, usually in per cent:

$$\delta^3\text{He} = \left[\frac{(^3\text{He}/^4\text{He})_s}{(^3\text{He}/^4\text{He})_{air}} - 1 \right] \times 100\% \qquad (7.8)$$

Here $(^3\text{He}/^4\text{He})_s$ is the observed sample ratio. It should be noted that the normalization is the air ratio, not an 'expected' ratio. The distinction is significant for He dissolved in water, for which there is a perceptible isotopic fractionation (Weiss, 1970b), so that the 'null' value, for solubility equilibrium with air, is $\delta^3\text{He} \approx -1.4\%$.

Uncertainties in seawater analyses and solubilities, and in resolving effects which can lead to apparent saturation anomalies (Table 7.5) make it difficult to identify small contributions of juvenile radiogenic ^{4}He by way of saturation anomalies. None of these impediments affects the ^{3}He/^{4}He ratio, however, but the addition of radiogenic ^{4}He nearly (Section 6.2) devoid of ^{3}He will create a compensating negative ratio anomaly, $\delta^3\text{He} \approx -\Delta_E{}^4\text{He}$ (for small variations), where Δ_E designates the excess ($=$ nonatmospheric $=$ juvenile) saturation anomaly.

In the expectation of using this effect to verify and more accurately quantify the presence of juvenile radiogenic ^{4}He, Clarke *et al.* (1969) measured ^{3}He/^{4}He ratios in South Pacific (Kermadec Trench) deep waters. Instead of negative anomalies they found positive anomalies,* δ^3He averaging around +10% and ranging up to +22% (Fig. 7.3). Thus the ^{3}He saturation anomaly, Δ^3He, was found to be even larger than Δ^4He, and juvenile He to contain ^{3}He as well as ^{4}He, with ^{3}He/^{4}He greater than the air ratio. In the absence of alternative sources of ^{3}He, Clarke *et al.* (1969) concluded that the ^{3}He is primordial, and that the solid earth is still degassing into the atmosphere volatiles it incorporated

> * The report by Clarke *et al.* (1969) is an interesting example of the utility of the δ-value normalization. They identified *ratio anomalies* at the 2% level of precision, even though their corrected (for instrumental discrimination) *absolute* values were too low by 25%. Their uncorrected absolute ratios were much closer to presently accepted values.

Fig. 7.3. Excess ^{3}He profiles at various locations in the Pacific Ocean. NOVA (Kermadec Trench) data from Clarke *et al.* (1969), GEOSECS data from Clarke *et al.* (1970), SCAN data from Craig *et al.* (1975). SCAN station 38 (6°30′ S, 107°24′ W) is at the crest of the East Pacific Rise; nearby stations 35 and 41 are on opposite flanks. Reproduced from Craig *et al.* (1975).

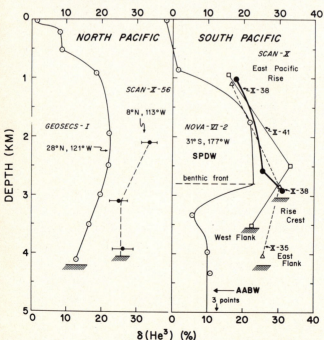

at its formation. Subsequent investigations have amply confirmed the existence
of the effect, extended the range of observation, and supported this interpretation.

Since radiogenic He has a $^3He/^4He$ ratio less than the air value, the only
serious candidate for an alternative to primordial ^3He is production by decay
of tritium (^3H). Cosmic-ray interactions in the atmosphere are a well-known source
of ^3He (Section 12.3), and some of this is channeled through tritium, which will
enter surface water and to some extent be entrained in deep water before its
decay. Craig & Clarke (1970), however, have shown that this is only a minor
perturbation, accounting for no more than about 2% of the primordial ^3He effect
in deep waters. In surface waters ($\leqslant 1$ km) on the other hand, anthropogenic
(nuclear weapons tests) tritium enrichments are large, and decay to ^3He can and
does produce δ^3He ratio anomalies up to a few per cent. This situation can even
be turned to advantage, and be used as a chronometer for isolation (from air) on
a timescale comparable to the 12 yr half-life of tritium (cf. Jenkins & Clarke,
1976).

The distribution of excess primordial ^3He in deep waters has been found to
exhibit considerable structure in both vertical and horizontal directions, indicating
that the sources of ^3He are not diffusely distributed about the ocean bottom but
are concentrated in localized areas. The logical expectation is that the sources
should be associated with accretional plate margins, where mantle material is
brought to the surface. In the discovery report, Clarke *et al.* (1969) observed
a mid-depth maximum in δ^3He and suggested its origin to be injection at the
East Pacific Rise. Although this particular mid-depth maximum was later found
to reflect a transition from one water mass to another and not association with
the East Pacific Rise, the general idea was nevertheless found to be correct.
Further work in the Pacific (see Fig. 7.3) showed that δ^3He is indeed characteristically at a maximum at mid-depth rather than in bottom waters (Clarke
et al., 1970), and can clearly be associated with the crest of the East Pacific
Rise (Craig *et al.*, 1975). The maximum anomaly found by Craig *et al.* (1975)
in the vicinity of the Pacific–Cocos–Nazca triple junction is $\delta^3He = 33\%$, and
since the same water exhibits a total saturation anomaly $\Delta He\ (= \Delta^4He) = 15\%$,
the corresponding saturation anomaly for ^3He is $\Delta^3He = 55\%$. In a similar study,
Lupton (1979) found higher (up to $\delta^3He = 68\%$, $\Delta^3He = 92\%$) and more localized
anomalies in the Gulf of California.

In the Atlantic, deep water anomalies in ^3He were found to be considerably
less than in the Pacific, in accord with the conclusions based on total saturation
anomalies (Section 7.3), but nevertheless quite definitely present in a characteristic
level $\delta^3He \approx 5\%$ (Jenkins *et al.*, 1972). Albeit at a lower level than in the Pacific,
the deep Atlantic ^3He excesses also show considerable structure; in a detailed
study of the western Atlantic Jenkins & Clarke (1976) observed a maximum δ^3He

of 13% and identified a localized source in the Gibbs Fracture Zone southwest of Iceland. To the south (at about $30°$ N), a section across the Mid-Atlantic Ridge shows no perceptible influence of the ridge itself on δ^3He, (Lupton, 1976), a result in marked contrast to the comparable data for the East Pacific Rise (Fig. 7.3).

Primordial ^{3}He concentrations are systematically lower in the Atlantic than in the Pacific (there are no data for the Indian Ocean). Part of this difference can be attributed to the more rapid flushing of Atlantic waters, and part to the circumstance that deep waters entering the Pacific already have positive anomalies (from the Atlantic). Part of the difference is also evidently due simply to weaker sources in the Atlantic than in the Pacific, as would be expected on the basis of relative rates of crustal formation.

Away from sources, excess ^{3}He will be conservative and so can be used for tracing movement and mixing of different water masses. Although experimental uncertainties are relatively high (about 1% in δ^3He), so too are the absolute magnitudes of the variations. Excess ^{3}He can thus be a useful adjunct to techniques employing temperature, salinity or other dissolved species (cf. Jenkins & Clarke, 1976). A dramatic illustration of both the locational nature of juvenile ^{3}He sources and its application to tracing of water mass movements is shown in Fig. 7.6 (Lupton & Craig, 1981).

An extremely fine localization of primordial ^{3}He injection, on a ten-meter scale, has also been observed. It has been suggested that lower-than-expected conductive heat flow at oceanic ridges could be due to significant heat transport by hydrothermal circulation (e.g. Talwani *et al.*, 1971), in which recently emplaced hot rock drives convection of local seawater. On the basis of temperature–salinity relationships, Weiss *et al.* (1977) made the first identification of hydrothermal circulation in the open ocean, observing several 'plumes' (temperature differential $\leqslant 0.2$ °C) above axial fissures in the Galapagos Rift (but not at a nearby East Pacific Rise site). Lupton *et al.* (1977b) analyzed water sampled in the plumes and found ^{3}He (and Rn) excesses (up to δ^3He $= 99\%$) relative to background water already rich in ^{3}He (δ^3He $= 30\%$). The primordial identification of ^{3}He provides strong support for the thesis that these plumes are indeed associated with recent emplacement of materials from the upper mantle.

More detailed examination and sampling allows association of hydrothermal circulation with specific vent fields. In such waters sampled in the Galapagos Rift by the Alvin deep submersible, Jenkins *et al.* (1978) report juvenile He enrichments which dwarf the normal saturation concentrations by factors up to 11 for ^{4}He and 60 for ^{3}He (Fig. 7.4). A particularly significant feature of this report is that added He occurs roughly in proportion to added heat (ΔT up to 12 °C in sampled water), corresponding to about 7.6×10^{-8} cal/atom of ^{3}He (Fig. 7.5). Jenkins *et al.* note that if this value is representative, hydrothermal circulation may indeed account for the depression of conductive heat flow relative to models for total heat flux.

An extreme example of hydrothermal circulation at an accretional plate margin (which was known first and served as an analog for the apparently more general case) occurs in the Red Sea (Craig, 1966, 1969). There the circulating water passes through evaporites, becoming brine ($S \approx 250^0/_{00}$) and is thus stabilized against convective mixing and dilution in spite of being hot (around 50 °C). The brines show striking enrichments of a number of elements, including He (Table 7.6): about 380 times saturation for ^{4}He, and a spectacular enrichment of ^{3}He, up to 3400 times saturation (Lupton *et al.*, 1977a). While the other enriched elements evidently come from the evaporites, the He must originate in the ridge basalts, the only plausible source for such a high ^{3}He/^{4}He ratio.

Fig. 7.4. Large excesses of juvenile He in hydrothermal waters sampled at specific vent fields in the Galapagos Rift. Note ambient seawater concentration near origin (cf. Table 7.6). Reproduced from Jenkins *et al.* (1978).

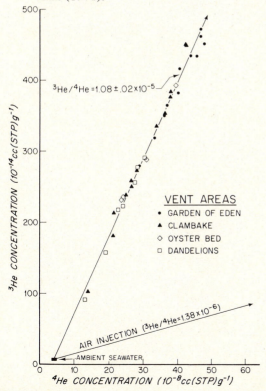

The composition of juvenile He is obviously a parameter of considerable interest. In normal waters, the limitation on the calculation is identification of the excess ^{4}He, since the fractional ^{4}He excess is smaller than that of ^{3}He, and must be identified by an absolute concentration excess which must be resolved from larger effects due to air injection and temperature/pressure variations. Combining their ^{3}He data with the He, Ne, and Ar data of Bieri & Koide (1972), Craig *et al.* (1975) adopted, as characterizing the waters in the Pacific–Cocos–Nazca triple junction region, ΔHe = 12%, Δ_EHe = (3.5 ± 0.5)%, and δ^3He = (32.0 ± 0.5)%, which leads to ^{3}He/^{4}He = (16 ± 3) × 10^{-6} for the juvenile He, eleven times the air ratio. The equivalent calculations can be made only with large uncertainties for the other Pacific or the Atlantic locations, where the anomalies are smaller.

Fig. 7.5. Relationship between ^{3}He and water temperature in hydro-thermal waters sampled in specific vent fields in the Galapagos Rift (cf. Fig. 7.4). The correlation suggests that both heat and juvenile He are brought to the surface by the same volcanic process, in the ratio 7.6 × 10^{-8} cal/atom (of ^{3}He). Reproduced from Jenkins *et al.* (1978).

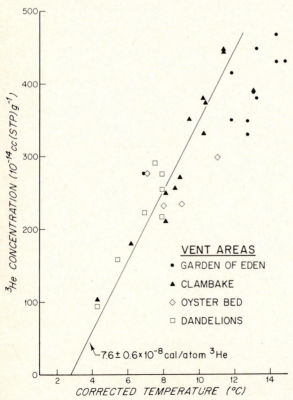

Identification of juvenile He composition in the more localized and more enriched waters is readily made. Compositions are apparently only narrowly variable at each location. Added $^3He/^4He$ is about 10×10^{-6} (cf. Fig. 7.4) in both Galapagos Rift data sets and also for the Gulf of California (Lupton *et al.*, 1977b; Jenkins *et al.*, 1978; Lupton, 1979). The Craig *et al.* (1975) value above, which presumably provides a more broadly based average, is higher, but in view of its relatively large uncertainty can be imagined consistent with this composition. The Red Sea value (Lupton *et al.*, 1977a) is 12×10^{-6}, significantly higher. It is interesting to compare these compositions to juvenile He in the rocks from which the water presumably extracts its excess He. The typical rock value is $^3He/^4He \approx 13 \times 10^{-6}$ (Section 9.6), quite in accord with the East Pacific Rise and Red Sea compositions. The 'mantle component' He observed

Fig. 7.6. A section of ^{3}He concentrations in seawater over the East Pacific Rise at $15°$ S. The contour labels are ^{3}He ratio anomalies (eq. (7.7)) in per cent. Reproduced from Lupton & Craig (1981).

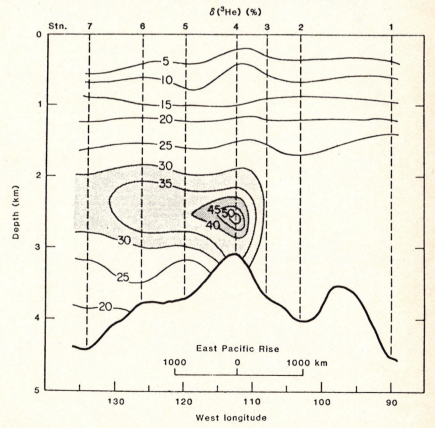

in oceanic rocks is perhaps not as uniform as was once thought (Section 11.4), however, so pending further data it is probably possible to reconcile the lower Galapagos and Gulf of California water ratios with local rocks without invoking an alternative or additional source. (Lupton *et al.* (1977b) note that the local rocks are too young for their *in situ* radiogenic He to significantly depress their inherited ^{3}He/^{4}He ratios.)

It is of considerable interest to determine the flux of juvenile He from the oceans into the atmosphere. This is a hazardous undertaking, since it requires making assumptions about the excess He distribution in the whole world ocean, mostly unsurveyed in this respect, and combining with not very well constrained estimates for the rate at which it is transported from deep water to the surface. Craig *et al.* (1975) calculate average fluxes of 3×10^5 atoms cm^{-2} sec^{-1} of ^{4}He and 4 atoms cm^{-2} sec^{-1} of ^{3}He for the oceanic input of juvenile He into the air (the cm^2 normalization is for the whole earth, not just the oceans), estimating a likely 50% uncertainty. Uncertain though they are, these fluxes are important because they are the only globally significant volatile fluxes from the solid earth which can be calculated on the basis of actual observations. Their role in budget and inventory calculations is considered in Chapter 12.

8 Sedimentary rocks

8.1 Introduction

The trapped noble gases in sedimentary rocks can be inferred to be principally atmospheric rather than juvenile. This will surely be the case for authigenic minerals, those whose microscopic structures are assembled at the surface of the earth. The trapped gases might be acquired directly from air; more likely, trapped gases will have been acquired from water, either seawater or groundwater. Even in the latter case, of course, the gases are also mostly or entirely atmospheric, since gases in water are usually nearly in equilibrium with air (Chapter 7). Detrital mineral grains or fragments, those whose microscopic structures are preserved from pre-existing igneous or metamorphic rock, also have an opportunity to trap atmospheric gases, but it is also possible that they retain juvenile gases. Actually, it is often not clear whether the trapped gases in igneous rocks are juvenile or atmospheric (Chapter 9). In any case, many sedimentary rocks have trapped noble gas contents very much higher than total gas contents characteristic of igneous rocks; in these cases at least, any juvenile gases which might be carried in detrital fragments will be quite minor, and most of the gas is atmospheric. How and when atmospheric gases are trapped are considered elsewhere in this chapter.

Some of the ^{4}He and ^{40}Ar in sedimentary rocks will be atmospheric, trapped along with the nonradiogenic gases. Some will have been produced by radioactive decay during the geological age, i.e. the time since sedimentation. If detrital minerals are present, some ^{4}He and ^{40}Ar may be retained from their pre-sedimentation history. In general, it is difficult to separate these contributions. Also, sedimentary rocks often lose juvenile radiogenic gases rather readily in comparison with igneous and metamorphic rocks. For these reasons, as is well known, it is rather difficult to apply the normal radiometric geochronology techniques, especially those based on ^{4}He and ^{40}Ar but also the others, to sedimentary rocks.

8.2 Elemental abundances

As far as noble gas contents are concerned, sedimentary rocks are not very well surveyed. The total number of published analyses is of the order of three score, of which many do not include all the gases and of which many represent replicate analyses of the same rock or the same formation. For many classes of sedimentary rock, no data at all are available. In part, this reflects merely the fashion in which the literature has grown and the absence of attempts at systematic surveys. In part, however, this also reflects technical difficulties. Sedimentary rocks are often the 'dirtiest' samples encountered in normal laboratory practice, containing large amounts of many volatile species which create problems in the cleanup procedures by which the noble gases are isolated for analysis.

Histograms displaying the available data for individual elements are presented in Fig. 8.1. Representative data are listed in Table 8.1 and elemental abundance

Table 8.1. *Selected data for noble gases in sedimentary rocks*

Sample	Gas concentrations (cm^3 STP/g)				Ref.[a]
	^{20}Ne ($\times 10^{-8}$)	^{36}Ar ($\times 10^{-8}$)	^{84}Kr ($\times 10^{-10}$)	^{130}Xe ($\times 10^{-12}$)	
Shales and related sedimentary rocks					
Wewoka E-22	<0.07	0.34	3.3	4.6	1
Larsh-Burroak Grn-3	1.3	1.1	10	139	1
ASZ II	–	2.1	43	20 000	2
Kibushi (clay)	0.65	0.11	1.1	≤1.1	1
Uncompaghre (slate)	<0.02	≤0.19	1.7	11	1
KH77-1-7[b]	0.65	13	355	2 674	1
Thucholite[c]					
Besner Mine #3	682	44	6 500	84 000	3
Kerogen[d]					
Swartkoppie	4.9	0.60	275	2 970	4
Shungite[e]					
Shungite I	27	2 237	9 006	1 129	5
Shungite V	0.91	3	38	104	5
'Air concentration'[f]					
Air	1.091	2.083	4.307	2.347	–

[a] Data references: 1, Podosek *et al.* (1980); 2, Phinney (1972); 3. Bogard *et al.* (1965); 4, Frick & Chang (1977); 5, Rison (1980b).
[b] Argillaceous marine sediment (Japan Trench).
[c] A hydrocarbon 'assemblage' rich in U and Th.
[d] A highly condensed (low H/C) carbon phase extracted from shale.
[e] A Precambrian sedimentary rock rich in amorphous organic carbon.
[f] Data from Table 2.3.

Fig. 8.1. Histograms illustrating frequency of observation of noble gas concentrations in sedimentary rocks. A more detailed illustration for Xe is given in Fig. 8.7. For each gas the lower abscissa scale is the normalized abundance, i.e. the ratio of observed concentration to the reference air 'concentration' (Table 8.1); this scale, labeled outside the frames, is the same for all gases. Within each frame, the upper abscissa scale gives the corresponding absolute concentration in cm^3 STP/g. Data are from Bogard *et al.* (1965), Canalas *et al.* (1968), Phinney (1972), Frick & Chang (1977), Kuroda & Sherrill (1977), Rison (1980b), and Podosek *et al.* (1980).

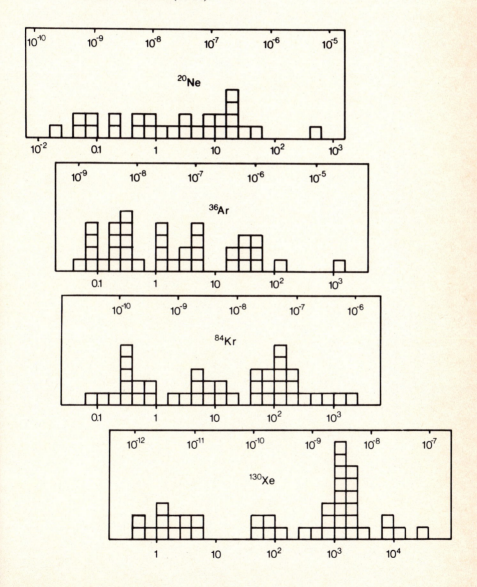

patterns shown in Figs. 8.2–8.4; these are selected to illustrate the range of noble gas characteristics. It should be noted explicitly that these data cannot be considered quantitatively representative of sedimentary rocks in general: the available data are strongly biased toward unusual rocks which for various reasons have been considered interesting.

In Figs. 8.1–8.4, data are displayed in terms of normalized abundances: for each gas the ratio of observed concentration to the 'air concentration' (Table 8.1). Assuming that trapped gases in sedimentary rocks are indeed acquired from air, this normalization provides a quantitative measurement of the efficiency of the trapping process.

Fig. 8.2. Display of noble gas abundances in sedimentary rocks. In Figs. 8.2, 8.3, and 8.4 all data are from Table 8.1 and the same format and ordinate scale are used (but note shifts in origin). Open circles with downward arrows indicate upper limits. The ordinate is normalized abundance (cf. Fig. 8.1); the horizontal dashed line at unity on the ordinate represents the air concentration normalization (Table 8.1).

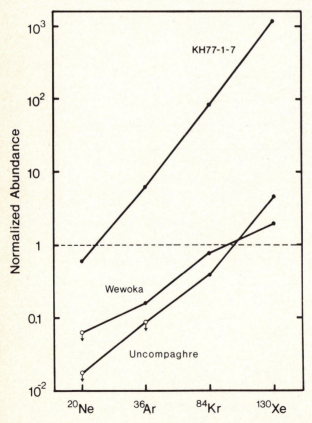

Some generalizations about trapped noble gases in sedimentary rocks are described below, but it should be noted that these are phenomenological. Physical mechanisms possibly involved in the trapping process are considered in Section 8.4; at this point, it seems fair to say that the problem is not well understood, and a genetic classification of the data is accordingly impossible.

In general, sedimentary rocks tend to have high gas concentrations, at least in comparison with igneous rocks (Chapter 9). The range of concentrations, however, is quite broad, for each gas spanning more than four orders of magnitude (Fig. 8.1). Concentrations are not very predictable: gas abundances in the same basic type of sediment or the same geological formation can vary by more than an order of magnitude (cf. Fig. 8.7) and abundances in specimens separated by a few millimeters can vary by a factor of two.

Elemental compositions are also quite variable, and in general not atmospheric. Assuming that air is the source of the trapped gases, nonatmospheric elemental

Fig. 8.3. Display of noble gas abundances in sedimentary rocks; cf. Fig. 8.2.

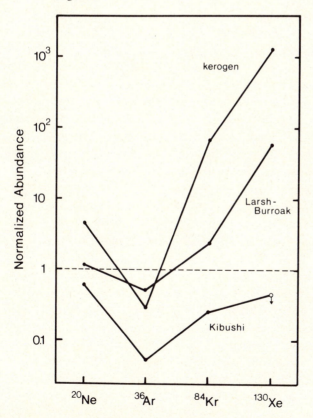

ratios thus indicate elemental fractionation to an extent given by the ratio of normalized abundances. As can be seen in Figs. 8.2–8.4, this fractionation can be quite large, up to four orders of magnitude.

Many sedimentary rocks have elemental compositions qualitatively similar to those illustrated in Fig. 8.2. This pattern is often considered normal for sedimentary rocks, and also has been designated a Type I pattern by Ozima & Alexander (1976) (but cf. Section 8.4). Its defining feature is progressively greater normalized abundances for the heavier gases. It occurs in samples whose overall gas contents are both relatively high and relatively low, and the magnitude of the elemental fractionation can vary significantly.

A second type of elemental composition pattern is illustrated in Fig. 8.3. This appears to be a modification of the normal pattern: progressively greater normalized abundances for the heavier gases, except for a much higher Ne abundance than would be expected for this trend. Ozima & Alexander designated this a Type III pattern, and considered it anomalous and rare. Podosek *et al.* (1980), however, note that it is not rare; it characterizes about half the samples for which an evaluation can be made, and the earlier assessment of scarcity can be seen to be a result of the scarcity of data. This pattern also occurs at all levels of overall gas content, and the magnitude of the 'anomalous' Ne enrichment is also quite variable. The effect can be present in some and absent in other samples of the same rock type, geological formation, or even a single hand specimen. All in all, it therefore appears that the enhanced Ne represents a separate component superposed in varying proportions on a 'normal' component (or components). If so, it is probably present to some extent in all or most samples, but noticeable as such only in those cases in which it is sufficiently prominent that the normalized abundance of Ne is higher than that of Ar. Presumably, this separate component whose low abundance of Ar and the heavier gases is insignificant in comparison with the normal heavy gas component.

Fig. 8.4 illustrates additional examples of both the types described above. Essentially all the available data can be fitted into one of these two molds. It should be re-emphasized, however, that these types are phenomenological and probably not genetic. The quantitative variations within each qualitative type are sufficiency large as to suggest that more than a single mechanism is involved. Furthermore, the qualitative types are not totally exhaustive. An interesting example is the shungite I composition (Fig. 8.4), the only known case of normalized Xe abundance lower than that of Kr (and Ar).

The trapped noble gases in sedimentary rocks presumably reflect conditions of sedimentation, the nature of the constituent mineralogy, and the subsequent history of diagenesis and metamorphism. Unfortunately, no systematic study of the relationship between trapped noble gases and these factors has yet been undertaken.

The prior discussion in this section has ignored He. This is because it is generally not possible to distinguish trapped He from the radiogenic He produced in the geological age of the sediment. The radiogenic ^{4}He is evidently lost with considerable ease, since observed ^{4}He is typically very much less than expected radiogenic ^{4}He (cf. Podosek *et al.*, 1980). In some cases, it is not even clear that the small amounts of ^{4}He observed are indeed primarily radiogenic. Thus, if the high Ne abundances (Fig. 8.3) actually reflect a light gas component, including He as well as Ne, there are some cases where the normalized abundance of trapped He need not be dramatically higher than that of Ne in order to allow the interpretation that most of the total ^{4}He is trapped rather than radiogenic. Resolution of trapped and radiogenic He might be effected on the basis of the ^{3}He/^{4}He ratio, but the matter is somewhat academic since there are few ^{3}He data for sedimentary rocks. There might also be difficulty with such an attempted resolution because ^{3}He is present in variable proportion in radiogenic He (Section 6.2) and because the composition of atmospheric He is likely to be significantly variable over geologic time (Section 12.5). Actually, the problem

Fig. 8.4. Display of noble gas abundances in sedimentary rocks; cf. Fig. 8.2.

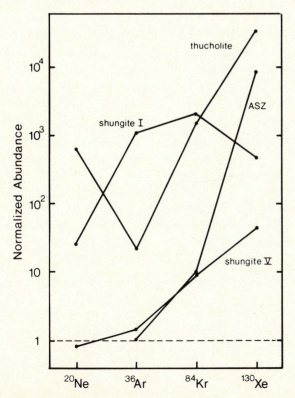

seems more likely to be interesting in the converse: if adequate means for
distinguishing radiogenic He from trapped He were available, and if ^{3}He data
were available, this would constitute the most promising avenue for studying
variations in atmospheric composition in the past (cf. Section 8.6).

8.3 Isotopery

Isotopic data for the noble gases in sedimentary rocks are, in general,
interpretable in fairly simple terms in a manner consistent with expectations.
It is expected that gases in sedimentary rocks will be trapped atmospheric gases,
so the isotopic compositions should be those of air. To first order, that is found
to be so. More generally, observed isotopic compositions can be understood to
be those of air, modified to varying degree by two effects: mass-dependent
isotopic fractionation and the addition of specific nuclear components.

In general, the nuclear components can be attributed to known nuclear
processes and are found in quantities reasonable for the occurrence of these
processes within the rock in question (as distinct from occurrence elsewhere
followed by transport of the products). The most common nuclear components,
of course, are radiogenic ^{4}He and ^{40}Ar, which are present in observable, usually
dominant quantities in most rocks; as discussed below, other components are
occasionally observed also. The ^{40}Ar produced by *in situ* decay of ^{40}K is sometimes nearly quantitatively retained, but it is also often lost to a significant or
extensive degree, so that the K–Ar method is not useful for sedimentary rock geochronology. The ^{4}He produced *in situ* is characteristically lost quite extensively.

Some further qualification is necessary for ^{4}He in sedimentary rocks. In many
cases ^{4}He is present in quantities much lower than calculated for *in situ* production and perhaps not implausible for trapping of atmospheric He (Section 8.2);
in such cases, either mechanism for the origin of the ^{4}He must be considered
viable and there are no convincing grounds for favoring one over the other. In
such cases data for ^{3}He might distinguish between a radiogenic and an atmospheric
origin for the He. Unfortunately, the ^{3}He/^{4}He in radiogenic He is variable
(Section 6.2), and it probably is variable also in atmospheric He (Section 12.5).
Also unfortunately, data for ^{3}He in sedimentary rocks are very sparse. Thus, in
many cases, whether the observed ^{4}He is radiogenic or atmospheric is not known.

In some unusual cases nuclear components besides ^{4}He and ^{40}Ar can be
identified. In the U- and Th-rich rock thucholite, for example, Bogard *et al*.
(1965) found not only a prominent Xe component from ^{238}U fission but other
radiogenic components (Section 6.2) as well. Two extreme cases are worthy of
note. Drozd *et al*. (1974) found a unique assemblage of nuclear components in
rocks from the Oklo natural fission chain reactor. Because of the high target/
product ratio in sedimentary barite($BaSO_4$), Srinivasan (1976) found not only

'radiogenic' ^{131}Xe (from ^{130}Ba (n, γ)) but also cosmic-ray spallation Xe (cf. Section 6.3).

Isotopic fractionation effects are not uncommon. It might be supposed that in general fractionation would be most prominent for the lightest gases He and Ne. This is perhaps so, but it is often difficult to tell. In the case of He, for which fractionation is probably most important, it is usually very difficult to identify fractionation in a two-isotope system subject to several potentially large nuclear effects. Ne has three isotopes, but the abundance of ^{21}Ne is low and so often not measured very precisely, and nuclear effects in Ne are also sometimes significant. Fractionation effects are thus generally more readily evident in the heavier gases: in Ar on the basis of the ^{38}Ar/^{36}Ar ratio because of the lesser importance of nuclear effects and in Kr and Xe because of the large number of isotopes available to distinguish fractionation from specific isotope effects.

Isotopic fractionation effects, such as those illustrated in Fig. 8.5, are often easily identifiable as such, but are not easy to explain, and no serious quantitative attempts have been made to account for them.

8.4   Origin of trapped gases

While it iseems clear that in general the trapped noble gases in sedimentary rocks are atmospheric, it is not at all clear by what mechanisms the gases

Fig. 8.5. Xe isotopic compositions in sedimentary rocks, displayed as per mil deviations from air composition (normalized at ^{132}Xe). Reproduced from Phinney (1972).

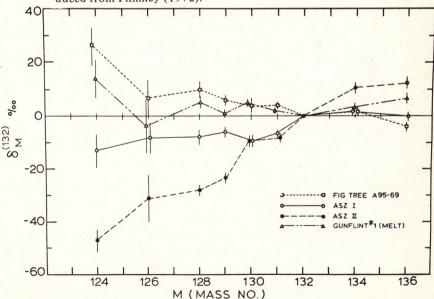

were trapped. Indeed, there has been relatively little discussion of this topic. The problem of noble gas trapping in general was raised in Section 4.5, and some of the data potentially relevant to sedimentary rocks are illustrated in Fig. 8.6.

An obvious candidate mechanism is solution. Noble gases may achieve solubility equilibrium during growth of authigenic minerals and become trapped by further growth (Section 4.5). The most relevant solubility data are for serpentine (Zaikowski & Schaeffer, 1979), which are illustrated in Fig. 8.6. Serpentine solubility is indeed relatively high (compared to igneous minerals) and fractionates in favor of the heavy gases, and if this is representative of sedimentary rock minerals then solubility may be important in cases of relatively low abundances and shallow fractionation patterns. By the same token, solubility appears inadequate to account for cases of much higher abundances and much more steeply fractionated compositions. It should be noted, however, that the serpentine data are for 340 °C, and if the heats of solution are negative the solubilities would be higher and probably more steeply fractionated at surface temperatures. Magnetite solubilities (Table 4.3) are much higher but with a fractionation pattern totally unlike those found in sedimentary rocks. Solubility seems a potentially viable mechanism for producing some of the trapped gases in sedimentary rocks, but there is very little basis for quantitative evaluation and until additional relevant solubility data are available the role of solubility will remain essentially hypothetical.

In using their 'Type I' designation for fractionation patterns progressively favoring the heavier gases (Figure 8.2) Ozima & Alexander (1976) suggested an association with water, in which dissolved gases also follow such a pattern (Fig. 8.6). It is plausible that most authigenic sedimentary minerals form in water, in which case the immediate source of the gases is water, and it may also be that hydrous minerals are more efficient than other minerals in noble gas trapping, but the relevance of water solubilities seems questionable. The solubilities would be important only if, by whatever means, water and its dissolved gases were assimilated into the rocks rather than exchanged with some larger reservoir. Assimilation, if it happens, might account for some of the cases of low abundance and little fractionation, but in general the gases dissolved in water are considerably lower in absolute concentration and considerably less fractionated than most of the observed trapped gases.

The only other mechanism which has been proposed is adsorption (Fanale & Cannon, 1971a; Podosek *et al.*, 1981), which has the immediately obvious virtues of producing very high concentrations and very strong fractionation (Fig. 8.6) at typical surface temperatures. Indeed, elemental fractionation by adsorption is so strong that it appears that it would be incapable of producing the right abundances of the lighter gases. The case for adsorption is also only hypothetical, but somewhat more data are available for comparison than in the case of solution. Podosek *et al.* (1981) noted that in the six samples for which comparison could be made the concentrations for equilibrium adsorption at air pressures were generally substantially higher than observed trapped gases in the same rocks and that there was no clear relationship between trapped gas contents and adsorption Henry constants, so that if adsorption is involved in producing the trapped gases so too were other factors, and the relationship is not straightforward.

A hypothesis for adsorption as a trapping mechanism is incomplete in that adsorbed gas is not really trapped and an auxiliary fixation mechanism is also needed. This is perhaps not so much of a problem. One possibility is that

Fig. 8.6. Noble gas abundances and compositions for means of incorporating gases into solids that may be relevant to the origin of trapped gases in sedimentary rocks. Abundances are those appropriate for atmospheric partial pressures (data from Table 4.6). Format is same as in Figs. 8.2–8.5.

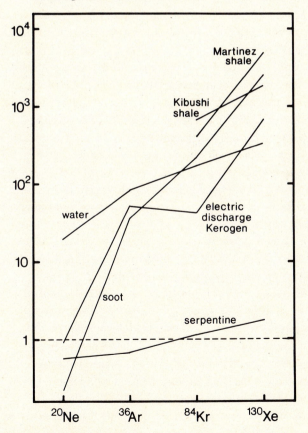

adsorbed gases could be trapped when covered over by grain growth, as also suggested for solution. Questions of mechanism aside, Honda *et al.* (1979) noted that so simple a process as static compression can act as a fixation agent, as demonstrated by an upward shift in characteristic degassing temperature. Podosek *et al.* (1981) suggested that at least in some cases no specific trapping mechanism at all is needed: gases in interior spaces of a sample must execute a random walk, making many collisions on surfaces, before escaping, and adsorption on these surfaces will slow down the process. Podosek *et al.* noted that even though the sticking time in an individual adsorption event is short (Section 4.2) the characteristic time for removal of adsorbed gas could be long on a laboratory timescale (cf. Section 4.5) if the total path length for escape is long, a few to several millimeters, and the individual step length in the random walk is short, 10–20 Å (a characteristic distance for interlaminar spacing in clays). Phinney (1972) noted that his ASZ shale acted as a long-lived gas source in the laboratory vacuum system and suggested that it was desorbing only slowly, an effect which could be accounted for by these considerations; Phinney also suggested that the difference in times of laboratory vacuum exposure could contribute to the order of magnitude difference in 'trapped' Xe contents in his two ASZ shale samples (Fig. 8.7).

Recent research on the localization of planetary trapped gases in meteorites (Section 5.2) suggests a particular relationship with the element C and has stimulated study of noble gases in various forms of C in terrestrial samples and laboratory synthesis (Frick & Chang, 1977; Frick *et al.*, 1979; Kothari *et al.*, 1979). Although the actual physical mechanisms involved can only be speculated upon, C does indeed tend to have high gas contents in natural samples (Fig. 8.3 and 8.4) and high distribution coefficients in laboratory synthesis (Fig. 8.6 and Tables 4.4 and 4.5), both with the steep elemental fractionation characteristic of sedimentary rocks. It is clear that C cannot be implicated for sedimentary rocks in general, but it may nevertheless play an especially significant role in some cases. (It is implicitly understood that it is reduced or organic C which has these features, rather than oxidized C; nevertheless, there are no data at all, natural or laboratory, for noble gases in carbonates.)

None of the possibilities described above are mutually exclusive, and all may be significant. Podosek *et al.* (1980) have suggested that the wide variety of abundances and compositions of trapped noble gases in sedimentary rocks almost demands that a multiplicity of trapping mechanisms be involved.

The list of relevant mechanisms presumably includes some not yet thought of. Particularly noteworthy is the need for some efficient mechanism which will strongly favor the lighter gases over the heavy, as illustrated by the strong Ne enhancements in a number of cases (Figs. 8.3 and 8.4). No such mechanism has even been proposed. In the absence of a plausible physical mechanism the possibility of experimental artifact must be considered seriously; it is also unclear, however, how such an effect might arise as an artifact. Furthermore, the Ne enhancement is not uncommon, and has been found in a variety of different samples and in a variety of different research laboratories.

8.5   The sedimentary xenon inventory

Noble gases now in sedimentary rocks, like noble gases dissolved in surface waters, are part of the generalized atmosphere (Section 12.1), and it is of interest to consider what contribution they make to the atmospheric inventory.

Fig. 8.7. Detail of distribution of Xe concentrations in sedimentary rocks (cf. Fig. 8.1). The line segment at normalized abundance 5976 shows average concentration needed for 10^{24} g to contain as much Xe as air. Codes inside boxes designate types: sh = shale, ch = chert, k = kerogen, ms = marine sediment, th = thucholite, c = coal, sg = shungite, ba = barite.

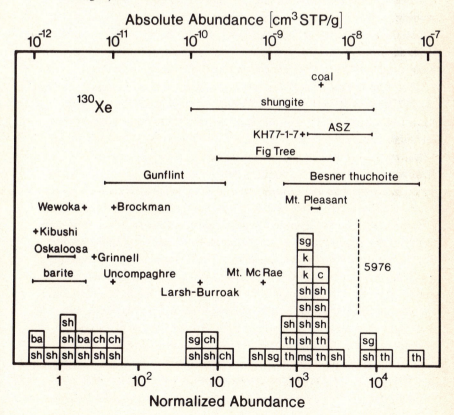

Unfortunately, assessing this contribution is much more difficult for sedimentary rocks than it is for water (Section 7.3), since gas concentrations in sedimentary rocks are far from uniform (Fig. 8.1) and the amount of sedimentary rock now extant (or ever extant) is subject to substantial uncertainty. An adequate inventory, one which would consider how much gas is present in all rock types likely to be important, is not possible at present. Nevertheless, some interesting generalizations and semiquantitative observations can be made.

As a scale marker, it is useful to inquire at what average concentration level the amount of gas in sedimentary rocks just matches the amount in air. For this purpose we will adopt a trial figure of 1×10^{24} g as a characteristic figure for sedimentary rock quantity; in particular this is evidently the best available estimate for the mass of shale extant (Wedepohl, 1969). This is 1/5976 the mass of the earth. Thus, in order that the sedimentary inventory of any gas match the air inventory, in 10^{24} g, its average concentration must correspond to a normalized abundance of 5976.

For all practical purposes, we may restrict further attention to the special case Xe. In all known cases but one, Xe has a higher normalized abundance than any of the lighter gases (Section 8.2 and Figs. 8.2–8.4); thus, in general, the sedimentary inventory will be larger in relation to the air inventory for Xe than it will be for any other gas. In particular, no sedimentary rock is known which has a normalized abundance greater than 5976 for any gas other than Xe (cf. Fig. 8.1). Unless an abundant but heretofore unsurveyed sedimentary rock type is found to be extremely rich in noble gases, the sedimentary contribution is apparently quite small, ≪1% of the air inventory, for Ne and Ar; the sedimentary contribution of Kr *might* be nontrivial, perhaps 1% or so of the air inventory. It is nevertheless clear that the sedimentary contribution will be most important for Xe.

Xe is also the case of greatest interest in a cosmochemical context. As discussed in Section 5.3, it is reasonable to expect that the total noble gas inventory of the earth should be 'planetary' in elemental composition. If the atmosphere originated by the essentially complete degassing of part or all of the earth (Chapters 12 and 13), the atmosphere should also have planetary composition. The relative abundances of Ne, Ar, and Kr in air conform remarkably well to this expectation, but Xe does not: its abundance in air is a factor of 24 lower than thus expected (Fig. 5.5). One interpretation of this situation is that both the atmosphere and the earth as a whole indeed have a planetary Xe abundance, but that the Xe is present primarily in an atmospheric reservoir other than air (or seawater), i.e. that it resides mostly in sedimentary rocks. Specifically, the candidate rocks are shales, and we will designate this interpretation the 'shale hypothesis'. This interpretation is generally favored in preference to its alternatives: that terrestrial Xe is planetary but that most of it has never entered the atmosphere, or that terrestrial Xe is simply not planetary. The shale hypothesis was first suggested by Canalas *et al.* (1968), and vigorously supported by Fanale & Cannon (1971a).

The argument of Canalas *et al.* was based on an average Xe content in their samples (normalized abundance 1700), in the extreme case corresponding to nine times more Xe in sediments than in air. This extreme case was based on an extreme estimate of 3×10^{25} g of shale, however; a more moderate 28% of the air inventory (for this average abundance) results from 1×10^{24} g.

Fanale & Cannon argued for the shale hypothesis in conjunction with their advocacy of adsorption as the relevant factor in noble gas trapping in sedimentary rocks (Section 8.4). Their data for 0 °C adsorption on Martinez shale (Table 4.2, Fig. 8.6) correspond to a normalized Xe abundance of about 5000. For their preferred figure of 3×10^{24} g of shale, this leads to a sedimentary Xe inventory three times that of air. This evaluation involves parameter choices favorable to the shale hypothesis, but even so falls nearly an order of magnitude short of planetary Xe abundance. One such choice is that of 3×10^{24} g; Fanale & Cannon prefer this to 1×10^{24} g as representing the total amount of shale ever produced rather than the amount presently extant, the difference having been consumed in tectonic recycling. If so, it must be further assumed that the recycling process does not return the Xe to the atmosphere. Furthermore, their calculations are based on equilibrium sorption, not on any observed gas contents in shales; the principal data base available to them was the work of Canalas *et al.*, in which the average Xe abundance was about a factor of three less than the adopted value for 0 °C sorption on Martinez shale.

Podosek *et al.* (1980) have argued that, irrespective of the trapping mechanism, the available data for trapped Xe in sedimentary rocks cannot be considered to support the shale hypothesis. A more detailed illustration of Xe abundances in sedimentary rocks is presented in Fig. 8.7. Only four samples have normalized abundances higher than 5976. Of these, three are thucholite or shungite, relatively scarce types which are presumably unimportant in the total inventory, and only one is a shale (ASZ II in Table 8.1 and Fig. 8.4). Most shales have lower concentrations, many of them very much lower. No terrestrial sample has been found to contain a normalized abundance as high as 1.4×10^5, the average abundance necessary if 10^{24} g of sedimentary rock accounts for the missing planetary Xe. If we arbitrarily adopt a normalized abundance of 1500 for Xe in sedimentary rock (cf. Fig. 8.7) as a reasonable upper limit, the corresponding sedimentary Xe inventory is 25% of the air inventory. This is certainly non-trivial but still a factor of 10^2 below accounting for planetary abundance of Xe.

It should again be stressed that the considerations above are only semi-quantitative and tentative. Many of the relevant input parameters are known only poorly, and there are insufficient data to construct an adequate inventory of noble gases in shales. The shale hypothesis is also easily generalized to include other 'sedimentary' reservoirs of atmospheric Xe, some of which might be important but which are presently unsurveyed.

In the arguments above, it has been assumed that gases 'in' a sedimentary rock are those normally reported as measured, i.e. trapped gases. Podosek *et al.* (1980) suggested a further generalization of the shale hypothesis to include gases actually adsorbed on sediments in equilibrium with the present atmosphere (an equivalent suggestion for Mars was made by Fanale & Cannon, 1971b, and Fanale *et al.*, 1978). This is distinct from the case for trapped gases, since adsorbed gas would be removed in the laboratory vacuum exposure involved in all trapped gas measurements. For this inventory, the relevant parameters are the amount of shale presently extant and the present temperature distribution, not the amount ever produced or the temperature of formation. For the amount adsorbed on 10^{24} g to equal the amount in air a mean Henry constant of $4 \, cm^3 STP \, g^{-1} \, atm^{-1}$ is required. Podosek *et al.* (1981) suggested that $1 \, cm^3 STP \, g^{-1} \, atm^{-1}$ is a more appropriate mean for Xe adsorption at surface conditions (cf. Table 4.2), which would correspond to adsorbed Xe being 25% of air Xe. They also cited factors which might change this estimate substantially, including potentially greater adsorption over geological timescales and less adsorption because of higher temperatures and pore closure in subsurface burial.

In summary, it seems plausible that Xe trapped in and/or adsorbed on sedimentary rocks accounts for a significant fraction of atmospheric Xe, perhaps of the order of 25% of the amount in air. This figure is very difficult to constrain, however, and the actual value might easily be substantially more or less. Still, this nominal value is some two orders of magnitude short of accounting for the putatively missing planetary Xe. These considerations certainly do not unambiguously *rule out* the hypothesis that sedimentary rocks contain the missing Xe, but it would be unwise to argue that the presently available sedimentary rock data support it.

8.6 Paleoatmospheric noble gases?

As noted in Section 8.1, the noble gases in sedimentary rocks are evidently dominantly atmospheric rather than juvenile. It is interesting and important to consider *when* these atmospheric noble gases were acquired. It is possible that these gases were acquired at or before the time of sedimentation, in which case sedimentary rocks preserve samples of the ancient atmosphere. This possibility is one of the principal reasons for interest in the noble gases in sedimentary rocks.

The alternative is that the trapped gases in sedimentary rocks are samples of the modern atmosphere, i.e. that any gases acquired at the time of sedimentation were lost during diagenesis or were exchanged with ambient atmosphere, continuously throughout geologic time or recently during weathering or in the laboratory. A middle ground is obviously possible also and the gases might be partly ancient and partly modern air. Literature discussions of noble gases in sedimentary rocks often implicitly assume that they are ancient rather than modern, but it is very difficult to bring any convincing explicit arguments to bear on this point. We feel that the present state of understanding of the mechanism of noble gas trapping in sedimentary rocks is so poor that no cogent arguments on the time of trapping are forthcoming from this direction. In this circumstance, appeal must be made to the character of the trapped gases themselves.

Considering the wide variety of absolute and relative elemental abundances displayed by noble gases (and other atmospheric species) in sedimentary rocks, and the lack of any theory to predict them quantitatively, it must be deemed hopeless (at least at present) to attempt identification of atmospheric elemental composition or absolute partial pressures on the basis of elemental abundances in sedimentary rocks. Isotopic effects offer more promise. Two classes of effect can be distinguished, mass-dependent fractionation and specific isotope variations. It would be difficult and perhaps impossible, however, to distinguish between the case of a rock sampling air by an isotopically biasing trapping mechanism (more likely) and the case in which ancient air is itself fractionated with respect to modern air, so there seem few prospects in this direction.

The field is thus narrowed to specific isotope effects. There are a number of candidates, all of them noble gases, in which radiogenic production over the course of geologic time is large enough to produce an observable compositional change in the total atmospheric inventory (Section 6.2). Specific models for plausible evolutionary paths are described in Chapter 13.

The greatest attention has been given to Ar because the possible effects are so large, and therefore potentially more easily detectable. The atmospheric $^{40}Ar/^{36}Ar$ ratio has presumably evolved from essentially zero to its present value of 296. Since the lifetime of ^{40}K is relatively large, this growth may have been stretched out over the age of the earth, and different models make quite different predictions for the evolution of atmospheric $^{40}Ar/^{36}Ar$. The difficulty, of course, is that K is an abundant element, and even in sediments selected for high $^{36}Ar/K$ ratio, the generation of radiogenic ^{40}Ar *in situ* during the geological age is enough to make a significant perturbation to trapped ^{40}Ar. In principle, if the geological age is known independently, correction for radiogenic production can be made; in practice, ambiguity about the possible escape of this radiogenic production has prohibited achievement of any definitive results.

A possible circumvention of this problem is available in the ^{40}Ar-^{39}Ar technique. In favorable cases, the isotopic correlations used in this method allow identification of K-Ar age and trapped Ar composition in spite of partial loss of radiogenic ^{40}Ar. The most extensive application of this approach to sedimentary rocks was made by Alexander (1975); unfortunately, his results are inconclusive. By the same technique, Cadogan (1977) asserts identification of trapped ^{40}Ar/^{36}Ar = 291 ± 1 in a Devonian (380 Ma) chert. We consider the data to be marginal, however, and do not feel that they justify the conclusion that the trapped Ar is different from modern air composition. Even if real, Cadogan's value is well within the range plausibly attributable to fractionation of modern air (cf. Section 6.4).

He is another case where variations in atmospheric composition are apt to be large, for essentially the same reasons as for Ar: radiogenic ^{4}He dominates atmospheric ^{4}He, and the parents are long-lived. The evolution of atmospheric ^{4}He/^{3}He is not closely tied to geologic sources (Section 12.5), however, so variations would probably be erratic and difficult to interpret, although interesting nonetheless. The experimental problem of ^{4}He production from U and Th during the geological age is even more formidable than in the case of ^{40}Ar, so the prospects for observing paleoatmospheric ^{4}He/^{3}He seem not particularly bright. In fact, it is not even clear that trapped atmospheric He has been observed at all in any sedimentary rock (Section 8.2).

One other candidate that seems not very promising is ^{21}Ne. The contribution of radiogenic ^{21}Ne is probably small ($\approx$4%; cf. Section 6.2), measurement uncertainties tend to be relatively large ($\approx$1%), and there are other potentially complicating effects (Section 6.5) that make isotopic resolution difficult for an element with only three isotopes.

The two remaining candidates are Xe isotopes, ^{129}Xe and ^{136}Xe. Although infrequently recognized as such, Xe is an important case (Bernatowicz & Podosek, 1978). While the likely radiogenic contributions to atmospheric ^{129}Xe and ^{136}Xe are relatively small, only a few per cent (Table 6.1), they are still much larger than available experimental precision. More important, Xe has two features that simplify the problem considerably. One is that Xe has several isotopes, so that it is easy to separate specific isotope effects from mass-dependent fractionation. The second is that both parents (^{129}I and ^{244}Pu) are now extinct, and are short-lived enough that they would not have been present in significant quantity in any known sedimentary rocks. Thus, no ambiguity arises because of radiogenic production during the geological age (^{238}U spontaneous fission is generally negligible). No effects of this kind are known; Phinney (1972), for example, reports Xe compositions in Precambrian sediments, but his data show no specific isotope variations from modern air, even in samples of the Fig Tree formation (Fig. 8.5), the oldest sedimentary rocks known.

We can summarize the discussion of this section fairly succinctly. As far as we are aware, every unambiguously identified trapped noble gas composition in sedimentary rocks has been the isotopic composition of modern air (or fractionated air). This applies to the isotopes discussed above and to all others as well. Unless exceptions appear, we must select from two alternative interpretations: (i) the observed gases actually *are* modern air, or (ii) to the extent that it has been sampled, ancient air has the same composition as modern air. If the first alternative turns out to be true, it would be informative because it would tell us about the origin of trapped gases in sedimentary rocks; it would nevertheless be disappointing. The second alternative is not implausible; essentially, it would correspond to formation of the atmosphere on a rapid timescale early in earth history (Chapter 13), models for which are currently popular for other reasons anyway. It would be important if this second alternative could be established as true, since competing models could then be eliminated, but it is not easy to imagine how this might be accomplished in satisfying fashion. A single unambiguous specific isotope effect distinguishing a paleoatmosphere from the modern atmosphere would clarify the picture considerably.

9   Igneous rocks

9.1   Introduction

In this chapter, we present a description of available information on the noble gas contents of igneous rocks. The data are important not only in the sense of a basic survey, but also because of their relation to the processes which produce igneous rocks and to the materials from which they are formed.

In the melting which forms a magma, some or all of the noble gases present in the parent solid will be partitioned into the melt. This partitioning will depend on the temperature and pressure environment, as well as on the parent composition. Nonequilibrium partitioning effects may be important. Noble gases in the magma are not necessarily preserved in the resultant igneous rocks. When the magma is transported, it will come into contact with rocks different from its source, and noble gas contents may change accordingly. Upon intrusion near the surface or actual surface extrusion, magmatic gases not only can be lost to the atmosphere (Chapter 10), but also can be gained from the atmosphere. Such effects are particularly evident in the comparison of other rocks with the rapidly-chilled glassy margins of submarine basalts (see Section 9.6).

One potentially significant exploitation of noble gas data for igneous rocks is the study of magmatic processes in general, particularly volatile transport of course. Actually, this area is not at all well developed, and efforts in this direction (e.g. Batiza *et al.*, 1979) are only primitive. The basic problem is lack of data, not only comprehensive survey data, especially for specific minerals, but also the relevant thermodynamic data (cf Section 4.3).

The complementary area of study, that of assessing the noble gas state of the source of the rocks, the earth's mantle in particular, has attracted keen attention. This subject is considered here, and in later chapters, particularly in Chapter 11.

9.2   Elemental abundances

Noble gas concentration data for igneous rocks are less scarce than data for sedimentary rocks, but are still far from plentiful. As is also the case for sedimentary rocks, analyses are concentrated on relatively few rock types, and there is not a sufficient data base for a comprehensive characterization of noble gas contents according to rock type or geological occurrence. Available abundance data are summarized in Figs. 9.1–9.6; selected data are presented in Table 9.1.

Noble gas concentrations in igneous rocks are variable across several orders of magnitude (Figs. 9.1–9.6) but, in general terms, can be described as low. Igneous rock concentration ranges are, for example, characteristically one to (more nearly) two orders of magnitude lower than the corresponding ranges for sedimentary rocks (cf. Figs. 8.1 and 9.1). They are also low in an experimental sense, in that there are often instrumental or procedural blank problems in their analysis and also in that some doubts can be raised about whether observed gases are actually indigenous to the samples (Section 9.5).

Concentrations are also low in the important sense of normalized abundance: rock concentration expressed as a ratio to air inventory divided by the mass of the earth. Average normalized abundances are less than unity, by about an order of magnitude or more (Fig. 9.1). At face value, and if the available data are representative of the solid earth in general, it is thus suggested that the entire inventory of noble gases in the solid earth is only a small fraction (a few per cent) of the inventory in air, i.e. that the earth is nearly fully degassed. This conclusion is even stronger if, as is usually expected (Section 4.4), noble gases behave as incompatible elements whose concentrations in melts are at least as great as (and perhaps substantially greater than) the concentration in the source rocks of the melts. This conclusion is also even stronger if observed gases are not all juvenile (Section 9.5). The conclusion is, however, weakened if it is supposed that igneous rocks lose gases when they come to the surface. A major qualification is whether the sampling is biased: it can be argued that there may be an 'undepleted' (of noble gases, among other elements) mantle reservoir which is grossly underrepresented by samples accessible on the surface (Section 11.2).

The low normalized abundances referred to above are for the primordial gases. The situation is less clear for the major radiogenic isotopes ^{40}Ar and ^{4}He. Trapped ^{40}Ar/^{36}Ar ratios in igneous rocks are often much higher than the air value (Fig. 11.1), so the average normalized abundance of ^{40}Ar will be higher than the average normalized abundance of ^{36}Ar. Whether it is sufficiently higher to reverse the sense of the tentative conclusion of the previous paragraph cannot be determined without a better inventory than is presently possible. Similar remarks presumably apply to ^{4}H as well; for ^{4}He, however, the very high average normalized abundance is essentially a reflection of the scarcity of He in air, i.e. a reflection of the circumstance that He, unlike the other gases, does not accumulate in the atmosphere.

The elemental abundance patterns illustrated in Figs. 9.9, 9.11–9.13, and 9.15 are normalized to 'air' concentrations: the air inventory divided by the mass of the earth. Such a normalization is inappropriate for He. Instead, we take

Table 9.1. Selected data for noble gases in igneous rocks

Sample	Concentration (cm³ STP/g)					Composition		
	^{3}He ($\times 10^{-12}$)	^{20}Ne ($\times 10^{-10}$)	^{36}Ar ($\times 10^{-10}$)	^{84}Kr ($\times 10^{-12}$)	^{130}Xe ($\times 10^{-14}$)	^{3}He/^{4}He ($\times 10^{-6}$)	^{40}Ar/^{36}Ar	^{129}Xe/^{130}Xe[a]
Submarine volcanics[b]								
MAR (TW 4-15)[c]	6[d]	447	24	62	64	12[d]	—	—
MAR (TW 4-15)[c]	—	26	85	210	131	—	—	—
MAR (TW 9-35)[c]	78[d]	273	5	9	20	12[d]	—	—
MAR (TW 4-118)[c]	—	7	89	458	794	—	—	—
EPR (HRX-6BZ)[e]	34	4	11	72	440	14	526	Air
EPR (735-1)[e]	56	1	2	6	7	14	4 188	Air
Hawaii (ERZ 1697)[e]	5	13	21	47	37	17	1 175	Air
Subareal volcanics								
New Mexico[f]	—	251	30	22	35	—	333	Air
Nigeria[g]	—	—	11	197	326	—	—	—
Gulf of California[h]	—	58	85	66	300	—	303	Air
Japan[i]	—	261	210	300	148	—	296	—
Hawaii[j]	2	12	10	82	153	48	311	Air
Xenoliths								
Kaersutite (New Zealand)[k]	187	34	140	436	4 600	49	400	Air
Spinel lherzolite (Hawaii)[l]	3	1	2	14	23	11	8 054	Air
Dunite (Hawaii)[m]	4	2	3	11	24	12	1 432	6.64
Dunite (Hawaii)[n]	—	7	4	6	13	—	4 437	6.77

Diamonds								
Batch 1 (<10% inclusions)^o	29	(0.2)	4	4	14	8	436	Air
Batch 2 (<1% inclusions)^o	18	(0.1)	0.7	2	10	20	1 121	Air
Plutonic rocks								
Skaergaard (upper zone)^p	—	2	24	133	800	—	520	Air
Skaergaard (lower zone)^p	—	34	48	108	95	—	3 800	6.74
Westerly granite^q	—	163	102	256	99	—	—	Air
Red Rock granite^q	—	273	327	268	5 320	—	—	Air
Air normalization^r								
Air	343^r	109	208	431	235	1.4	296	6.50

^{a} 'Air' designates ratios equal, within error limits, to the air ratio; blank spaces indicate no data.
^{b} The second and fourth samples are holocrystalline, all others are glasses; MAR = Mid-Atlantic Ridge, EPR = East Pacific Rise.
^{c} Dymond & Hogan (1973).
^{d} ^{3}He/^{4}He ratio not measured, here assumed to be a representative value for quenched glass; ^{3}He concentration calculated from assumed ^{3}He/^{4}He ratio and measured ^{4}He abundance.
^{e} Rison (1980a).
^{f} Hennecke & Manuel (1975a).
^{g} Fisher (1974).
^{h} Batiza et al. (1979) (Tortuga T-2-56).
^{i} Kaneoka (1980) (Mt Usu pumice).
^{j} Kaneoka & Takaoka (1980) (HA a Cpx, augite phenocrysts, Haleakala, Maui).
^{k} Saito et al. (1978).
^{l} Kaneoka & Takaoka (1980) (SLC-52).
^{m} Kaneoka & Takaoka (1978) (Hualalai 1801).
^{n} Hennecke & Manuel (1975b).
^{o} Takaoka & Ozima (1978); Ne values are upper limits.
^{p} Smith (1978); ^{40}Ar/^{36}Ar corrected for in situ ^{40}Ar.
^{q} Kuroda et al. (1977).
^{r} Composition is that of air (Table 2.2); concentration is air inventory divided by mass of the earth (Table 2.3), except for ^{3}He, which is a reference concentration chosen so that the ^{3}He/^{20}Ne ratio is the nominal planetary value (Tables 5.3 and 5.4).

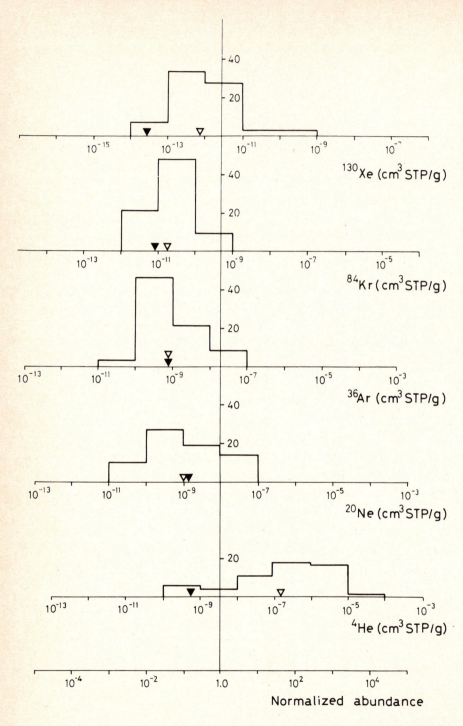

Normalized abundance

as a normalizing concentration for ^3He the value such that the $^3He/^{20}Ne$ value is the planetary value: this is intended as an approximation to the amount of ^3He which would be in air if He did not escape and if the planetary gas analogy (Section 5.3) were valid for He. Note that all the isotopes used for elemental abundance illustrations (Figs. 9.11–9.15) are primordial (rather than radiogenic).

A great deal of attention has been given to elemental abundance patterns, and many attempts have been made to classify them, explain their origins, or establish correlations between compositions, absolute concentrations, isotopic structures, rock type, location, and so on. In many cases elemental patterns have been used to argue that observed gases are not even juvenile, but only atmospheric contamination (cf. Section 9.5); this is probably correct in some instances and at least questionable in others. In such cases it is usually considered that the contamination is trivial in that it occurred in surface exposure to air or water, but some investigators suggest a more profound contamination of the magma source by recycling of atmospheric volatiles into the mantle (cf. Section 12.4). A persistent theme is the attempt to identify some elemental composition which is characteristic of the mantle as a whole and interpret other compositions as modifications of it, either by igneous processes or by contamination with atmospheric gases. Often an inferred mantle composition is identified with one of the two well-defined extraterrestrial compositions known, i.e. solar or planetary gases. For comparison, Fig. 9.7 illustrates solar and planetary compositions in the same format as Figs. 9.11–9.13 and 9.15.

Dymond & Hogan (1973), for example, noted that gas abundances in the quenched glass rims of marine basalts displayed a characteristic abundance pattern (Fig. 9.11) which they viewed as very similar to the solar pattern (cf. Fig. 9.7); they thus considered that the quenched basalts contained 'primordial' gases, which were different from atmospheric gases. It is clear, however, that the similarity is only in the Ne/Ar ratio, and the other gases do not conform to the expected abundance pattern. Craig & Lupton (1976) have also argued for the presence of 'primordial' (meaning 'solar') gas on the basis of Ne isotopic composition, but we disagree with this interpretation (Section 6.5).

Fig. 9.1. Histogram of available noble gas elemental abundance data for igneous rocks (cf. Fig. 8.1). More detailed displays for individual elements are given in Figs. 9.2–9.6. Absolute concentration scales are shown for each gas. The normalized abundance scale (bottom) is the same for all the gases; the normalization for each is the amount in air divided by the mass of the earth (Table 9.1). The solid triangles indicate concentrations corresponding to igneous melt solubility equilibrium (Kirsten, 1968; Fisher, 1970) at atmospheric partial pressures (also cf. Fig. 9.8). The open triangles indicate geometric mean concentrations.

Fig. 9.2. Histogram of [4]He concentrations in igneous rocks (cf. Fig. 9.1).

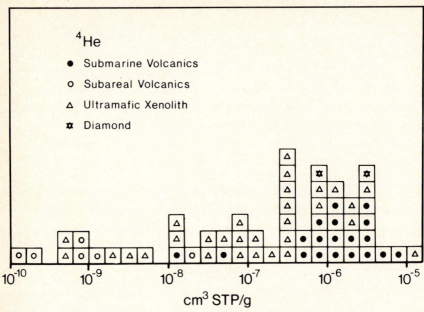

Fig. 9.3. Histogram of [20]Ne concentrations in igneous rocks (cf. Fig. 9.1).

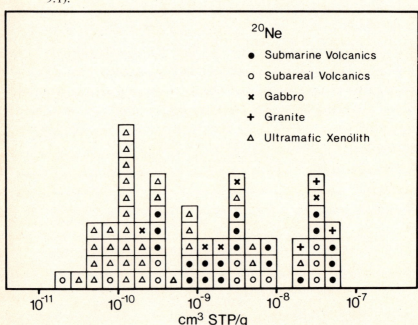

Fig. 9.4. Histogram of ^{36}Ar concentrations in igneous rocks (cf. Fig. 9.1, also Fig. 11.1).

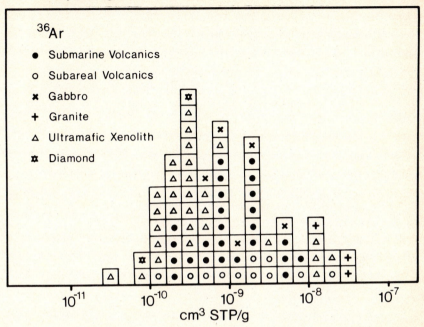

Fig. 9.5. Histogram of ^{84}Kr concentrations in igneous rocks (cf. Fig. 9.1).

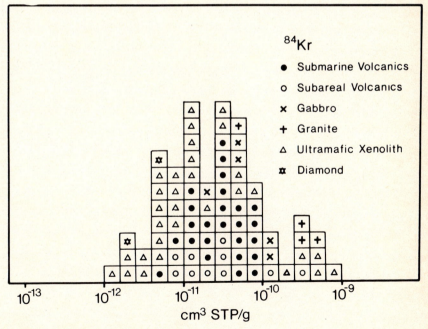

Any model which considers that gases in the earth's interior are solar in origin must also stipulate either that the atmosphere has a separate origin or that the present atmosphere has been greatly modified after degassing of such solar gases. This seems too complicated a hypothesis, since we see no evidence, either in elemental abundance patterns or isotopic structures, that gases in the interior of the earth are any more solar than gases in the atmosphere. The isotopic structure of Ne indeed does suggest an added solar component (Section 6.5), but only a relatively small contribution which would not much affect elemental abundances of the light gases and which would be irrelevant to the heavy gases, and in any case there is no evidence that such solar gas is more prominent in the interior than in the atmosphere. In the absence of any positive evidence, the only reason for considering a solar gas pattern in the interior would seem to be expectation based on analogy with extraterrestrial materials. We feel, however, that the analogy with planetary gases in extraterrestrial material is a considerably better one, in terms of both expectations and observations.

Indeed, the planetary gas analogy is more frequently invoked than is the solar. As an example, the same pattern that Dymond & Hogan (1973) considered to resemble solar gases, Fisher (1974) considered to resemble planetary gases.

Fig. 9.6. Histogram of ^{130}Xe concentrations in igneous rocks (cf. Fig. 9.1).

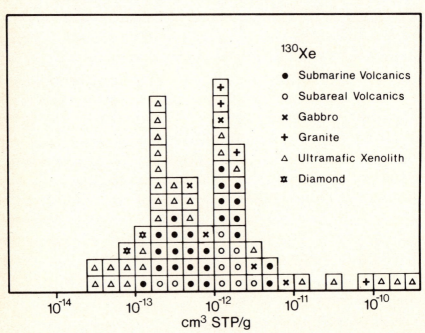

The latter view is tenable, however, only if the Ne/Ar ratio is ignored. Ozima
& Alexander (1976) proposed a classification of abundance patterns in which
their Type II, their basic (and only) juvenile gas pattern, was defined by the
qualitative features of the basalt glass pattern (Fig. 9.11): nearly atmospheric
Kr/Ar ratio and, relative to Ar and Kr, higher normalized abundances of both Xe

Fig. 9.7. The solar and planetary elemental abundance patterns normalized to the air abundance patterns (data from Tables 5.3 and 5.4). The
format is the same as in Figs. 9.9 and 9.11–9.13, and tickmarks on the
ordinate represent factors of ten, but only the shapes of the patterns
should be compared; both the absolute or relative ordinate positions
are arbitrary so no ordinate scale is given.

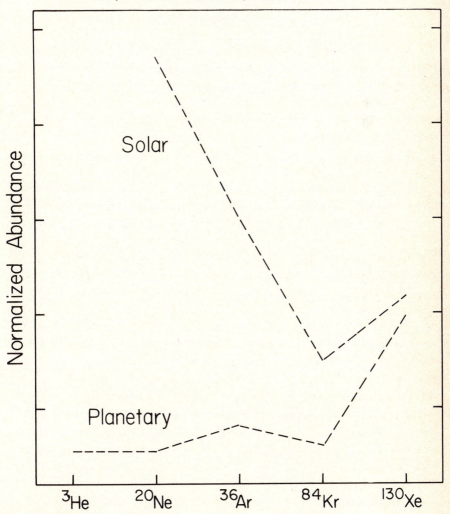

and Ne. Their interpretation is that this indeed represents primordial planetary gas remaining in the interior, which accounts for the Xe enhancement (cf. Fig. 9.7). This requires the usual special explanation for atmospheric Xe (Sections 5.3 and 12.6). It also requires a special explanation for the Ne enhancement. Ozima & Alexander suggested preferential diffusion of Ne into the magma. More recently, Ozima & Zashu (1983a) suggested that the Type II pattern could result from two-stage extraction, i.e. partitioning from a reservoir (depleted mantle) from which gases, originally in atmospheric proportions, have already been extracted once. Either view is qualitatively plausible but not easy to defend quantitatively. Furthermore, glassy basalts have also been found to have ^{3}He/Ne ratios which are quite variable and characteristically lower than the planetary value, a feature which considerably complicates any attempt to offer simple interpretations.

There are also other samples whose abundance patterns suggest interpretation in terms of primordial planetary gas. A well-known example is the kaersutite (an amphibole xenolith) analysis of Saito *et al.* (1978), but there are others as well. The variations are still large, however, and there is usually no basis for correlation between petrological primitiveness and any inferred 'primitiveness' of noble gas compositions.

Overall, noble gas elemental abundance patterns observed in igneous rocks are both complex and diverse. This is not particularly surprising, given the number of effects which can modify compositions both in igneous environments and on the surface. There have been many complex and detailed explanations offered for both individual rock patterns and for perceived classes of patterns. Such interpretations, which must in many cases be considered overinterpretations, are typically plausible but also typically are either only qualitative or require one assumption for every quantitative feature explained. Even empirical classifications tend to be disputed; the characteristically high Ne abundances and Ne/Ar ratios thought characteristic of quenched marine glasses ('primordial' to Dymond & Hogan (1973), Type II to Ozima & Alexander (1976)), are much less prominent or absent (Fig. 9.11a) in some data (Craig & Lupton, 1976; Rison, 1980a) but quite clearly present and prominent (Fig. 9.11b) in more recently acquired data (Kirsten *at al.*, 1981; Ozima & Zashu, 1983a).

The important enterprise of trying to characterize the elemental abundance patterns of primordial noble gases in the sources of igneous rocks unfortunately, must be considered, so far, unsuccessful. We feel that no case for a solar pattern can be made, at least not on the basis of observations. A somewhat stronger case can be made for a planetary pattern in some instances, although only qualitatively but it seems nevertheless doubtful that a planetary pattern would be inferred from observations were it not for expectations. It seems a significant but little remarked circumstance that there are no igneous rocks whose noble gas elemental abundance patterns are particularly similar to air composition.

9.3   Isotopery

A frequent theme in terrestrial noble gas studies is the possibility that atmospheric gases are in some way unrepresentative of the earth as a whole, either a biased sampling of the total or original complement of noble gases or a modification beyond the expected addition of radiogenic gases. Concern for this possibility is presumably heightened by definite differences between atmospheric noble gas structures and those in extraterrestrial analogs (Chapters 5 and 6). Considerable attention has thus been given to the possibility of observing, in igneous rocks, evidence that noble gases in the earth's interior are different from noble gases in its atmosphere. Understandably, it is difficult to reach firm conclusions on the basis of elemental abundances (cf. Sections 5.3, 9.2 and 12.4). Isotopic comparisons offer the prospect of less ambiguous conclusions. It is appropriate to consider such comparisons explicitly here, and our evaluation can be stated succinctly: there are no such differences, i.e. noble gases in igneous rocks are the same as noble gases in air.

One class of isotopic differences between terrestrial and extraterrestrial gases is fractionation: there is a severe and all but certain fractionation in Xe (Sections 5.6 and 6.7), a smaller and perhaps less certain fractionation (in the opposite direction) in Kr (Sections 5.6 and 6.8), no discernible fractionation in Ar, and a large effect in Ne which is possibly fractionation but which we consider better interpreted in other terms (Sections 5.5 and 6.5). Igneous rock gases frequently appear to differ from air gases by fractionation, but not nearly to the extent referred to above for Xe or Ne and the effects occur for all gases. All in all, we see no evidence which suggests that igneous rocks sample a primordial component significantly different (by fractionation) from air compositions.

The other class of differences is specific isotope effects. There are many such differences between air and igneous rock compositions, of course, but to our knowledge, all of them can be understood in terms of known radiogenic components generated within the earth: radiogenic ^{40}Ar and ^{4}He are common, and there are several other radiogenic gases (Section 6.2) which are less common. Potential exceptions (Section 6.10) seem highly improbable. Even a spallation component (Section 6.3), if it exists at all, exists in the same proportions in air and igneous rock gases.

He is an exception to this generalization – the only exception – in the sense that atmospheric He is not representative of He in the solid earth, now or in the past. It is important to appreciate how superficial this distinction actually is, however, particularly since there is sometimes a tendency to view atmospheric

He as the norm, compared to which He in igneous rocks, specifically 'mantle component He', looks unusual and demands a special explanation. 'Mantle' He, characterized by ^{3}He/^{4}He about an order of magnitude higher than the air value, is plausibly understood as a primordial He component (probably Helium A, possibly with admixed solar He (cf. 9.8), perhaps even with minor spallation He) which accounts for most of the ^{3}He, to which has been added radiogenic He, generated within the earth, which accounts for most of the ^{4}He (Section 6.6). Atmospheric He is basically different only because it does not accumulate; the atmospheric ^{3}He/^{4}He ratio has no fundamental geochemical significance and reflects only a probably fluctuating balance, achieved independently for each isotope, between sources and sinks (Section 12.5).

The differences between terrestrial and extraterrestrial noble gases must ultimately be accounted for in the formation of the earth and/or its planetary evolution. There is, however, no evidence that, in terms of fundamental isotopic structures, atmospheric gases are distorted or otherwise unrepresentative of the volatiles originally present in the earth, nor is there any basic mystery in how the atmosphere could have been derived from the interior. If the earth's noble gases were ever, in any fundamental sense, more primitive than they are now, the traces of these primitive gases have vanished or remain hidden from observation.

9.4   Excess Ar and He

The terms 'excess Ar' and 'excess He' refer to cases in which an igneous rock contains unsupported radiogenic gas, i.e. more ^{40}Ar and/or ^{4}He than can be attributed to ^{40}K and/or U/Th decay during its geological age. The effect is now recognized to occur widely in a variety of igneous and metamorphic rocks. The term originally arose in the context of geochronological studies and was used to contrast with the 'normal' case of partial loss of radiogenic gases.

In some cases the 'excess' ^{40}Ar could be recognized, on the basis of the ^{40}Ar/^{36}Ar ratio, to be due to an air Ar contamination (cf. Section 6.4). In current usage the term 'excess' is usually reserved for unsupported ^{40}Ar beyond what can be attributed to air on the basis of ^{36}Ar. In some cases allowance for modest isotopic fractionation of air Ar must also be made (cf. Krummenacher, 1970).

Excess ^{40}Ar or ^{4}He, at least in principle, might be due to preferential loss of K or U/Th. This cannot be the explanation in very young samples, however, nor does it seem very likely *a priori*; it seems doubtful that loss of radioactive parent is the cause of very many cases of excess gases, if any at all. The general case is evidently that the excess ^{40}Ar and ^{4}He are inherited, i.e. the melt from which the igneous rock formed contained radiogenic gas accumulated in its parent rocks There are also cases, however, in which rocks evidently acquired radiogenic gases after solidification, e.g. in metamorphism (see below). In either case, the excess gases are, of course, radiogenic and the essence of the effect is that they are *trapped* radiogenic gases rather than *in situ* radiogenic gases (see Section 1.4).

It should be noted that, from a geochemical rather than a geochronological perspective, excess gases should not be considered unusual or remarkable. Essentially all the ^{40}Ar (Section 6.4) and even most of the ^{4}He (Section 6.6) observed in any terrestrial environment is radiogenic. Thus, any juvenile (non-atmospheric) ^{40}Ar or ^{4}He which is trapped (not *in situ*) is, by definition, in excess.

It is also important to note that the excesses are identified when the nominal gas retention age is greater than the actual 'geologic age', i.e. the age of extrusion or intrusion. In most (not all) cases, there is no age incompatibility if allowance is made for generation of the radiogenic gases in the parent material of the magma; in such cases, the excess Ar or He effect can be viewed simply as a failure of the 'radioactive clock' to be completely reset. The effect is thus particularly striking in very young rocks in which *in situ* radiogenic production is essentially negligible (cf. Dalrymple, 1969).

The existence of excess ^{40}Ar and ^{4}He was first noted (Strutt, 1908; Damon & Kulp, 1958a) in the cyclosilicate minerals beryl, cordierite, and tourmaline (see Section 9.9). Hart & Dodd (1962) showed that the effect also occurred in common rock-forming minerals such as pyroxene and amphibole. Subsequently, the effect has become well known to characterize submarine pillow basalts (Funkhouser *et al.*, 1968; Dalrymple & Moore, 1968; Fisher *et al.*, 1968; Dymond, 1970); it also occurs in subareal basalts, however (McDougall *et al.*, 1969; Fisher, 1971). Excesses have also been found in gabbros and diorites (Civetta *et al.*, 1973; Hayatsu & Palmer, 1975), in ultramafic xenoliths (Lovering & Richards, 1964; McDougall & Green, 1964; Kirsten & Gentner, 1966; Kirsten & Müller, 1967; Kaneoka, 1974), and in a variety of specific minerals (Roddick & Farrar, 1971; Harper & Schamel, 1971; Takaoka & Ozima, 1978).

Inherited excess ^{40}Ar and ^{4}He in igneous rocks – that present in the magma from which the rocks formed – may be a nuisance from the viewpoint of geochronology, but it is an extremely important effect in a different perspective. Inherited trapped gases in mantle-derived igneous rocks are samples of noble gas in the mantle, and their characterization is an important subject pursued elsewhere in this chapter and in Chapter 11. The fact that the excess gases are isotopically distinct from air gases guarantees that they are not derived from air, i.e. that at least in such cases there are trapped gases in igneous rocks which are juvenile gases from the mantle. Such gases are often denoted by the informal term 'mantle component'. Some confusion arises in the use of this designation as opposed to 'excess' gases. We will use 'excess', as above, to denote unsupported radiogenic gases, not necessarily restricted to ^{40}Ar and ^{4}He (cf. Section 6.9); excess gases may be mantle-derived, but they need not be (see below). The mantle component includes not only excess radiogenic gases but also any trapped mantle-derived juvenile nonradiogenic (i.e. primordial) gases.

Not all excess radiogenic gases are mantle-derived, nor inherited in the sense described above. On the basis of a K–Ar isochron diagram (Fig. 9.8), for example, Civetta *et al.* (1973) infer excess ^{40}Ar in diorite minerals. Their interpretation is that in thermal metamorphism radiogenic ^{40}Ar was rehomogenized and trapped so that the whole rock and biotite and quartz–feldspar separates each contain about 4×10^{-6} cm^3 STP/g excess ^{40}Ar. They also note that muscovite in pegmatites in the same locality are deficient in ^{40}Ar by about the same amount that the biotite and quartz–feldspar are in excess, and suggest that the muscovites were the source of the redistributed ^{40}Ar. Similar results have also been reported by others, e.g. Roddick & Farrar (1971).

Metamorphism may cause loss as well as gain of excess gases, of course. Hebeda *et al.* (1980), for example, observed that unmetamorphosed relicts of a basic intrusive in an alpine orogen in southern Spain contained large excesses of ^{40}Ar, about 10^{-5} cm^3 STP/g, whereas relicts of metamorphosed rocks contained little excess ^{40}Ar. Their interpretation is that ^{40}Ar liberated from sediments diffused into the intrusion and remained trapped in the parts which were not

Fig. 9.8. K–Ar isochron diagram for whole rock, biotite, and quartz–feldspar separates (see text). The fitted line is interpreted as an isochron corresponding to the age shown; the nonzero ordinate intercept is interpreted as evidence for excess radiogenic ^{40}Ar. Reproduced from Civetta *et al.* (1973).

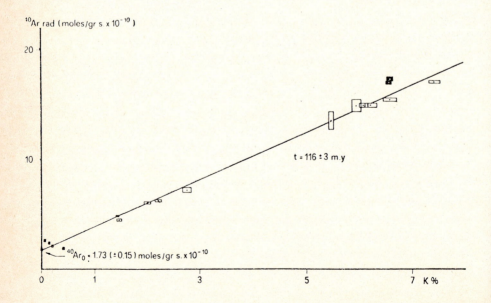

metamorphosed but were expelled during recrystallization of those which were metamorphosed.

9.5 Air contamination?

The trapped noble gases in sedimentary rocks are generally held to be of atmospheric origin; indeed, they are properly considered still to be part of the atmosphere (Section 12.1). In contrast, the usual principal interest in trapped noble gases in igneous rocks is in the characterization of juvenile gases and assessment of the noble gas state of the mantle. We have noted that noble gases in igneous rocks are fundamentally the same as those in the atmosphere (Section 9.3). It is legitimate to inquire whether this is due to the trivial reason that they actually are atmospheric gases, and there is thus considerable concern about whether the trapped gases in igneous rocks are indeed juvenile or only atmospheric. Unfortunately, it is often not easy to answer this question. A number of semi-quantitative considerations can be made for this problem, however; some of the relevant data are illustrated in Fig. 9.9.

An obvious comparison is with solubility equilibrium. Volcanics erupted on the surface of the earth, whether in air or under water, will be in an environment where the noble gas fugacities are those of air. If they equilibrate with this environment, they will have gas contents determined by air pressures and solubilities, and observed concentrations would tell us nothing about the original magmatic gases. The usual comparison is with the enstatite melt solubilities of Kirsten (1968); the corresponding gas levels are illustrated in Fig. 9.9. It should be noted that these solubilities are for 1500 °C, and lower solubilities are expected at lower temperatures; on the other hand, solubilities for less mafic melts, at a more typical igneous temperature, tend to be higher (Table 4.3). It is not at all clear whether lavas *should* equilibrate with air pressures, but it is often suspected that they do.

Equilibrium solubility levels (Kirsten, 1968; Fisher, 1970) are also indicated in Fig. 9.1. These concentrations are seen to be of the same general order of magnitude as the observed concentrations, although there is also a clear trend of a better match for the lighter gases than for the heavier. By the thesis that such a coincidence is no accident, it is thus suggested that typical trapped noble gas contents in rocks are indeed often nothing but dissolved air, especially the lighter gases Ne and (nonradiogenic) Ar. It should be noted, however, that an alternative hypothesis is possible, namely that noble gas fugacities in magmas (and thus the mantle?) are of the same order of magnitude as in the atmosphere, i.e. that as far as noble gases are concerned the mantle is not too far out of equilibrium with the atmosphere.

For the heavier gases typical concentrations tend to be higher than melt solubility concentrations, whence the suggestion that the observations reflect only dissolved air is less plausible, especially for Xe (Fig. 9.1).

In the paragraphs above the solubility considered was for melts. For the solid phases typical of igneous rocks gas solubilities are evidently lower than for lavas (Section 4.4). This may not be true for phases formed by surface alteration, however, especially hydrous minerals. Here the most relevant available data are for serpentine (Zaikowski & Schaeffer, 1979), and Fig. 9.9 illustrates gas contents for a hypothetical rock devoid of gases except for gases dissolved in a 5% serpen-

Fig. 9.9. Illustration of possible artifacts by atmospheric contamination. Concentrations are shown for a rock devoid of noble gases except for: gases dissolved in molten enstatite at air pressures, gases dissolved at air pressures in a sample containing 5% by weight serpentine, gases contained in a 2% by weight content of air-saturated seawater, and gases contained in 1% by weight of a material like marine sediment KH77-1-7. Normalization is the same as in Figs. 9.11–9.13. Data from Tables 4.3, 7.6, and 8.1.

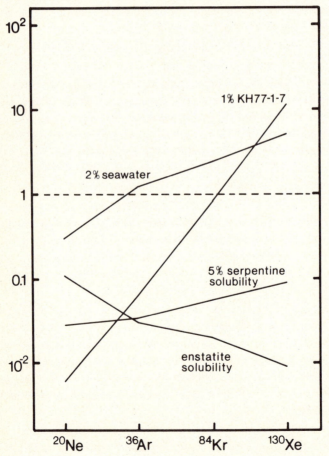

tine component. It is noteworthy that serpentine solubility, in contrast to igneous melt solubility, favors the heavy gases gases over the light, a feature which could be invoked to account for the general trend of higher normalized abundances for heavier gases (Fig. 9.1) in terms of solution of atmospheric gases.

Solution is not the only possible effect, of course. Fig. 9.9 also illustrates the gas contents an igneous rock would have if, by whatever process, it assimilated 2% seawater along with its dissolved gases or contained 1% material like marine sediment KH77-1-7. Both of these patterns also enhance the heavier gases over the lighter.

By comparison of various conceivable effects (Fig. 9.9) with the observations (Figs. 9.1–9.6) it can be seen that, at least as far as gas quantities are concerned, plausible mechanisms for atmospheric contamination, individually or in concert, could be invoked to account for trapped noble gases found in igneous rocks. Whether or not they actually do so is a different question entirely. In some instances a strong case can be made that observed trapped gases are indeed only air contamination; in others a strong case can be made that they are not. Unfortunately, generalization is impossible; individual cases must be considered individually, and even then there is usually substantial uncertainty.

Thus, for example, Fisher (1970) noted that if gas solubilities follow the model of Blander *et al.* (1959) (see Section 4.3), solubility ratios would be univariant; specifically a linear correlation between, say, $\log(Ar/Kr)$ and $\log(Kr/Xe)$ would be expected. Observing such a correlation (which passed through air composition) in a set of submarine basalts, Fisher asserted thereby that the trapped gases were atmospheric. Ozima & Alexander (1976), however, pointed out that Fisher's reasoning was incorrect: according to the model only part of the correlation line (heavy gases depleted, relative to air ratios) should be populated, and the data under consideration were on the wrong side of air composition (corresponding to a negative surface tension).

Several investigators have stressed the importance of the glassy rims of submarine pillow basalts (cf. Funkhouser *et al.*, 1968; Dymond & Hogan, 1978). These glasses, formed by rapid quenching of lavas extruded into seawater, are often considered the best candidates for minimal interaction with atmospheric gases and thus for the best preservation of juvenile noble gases and the least contamination with atmospheric noble gases. Dymond & Hogan (1978), for example, find (Fig. 9.10) that very near the outer rim (a few millimeters) of some pillow basalts the Ar contents are relatively high, as is the $^{40}Ar/^{36}Ar$ ratio, an indication of juvenile gas (excess ^{40}Ar). Farther in (about 1 cm) they find lower Ar concentrations, which they attribute to expulsion of gases from growing crystals. Still farther in (several centimeters), they find again higher Ar concentrations but lower, more nearly atmospheric, $^{40}Ar/^{36}Ar$, which they interpret as atmospheric contamination arising when seawater invades cracks formed in the more slowly cooling interior of the lavas. More generally, they state, holocrystalline submarine basalts have higher and more steeply fractionated (heavy gas enhancement) gases than do glassy basalts (Fig. 9.11). Both features are characteristic of sedimentary rocks (Figs. 8.2–8.4 and 9.9) so their interpretation is that such cases reflect atmospheric gases introduced during aqueous alteration.

The only truly unambiguous identification of juvenile gas, as opposed to atmospheric gas, is in such cases where trapped gases are *isotopically* distinct from atmospheric gas. Craig & Lupton (1976) applied this criterion in asserting identification of primordial solar Ne, for example, although in this case we feel

Fig. 9.10. Ar in marine basalts. The lines connect results obtained for the same rock at different depths (labeled in centimeters) from the rock–water interface. The trends are interpreted to indicate that only at rims do the trapped gases reflect magmatic gases (see text). Reproduced from Dymond & Hogan (1978).

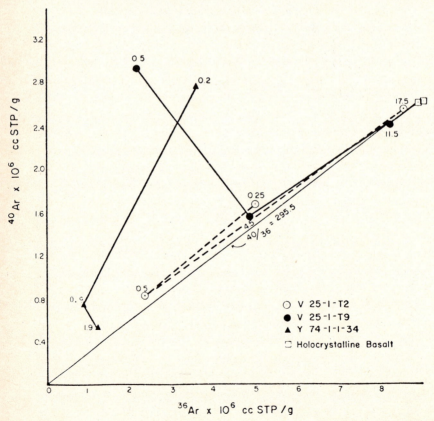

that a better interpretation can be made in terms of fractionated Ne of air composition (Section 6.5). More generally, the only clear cases of isotopic difference, other than fractionation, between rocks and air are cases of excess radiogenic isotopes in the rocks (Section 9.3). The phenomenon of radiogenic excesses is thus rather important in that it demonstrates that at least in some cases trapped gases are juvenile rather than atmospheric. Even in these cases, however, only the *excesses* are demonstrably juvenile: the same ambiguities arise for the other gases. The problem is analogous to that described in Section 8.6. Given that most trapped gases in igneous rocks are not isotopically distinct from atmospheric composition, there are two possibilities: they actually are atmospheric, or they are juvenile and juvenile gases have the same composition as air. Both possibilities are plausible and we usually have insufficient information to make a clear resolution.

Finally, it should be noted that the discussion of this section hardly applies to He, the one gas for which the likelihood of atmospheric contamination can often be dismissed out of hand. Equilibrium solubility concentration is quite low compared to observed concentrations (Fig. 9.1), for example, and the other effects illustrated in Fig. 9.9 also give low He contents. The basic reason is that He, relative to the other noble gases, is so scarce in the atmosphere. This is because He escapes from the top of the atmosphere, and its present atmospheric abundance is between three and four orders of magnitude lower than it would be if it accumulated (cf. Table 12.1). He accordingly has a special status as by far the most easily identifiable juvenile gas.

9.6   Volcanics

The observations that seawater contains He in excess of solubility equilibrium (Sections 7.5 and 7.6) were quickly interpreted as a juvenile flux associated with seafloor spreading, the quantitatively dominant volcanic process on the earth. An obvious suggestion is that the juvenile gases might be preserved in the volcanic rocks themselves; this is indeed the case, of course, as was strikingly verified by observation of ^{3}He in marine basalts by Lupton & Craig (1975) and Craig & Lupton (1976). That this is also the case for the radiogenic gases ^{4}He and ^{40}Ar was already well known (Section 9.4) at the time.

Characteristically, juvenile gases are best preserved in the quenched glass rims (not more than a few centimeters deep) of marine basalts extruded directly into seawater. It is often considered that these rapidly chilled glasses preserve, or at least come closest to preserving, the gas contents of their source magmas. Nonglassy samples, in contrast, characteristically have different and diverse elemental and isotopic patterns which are generally interpreted in terms of interaction with surrounding seawater, either loss to the water or contamination during alteration (Section 9.5, Fig. 9.11).

Juvenile gases in the marine glasses characteristically have high $^{40}Ar/^{36}Ar$ (Section 11.3), i.e. excess radiogenic ^{40}Ar, and high $^{3}He/^{4}He$ ratios. The high $^{3}He/^{4}He$ ratios unambiguously indicate that the ^{3}He is juvenile, not atmospheric, and the fact that ^{3}He is the only primordial (nonradiogenic) species for which so clear a demonstration is possible accounts for much of the interest in such observations. The $^{3}He/^{4}He$ ratio is high only in comparison with air He, however, and most of the ^{4}He is radiogenic (Section 6.6), i.e. like the ^{40}Ar it is an excess component.

For fresh glasses the $^{3}He/^{4}He$ ratio is variable only in a narrow range, about 1.1×10^{-5} to 1.4×10^{-5} (cf. Craig & Lupton, 1976; Rison, 1980a; Kurz & Jenkins, 1981), or about 8 to 10 times the air value. Possibly the real range is even narrower, as the reported spread may be due to systematic experimental uncertainty or interlaboratory bias. In any case, the glass compositions are the same as the compositions of excess He in seawater (Section 7.6), so both sets of observations can be taken to represent the same magmatic source of He.

The narrow variability (or constancy) of the $^{3}He/^{4}He$ ratio in marine glasses from all over the world ocean is rather remarkable in view of the nearly totally separate origins of the two isotopes: ^{3}He is primordial and ^{4}He is radiogenic. The nearly constant composition of the mantle He sometimes inspires the suggestion that it is itself a distinct primordial component, not a primordial/radiogenic mixture; we consider this untenable, however (Section 6.6). Rather, the constancy implies that the mantle source of the magmas is quite well mixed, at least so far as He and U/Th are concerned, on a global scale.

In view of the He (and Ar) isotopic data, the elemental abundance patterns of the glasses (Fig. 9.11) are sometimes also considered representative of juvenile gases (Section 9.2), although there is considerably more uncertainty about interpretation and representativeness. Unfortunately, it is difficult to make a direct comparison between ^{3}He and the heavier primordial gases, since the relevant data typically lack data for either ^{3}He or the gases heavier than Ne. Abundances for ^{3}He calculated by *assuming* representative $^{3}He/^{4}He$ values for glasses are given in Table 9.1; as illustrated in Fig. 9.11, ^{3}He abundances are often not as great as might be predicted on the basis of Ne. Craig & Lupton (1976) found $^{3}He/^{20}Ne$ ratios to vary by more than three orders of magnitude, mostly but not all below the planetary ratio.

It is worth an explicit remark that oceanic tholeiites, the dominant oceanic basalt type which most of the He data describe, are not where one might most logically expect to find primordial He. Geochemical and petrologic models generally stipulate that tholeiites are derived from mantle rocks which are already depleted in incompatible elements by at least one and perhaps multiple generations of magma extraction. The expectation would thus be that primordial He should also have been extracted. This does appear to be the case for ^{4}He: a 4.5 Ga age for a U content as low as 10 ppb (with Th/U = 3) corresponds to a ^{4}He content greater than 10^{-5} cm^3 STP/g in the source rock and presumably higher still in a derivative magma. This is considerably higher than most ^{4}He contents in glasses (Fig. 9.2), so it seems likely that ^{4}He has been extracted reasonably efficiently at least once. It may be that (primordial) He does not partition into a melt very efficiently (cf. Section 4.4) or perhaps that the source of tholeiites is resupplied with primordial He from other reservoirs.

Volcanics other than ocean floor basalts frequently have different noble gas characteristics (cf. Fig. 9.12). A noteworthy example is Hawaiian magmas, for which ^{3}He/^{4}He ratios are sometimes significantly higher than the typical oceanic value (Fig. 11.6), and there may be associated high ^{129}Xe and low ^{40}Ar/^{36}Ar values (Section 11.3). Such results are often interpreted in terms of a 'hot spot' which taps a different, more primitive, mantle source.

9.7   Xenoliths

Basaltic volcanics occasionally contain ultramafic xenoliths of unambiguous mantle origin. Noble gases in these xenoliths are thus studied in expectation of direct information about mantle gases. In this respect the xenoliths might have an advantage over basaltic volcanics also originating in the mantle, since they crystallized at depth and may be less susceptible to gas loss or air contamination on the surface. There is not a striking systematic difference between xenoliths and other igneous rocks in terms of total gas contents, however, so the noble gas contents in sources of the xenoliths and ordinary basaltic volcanics might be similar.

Noting that some ultramafics can be used as geobarometers (cf. MacGregor & Basu, 1974), Kaneoka *et al.* (1978) suggested that their study might yield information about vertical variation of noble gases in the mantle. They found that in a South African kimberlite an olivine megacryst has a lower ^{40}Ar/^{36}Ar ratio than a phlogopite-bearing peridotite. Since the former has an inferred deeper origin than the latter, they suggested a decrease of ^{40}Ar/^{36}Ar with depth in the mantle. Rison (1980a) reached a similar conclusion from study of three garnet lherzolites, also from South African kimberlite; in these samples he also found a systematic decrease in ^{40}Ar/^{36}Ar with depth of origin.

In order to interpret xenolith data it is important to consider the general question of how well they can be expected to preserve the record of their mantle environment. Studies of Sr, Pb, and Nd isotopic compositions in various xenoliths indicate that they have generally not equilibrated with their host lavas. Experimental studies of diffusion coefficients of Sr, Mg, Al, and other cations indicate that this is reasonable, and that isotopic disequilibrium on a centimeter scale could persist for periods at least of the order 10^8 to 10^9 yr in dry crystalline mantle. It is not clear to what extent this generalization extends to the noble gases, since they are expected to have higher diffusion coefficients, particularly the lighter gases. Bernatowicz (1981), for example, found the same gas compositions (except for plausibly *in situ* radiogenic ^{40}Ar) in xenolith minerals and in

Fig. 9.11. Noble gas concentrations in submarine volcanics, displayed as ratios of observed concentrations to the corresponding normalization concentration, the air inventory divided by the mass of the earth. (A) MAR = Mid-Atlantic Ridge; EPR = East Pacific Rise; samples designated (x) are crystalline, others are quenched glass. Dashed line segments designate data which are assumed rather than inferred (see text). Data from Table 9.1. (B) CYAMEX (East Pacific Rise) basaltic glasses. Reproduced from Ozima & Zashu (1983a).

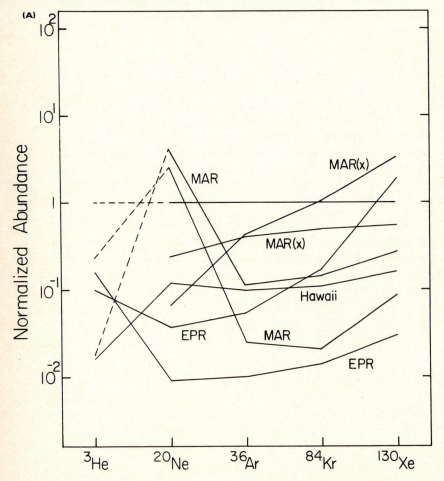

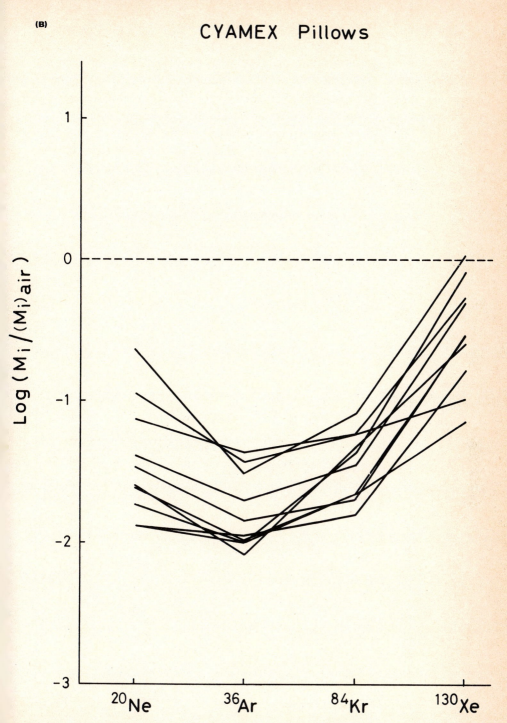

(B)

CYAMEX Pillows

host basalt; this might indicate that the sources of both had the same gas composition. There are, however, data indicating lack of isotopic equilibration. Funkhauser & Naughton (1968) and Gramlich & Naughton (1972), for example, found different $^{40}Ar/^{36}Ar$ ratios in different minerals in Hawaiian lherzolites. Rison (1980a) also found differences among the minerals in a South African kimberlite xenolith: $^{40}Ar/^{36}Ar$ ratios of 3894 in clinopyroxene, 1709 in orthopyroxene, and 482 in olivine. Rison interpreted these data in terms of an initially uniform composition and variable degree of atmospheric contamination; an alternative interpretation of variable degree of isotopic exchange with the host lava is also possible.

Fig. 9.12. Noble gases in subareal volcanics (cf. Fig. 9.11). Data from Table 9.1.

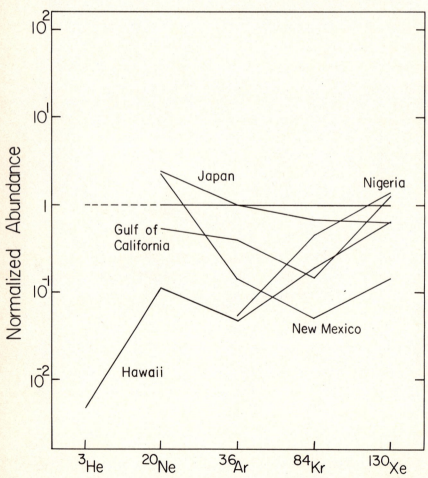

The data base for noble gases in xenoliths is still relatively small, and it is difficult to reach general conclusions on the degree to which xenolith noble gases do or do not approach equilibration with their host lavas or with the atmosphere. Until the problem is clarified assessments of mantle noble gases on the basis of xenolith noble gases must be made with some caution.

9.8   Diamonds

Diamond is a unique form of mantle xenolith: it is formed at high pressure and is a notably 'inert' mineral, resistant to chemical and thermal alteration, particularly at low oxygen fugacity such as prevails in the mantle. Diamond is thus a prime candidate for sampling noble gases in their source region, the mantle.

While diamond is a rather 'refractory' material, it is well known to contain volatiles; diamonds have been reported to contain O_2, H_2, CH_4, H_2O, CO, CO_2 and N_2 (e.g. Melton & Giardini, 1976). Noble gas contents are illustrated in Table 9.1 and in Fig. 9.13. It is often assumed that the noble gases, along with other volatiles, are in inclusions contained within the diamonds, rather than the actual diamond lattice. This would account for observation of Ar release by simple vacuum crushing (Melton & Giardini, 1980), for example. In the stepwise heating data of Takaoka & Ozima (1978) the gases were not released until graphitization at 2000 °C (Fig. 9.14), however, which favors lattice siting, but this might also be due to difficulty of escape from wholly contained inclusions without graphitization. In any case, there is no clear correlation between gas amounts and quantity of inclusions, nor a clear distinction between diamonds and other igneous rocks in terms of gas amounts (Figs. 9.1-9.6).

He and Ar compositions in diamond are of substantial interest as samples of mantle gases. Diamonds clearly contain mantle He, as evidenced by elevated $^3He/^4He$ ratios (Table 9.1), in some cases at least significantly higher than the normal abyssal basalt ratios (see Section 11.4). Some diamonds also have relatively low trapped $^{40}Ar/^{36}Ar$ ratios (Table 9.1), and Melton & Giardini (1980) even report a ratio significantly lower than atmospheric. Others, however, have much higher ratios (e.g. Ozima et al., 1982). The problem of whether the low values represent actual mantle gas or merely atmospheric contamination is important in assessing the state of the mantle (see Section 11.3).

Recently, Ozima & Zashu (1983a) found a wide range of $^3He/^4He$ ratios in a suite of South African diamonds, from less than 10^{-7} to as high as $(3.2 \pm 0.3) \times 10^{-4}$ (for this sample the concentration of ^3He is also remarkably high: 4×10^{-11} cm^3 STP/g). They suggested that the isotopic variations could be understood in terms of dilution of primordial He with variable amounts of radiogenic He and noted that preservation of such high $^3He/^4He$ ratios requires a thorough and very early isolation of He from U and Th. Perhaps more remarkable still, in at least two of the diamonds the *measured* $^3He/^4He$ ratio is higher than that of Helium A and nearly that of Helium B (Table 5.4); substantiation of these observations would require that prior to accretion earth materials received a significant component of solar noble gases.

9.9   Plutonic rocks

Granites and other igneous continental crustal rocks typically are older and richer in U, Th, and K than marine or mantle xenolith rocks, and thus generally contain large amounts of *in situ* radiogenic ^4He and ^{40}Ar. They have

Fig. 9.13. Noble gases in xenoliths and diamonds (cf. Fig. 9.11). Data from Table 9.1.

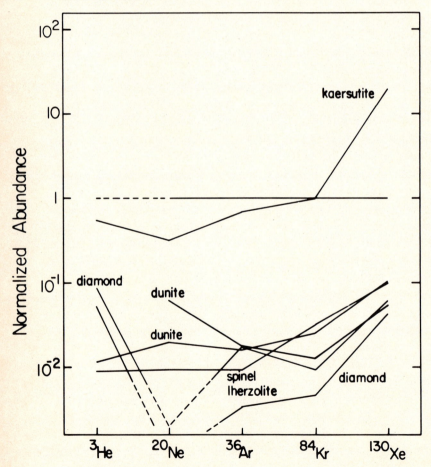

Fig. 9.14. Thermal release patterns for noble gases in diamonds (Batch 2 in Table 9.1). The principal gas release occurs in the 2000 °C step, in which the diamond is graphitized. Reproduced from Takaoka & Ozima (1978).

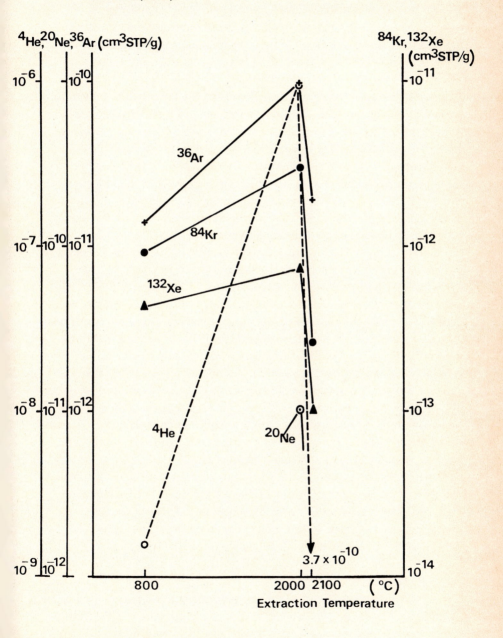

also been observed to contain ^{238}U-fission Xe and radiogenic ^{21}Ne, and their ^{3}He is presumably radiogenic also (cf. Section 6.2). Only the nonradiogenic isotopes are shown in Figs. 9.1–9.6 and 9.15.

Aside from geochronology studies, granites and related continental rocks have not attracted very much attention in noble gas investigations. The available data tend to be near the high end of the concentration range for igneous rocks in general (Figs. 9.1–9.6), but not outside it, but the data are so sparse that it would be hazardous to consider this a generalization.

It is also interesting to note that in terms of the relation between ^{130}Xe/^{36}Ar ratio and ^{130}Xe content (Fig. 12.4), the limited data available for granites are in the same general field as other igneous rocks and are not in the typical sedimentary rock field. The inference is that in their formation these granites did not assimilate large amounts of sedimentary rock (see Section 12.4), at least not without major loss of noble gases. The suggested implication is indeed tenuous, but if supported by future studies this observation may be relevant to the long-standing problem of whether granites form by differentiation of mantle magmas or by granitization of previously existing crustal rocks, including sediments.

Kuroda *et al.* (1977) have presented noble gas data for the Red Rock granite, a formation perhaps associated with the Sudbury structure, which is usually believed to have been created by a meteorite impact 1.7 Ga ago (cf. Dietz, 1964). Kuroda *et al.* assert, on the basis of elemental composition and Xe isotopic composition, that noble gases in the Red Rock granite contain a small but nontrivial fraction of meteoritic (i.e. planetary) noble gas from the impacting object. It is our opinion that, in view of the wide variety of ill-understood noble gas elemental fractionations evident in igneous rocks, no such argument based on elemental composition can be considered even strongly suggestive. Also, while the Xe isotopic data are consistent with trapped air Xe plus *in situ* fission Xe and minor admixture of planetary Xe, they are certainly not compelling. Kuroda *et al.* do not discuss transport mechanisms or mass balances. Altogether, we thus feel that their advocacy of this hypothesis is unfounded.

Smith (1978) calls attention to an interesting contrast in noble gas abundances in samples of the Skaergaard intrusive (Table 9.1 and Fig. 9.15). His interpretation is that gases in the lower zone sample reflect the original magmatic gases (this sample contains excess ^{129}Xe), while those in the upper zone sample are greatly modified by interaction with groundwater (cf. Section 9.5).

9.10 Cyclosilicates

The minerals beryl, cordierite, and tourmaline, collectively designated as cyclosilicates, have large open channels in their crystal structures. These channels are large enough to accommodate atoms and molecules extraneous to the structure, and these minerals are well known as hosts of such extraneous elements, including noble gases (Fig. 9.16). Beryl and tourmaline are characteristic minerals of pegmatites, and cordierite is a common metamorphic mineral.

These minerals, particularly beryl, typically contain large amounts of ^{4}He and ^{40}Ar, i.e. they contain excess ^{4}He and ^{40}Ar in the sense described in Section 9.4. Large ^{4}He excesses in beryl were first observed by Rayleigh (Strutt, 1908), and excesses of both ^{4}He and ^{40}Ar in all three minerals have been amply documented since (Aldrich & Nier, 1948; Damon & Kulp, 1958a; Gerling *et al.*, 1968; Ginzburg & Panteleyev, 1971). Similarly, excesses of radiogenic Xe and Ne (Section 6.2) have also been observed (Smith, 1978; Saito *et al.*, 1983).

Fig. 9.15. Noble gases in plutonic rocks (cf. Fig. 9.11). Data from Table 9.1.

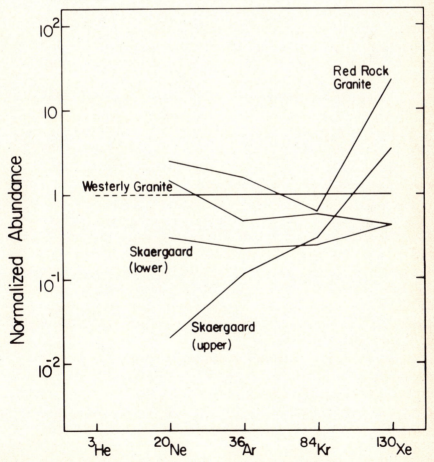

For both ^{4}He and ^{40}Ar there is an age effect, first noted by Rayleigh (1933): older samples contain more excess gases. Damon & Kulp (1958a), for example, found that beryls older than 2.5 Ga contain almost a factor of 10^2 more excess ^{40}Ar than those with ages younger than 1.0 Ga. In general, later investigations have supported the age effect, but it may still be possible that it is a sampling bias rather than a reflection of some geological process.

It is not clear whether the excess gases are trapped at the time of mineral formation or accumulated throughout the subsequent geological history. Noting that many minerals do not retain Ar well during a high-temperature period after formation, York & Farquhar (1972) argue for the latter possibility and propose that beryl absorbs radiogenic gases liberated from the major minerals throughout postformational history. The age effect is a natural consequence of this model. Damon & Kulp (1958a) however, argue that the gases are trapped only during crystallization. This would follow from the model that gases are trapped by water molecules or alkali ions, incorporated during crystallization, which plug the structural channels (Ginzburg & Panteleyev, 1971; Schreyer *et al.*, 1960; Smith & Schreyer, 1962). The age effect might then be due to generally higher levels of radiogenic ^{4}He and ^{40}Ar mantle fugacities the further in the past the minerals formed (Damon & Kulp, 1958a).

The cyclosilicates evidently impose a strong elemental fractionation in trapping. By comparing observed abundances with plausible radiogenic compositions (cf. Section 6.2), Saito *et al.* (1983) found that both beryl and cordierite trapped ^{4}He more efficiently than ^{40}Ar by about an order of magnitude or more. They also found that while beryl consistently trapped ^{40}Ar more efficiently than ^{136}Xe, cordierite generally favored ^{136}Xe over ^{40}Ar.

While pegmatites form from volatile-rich last-stage magmatic liquids, this cannot be the only reason for the concentration of gases in beryl and tourmaline, since other coexisting pegmatite minerals do not show such large excesses of ^{4}He and ^{40}Ar (Fig. 9.16). Indeed Saito *et al.* (1983) show that beryl did not sample the same gas reservoirs as the other minerals, since its ^{40}Ar/^{36}Ar ratios are much higher than the other minerals. Saito *et al.* interpret this in terms of different times of crystallization and changing fluid composition rather than postcrystallization transfer of radiogenic gas from other minerals to beryl, since such transfer would not produce higher ^{40}Ar/^{36}Ar in beryl.

It is noteworthy that the cyclosilicates and other minerals in pegmatites have nonradiogenic gas contents which are higher than typical igneous rocks, but certainly not strikingly so (cf. Fig. 9.16 and 9.1), and not at all comparable to the excess radiogenic gas enhancement. In beryl and tourmaline, the heavy gases are not particularly enriched in comparison with other minerals in the same pegmatites.

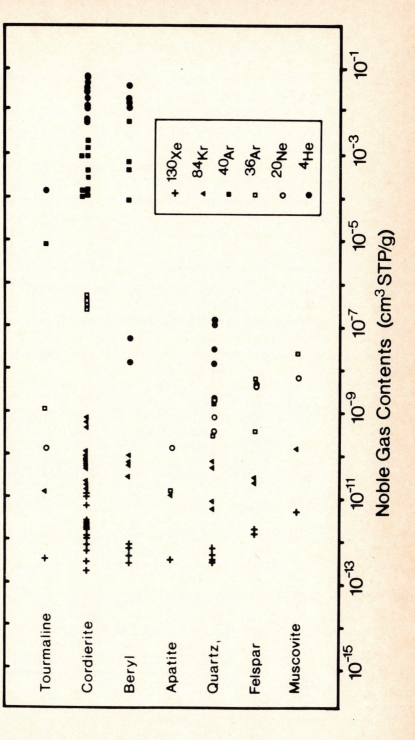

Fig. 9.16. Noble gases in cyclosilicates and associated pegmatite minerals. Data are from Saito et al. (1983).

9.11   Noble gas locations within rocks

As described elsewhere in this chapter, igneous rocks contain trapped noble gases from their sources. It is clearly important to consider where in the rocks these gases are located. Such information can help with the understanding of partitioning effects between magma and various solid phases and is necessary for proper consideration of magmatic noble gas transport processes and thus mantle degassing. Unfortunately, this is one of the most weakly understood areas of noble gas geochemistry. There are some data for specific and uncommon minerals such as diamond (Section 9.8) and cyclosilicates (Section 9.10), but even in these cases, it is unclear whether the gases are actually in the mineral lattice or in inclusions (diamonds) or when and how the gases are incorporated (cyclosilicates). Very little can be said definitively for the major constituents of common and important rock types.

A direct approach can be made by analysis of separated minerals. Such data are rare, and even for available data there are problems, as may be illustrated by comparison of two specific cases. Bernatowicz (1981) analyzed gases in a spinel lherzolite ultramafic inclusion (from Arizona, USA), and also in mineral separates of the major phases: clinopyroxene, orthopyroxene, olivine, and spinel. On the basis of these data (Table 9.2) and a modal analysis Bernatowicz found that the constituent minerals could not account for the noble gases in the bulk rock.

Table 9.2. Noble gas distribution in spinel lherzolites

	Concentrations (cm^3 STP/g)				
Sample	^{4}He ($\times 10^{-8}$)	^{20}Ne ($\times 10^{-10}$)	^{36}Ar ($\times 10^{-10}$)	^{84}Kr ($\times 10^{-12}$)	^{130}Xe ($\times 10^{-14}$)
Spinel lherzolite (Hawaii)[a]					
Bulk	2.4	0.40	2.4	7.1	22
Clinopyroxene	12	0.82	3.9	24	99
Orthopyroxene	1.9	0.42	3.0	11	41
Olivine	1.5	0.40	3.4	11	26
Spinel	<10	<4	<4	16	24
Spinel lherzolite (Arizona)[b]					
Bulk	—	3.0	0.36	2.2	12
Clinopyroxene	—	0.14	0.05	0.8	1.5
Orthopyroxene	—	<0.2	0.13	0.2	0.3
Olivine	—	0.011	0.15	0.7	1.9
Spinel	—	0.053	0.015	<0.01	0.03

[a] Rison (1980a).
[b] Bernatowicz (1981).

He concluded that a major fraction of the gases were lost in the mineral separation procedure and he suggested that they were present in inclusions, specifically the planar arrays of (CO_2?) bubbles which he observed in this rock and which would be likely fracture loci. Rison (1980a) also analyzed a spinel lherzolite from Salt Lake Crater, Hawaii) and its minerals. Although lack of a modal analysis prohibits a mass balance calculation for these data, the mineral gas contents in this case are capable of accounting for the whole rock data (Table 9.2); indeed, the minerals are mostly higher in gas content than the whole rock.

Indirect information about noble gas locations can also be obtained by stepwise heating, since the characteristic gas release temperature depends on mineral identity (Section 4.9 and Fig. 4.7). Feldspar, for example, loses its gases characteristically at about 1200 °C, while for biotite the release temperature is substantially lower, 800–900 °C. For most minerals the major gas release occurs just below the temperature of a major structural change such as melting or dehydration, e.g. graphitization of diamond (Fig. 9.14). Degassing from glasses occurs at relatively low temperatures, e.g. 800 °C in glassy submarine basalts (cf. Ozima & Takigami, 1980). Stepwise degassing is rather limited as a source of information on gas location, however, since gas release rates depend sensitively on grain size as well as mineral identity and, in practice, it is not really feasible to separate the contributions of more than two phases.

In volcanics which form under relatively low confining pressure, volatiles such as CO_2 and H_2O may become supersaturated and form a separate fluid phase, so that the resultant rocks are vesiculated. It is generally expected that noble gases would be strongly partitioned into such a fluid phase. Kurz & Jenkins (1981), for example, inferred that major fractions (up to 80% in some cases) of the He in a group of submarine basaltic glasses they examined were actually in vesicles.

10 Emanation

10.1   Introduction

The term 'emanation' is used to denote transfer of gases from the solid earth into the atmosphere. Usage is somewhat loose and it is difficult – and not particularly necessary – to specify a rigorous definition of what is or is not a form of emanation. The usual understanding is restricted to juvenile gases, whether primordial or radiogenic, i.e. cycling of atmospheric gases is excluded. Nevertheless, it is often difficult to determine whether gases are juvenile or atmospheric, e.g. gases vented in volcanic activity might be atmospheric (recycled after subduction). Radiogenic gases released into the atmosphere by weathering of surface rocks are certainly juvenile but not usually considered a form of emanation. As befits the ambiguity in which phenomena should be included under this rubric, various forms of emanation are described, not only in this chapter but also in appropriate context in other chapters. The very important special case of He in ocean water, for example, is treated in Sections 7.5 and 7.6.

It is convenient to distinguish two basic types of emanation, volcanic and nonvolcanic. The division is by no means strict, and there are cases when both volcanic and nonvolcanic processes are involved and others for which it is not clear what processes are involved.

In vulcanism, gases will be liberated from mantle solids and partitioned into the magma. Some gases remain in the resultant igneous rocks (Chapter 9), others will be released as gases into the atmosphere (i.e. emanation). Study of volcanic emanations is thus complementary to study of gases in igneous rocks in terms of providing information about igneous processes, particularly volatile transport, and in assessing the noble gas state of the mantle (Chapter 11).

Nonvolcanic emanation involves release of gases from solids from the crust, and probably also in the mantle, without first partitioning into a magma. Understandably, nonvolcanic emanation processes are more complicated and more poorly understood. In some cases emanation must be prompt, as inferred directly from the mere observation of emanation of short-lived Rn. In other cases long storage in some intermediate reservoir occurs, as inferred in the substantial He content in some well gases.

Mechanical disturbances such as fault movement and/or stress change appear to be important in nonvolcanic emanation. Noble gas emanation is thus currently a very active research area because of its possible utility in the study and prediction of earthquakes.

10.2   Volcanic emanation

It has long been known that volcanic emanations are relatively rich in elements such as C, Cl, S (along with water), i.e. those chemical species identified as excess volatiles and the constituents of the atmosphere (cf. Section 12.1). This observation provides strong support for the belief that volatile emanation from the solid earth is indeed the source of the atmosphere (cf. Section 12.3). The major volatiles are presumably accompanied by juvenile noble gases as well, but relevant data are relatively scarce. In part, this is attributable to their low absolute abundances. It is also partially attributable to the predominantly chemical identification of the volatiles, so that noble gases, if identified at all, are lumped together as inert gases or grouped with 'nitrogen'.

It is also well known that much or most of the volatiles in various forms of volcanic emanation are actually atmospheric rather than juvenile. Atmospheric noble gases, for example, can be recycled in volcanic emanations through direct infusion of air, either in the actual emanation or in the sampling, through solution in marine or meteoric water, through sediments, and perhaps even real magmatic gases may be partially atmospheric rather than juvenile (cf. Section 12.4). In general, it is very difficult, in nearly all cases impossible, to make a reliable identification of juvenile noble gases in volcanic emanation, except by isotopic arguments.

An important and well-known form of volcanic emanation of noble gases is that associated with submarine volcanism at oceanic ridges. As described in greater detail in Sections 7.5–7.6, juvenile He emanation is readily observed in seawater, particularly near active hydrothermal vents. Juvenile primordial ^{3}He is easily identified isotopically by a characteristically elevated (relative to air) ^{3}He/^{4}He ratio. Juvenile radiogenic ^{4}He (as well as the primordial ^{3}He) can also be identified independently on the basis of excesses relative to solubility equilibrium, especially at the vents. This is the most important (perhaps only) case in which juvenile emanation is unambiguously identified without resort to isotopic argument, and accordingly permits an independent and unambiguous determination of the isotopic composition (rather than just a limit or constraint) of the juvenile gas. The juvenile He has the characteristic composition ^{3}He/^{4}He $\approx 1.2 \times 10^{-5}$, about eight times the air ratio. The same composition is also characteristic of the associated igneous rocks (Section 9.6). He of this composition is widely referred to as 'mantle' He.

Juvenile He emanation is not restricted to oceanic ridges. Polak *et al*. (1975) reported that ^{3}He/^{4}He in most volcanic gas emanation is significantly higher than the air ratio, with many of the values as high as the 'mantle' ratio. Similar high ^{3}He/^{4}He ratios have also been reported for volcanic gases from Lassen Park and Yellowstone Park by Craig *et al*. (1978b) and for gases from various Japanese volcanoes by Nagao *et al*. (1981). Gaseous emanations as well as volcanic rocks from Hawaii sometimes have ^{3}He/^{4}He ratios substantially higher than the nominal mantle value, an observation which is sometimes interpreted in terms of a relatively less 'depleted' deep mantle source (cf. Section 11.4).

The majority of the world's presently active volcanoes are associated with subduction zones where oceanic crust is remixed into the mantle. It should be noted explicitly that gaseous emanations in subduction zones are characterized by juvenile He, as follows from observations of high ^{3}He/^{4}He ratios. He composition data for volcanic gases as well as igneous rocks are included in the compilation illustrated in Fig. 11.4, and it can be seen that the compositional range is large. However, the highest ^{3}He/^{4}He values are quite similar in many geographical locations and essentially identical to 'mantle' compositions.

^{3}He/^{4}He ratios ranging downwards from the nominal mantle value are generally interpreted as reflecting addition of radiogenic He. In subduction zones the radiogenic He might be that generated *in situ* in the subducted oceanic crust or subcrustal lithosphere, or it might be He extracted by ascending magma from older crust/lithosphere overlying the subducting slab. It is not possible to make an unequivocal choice of these alternatives on the basis of He composition alone, however. The latter possibility, extraction from overlying material, is often favored not only because of the greater amount of radiogenic He available in the older and more U-rich rock but also because it might be expected that magma generation would homogenize the He and produce a uniform characteristic He composition, contrary to the observations.

The wide compositional range which can be generated by mixing mantle and radiogenic He suggests interesting possible applications of He as a geochemical tracer. Craig *et al*. (1978a), for example, have commented on its relevance to models for magma generation in andesitic volcanism at convergent plate boundaries (cf. Ringwood, 1975). They note that for typical fresh oceanic basalt concentrations of 0.25 ppm U and 1.5×10^{-6} cm^{3}STP/g trapped ^{4}He (cf. Fig. 9.2), radiogenic He production in 100 Ma, a characteristic age of the western Pacific plate, will decrease ^{3}He/^{4}He from an assumed initial ratio 10 times atmospheric to 2.5 times atmospheric. Higher U concentrations, such as McDougall's (1977) 0.44 ppm mean for older basalts altered by seawater, will produce even lower ratios, as will any contamination with radiogenic He from older crust overlying the subduction zone. Nonetheless, ^{3}He/^{4}He ratios in western Pacific emanations are often substantially higher than 2.5 times atmospheric (Craig *et al.* (1978a) cite 5–7 times atmospheric in the Hakone (Japan) volcano and in the Marianas). If the assumptions and numerical values involved are approximately correct, the inference is that subducting oceanic crust is not the major source of the He, and presumably other volatiles as well, in these island-arc magmas; rather, the suggestion is that the source is the 'pristine' mantle wedge above the subduction zone, with any volatiles from the subducting crust being greatly diluted by those from the mantle wedge.

There is, of course, substantial interest in quantitative evaluation of the overall rate of volcanic discharge into the atmosphere. An estimate of submarine He emanation is possible (Section 7.6), but otherwise there have been many attempts at this problem without a great deal of success. The difficulties are particularly marked for subareal volcanism. Adequate sampling is an obvious and serious problem. The most severe problem, however, is variability: even in a single volcano both the discharge rate and composition of emanation vary greatly with time and it is almost impossible to derive any realistic estimates of average values.

The situation is illustrated by recent studies by Matsuo *et al.* (1978) on variations of N_2/Ar in fumarolic gases of Mt Showa-Shinzan (Japan). They found an irregular decline of N_2/Ar from more than 2000 to about 100 (close to the air value of 84) over a 10 year period (1954–1965); over this same period both the overall fumarolic activity and the discharge temperature declined as well. Matsuo *et al.* interpret these observations in terms of a greater influence of air contamination as activity decreased. This case also illustrates the difficulties involved in identifying juvenile components. Matsuo *et al.* noted that simple degassing models (cf. Chapter 13) predict a low value, about 12, for juvenile N_2/Ar in the mantle. Such low values are rather scarce; Shepherd (1938) observed the range 5–39 at Kilauea (Hawaii) but most values are substantially higher and Matsuo *et al.* concluded that in the typical case, including their study, the N_2 is derived primarily from organic materials in sediments, i.e. is atmospheric rather than juvenile. This leaves unanswered the question of how much of the Ar is juvenile. For Showa-Shinzan the Ar is evidently primarily but not entirely atmospheric, since $^{40}Ar/^{36}Ar$ ratios range from 1% to 9% higher than the air value.

As noted, there is also a considerable problem in sampling technique for obtaining representative volcanic gases. Recently, introduction of a new technique involving use of a Barringer correlation spectrometer has yielded some success in estimates of the discharge rate of SO_2 in volcanic gases (Stoiber & Jepsen, 1973). Compiling such data, Okita & Shimozuru (1975) estimated the global SO_2 discharge rate at 3×10^7 tons/yr. (It is interesting to note that at this rate accumulation in 4.5 Ga produces four times the nominal atmospheric inventory in Table 12.1.) There are rather few observations of both SO_2 and noble gases (essentially Ar); available data (Jaeger, 1940; Matsuo, 1961; 1979; White & Waring, 1963) indicate Ar/SO_2 values in the range 10^{-2} to 10^{-4} (by volume). A nominal Ar/SO_2 ratio of 10^{-3} combined with the estimated SO_2 flux above corresponds to ^{40}Ar emanation at the rate of 1.1×10^{13} cm^3 STP/yr. Accumulation of this flux for 4.5 Ga is 4.7×10^{22} cm^{3}STP, about 130% of the present atmospheric inventory. These figures are suggestive, but in view of the large uncertainties in both the SO_2 discharge and the Ar/SO_2 ratio, along with the difficulties in estimating how much of the Ar (or the SO_2) is actually juvenile, it is evident that such comparisons can have little more than qualitative significance.

It is interesting to make a similar comparison with ^4He emanation. The oceanic ^4He flux (Section 7.6) estimated by Craig *et al.* (1975) is 3×10^5 atoms cm^{-2} sec^{-1}, or 1.8×10^{12} cm^3 STP/yr. There are no observational data for the accompanying ^{40}Ar flux, but if we assume current production of $^4He/^{40}Ar = 6$ (Section 6.2), the corresponding ^{40}Ar flux is 3×10^{11} cm^3 STP/yr. Accumulation at this rate for 4.5 Ga produces 3.7% of the present atmospheric inventory. The ^{40}Ar discharge thus estimated is substantially lower than that estimated by relation to SO_2, but the discrepancy is easily attributable to large uncertainties in both calculations.

It is often considered that the principal channel for volatile emanation into the atmosphere is hot springs (cf. Rubey, 1951), because of their numerous and widespread occurrence. Hot spring volatiles consist mainly of water with minor amounts, generally less than 1%, of other species such as S, N, Cl, and noble gases. Very nearly all hot spring water is recycling groundwater, however, as Craig (1963) has convincingly shown on the basis of O and H isotopic compositions. It is not clear how much of the other volatiles as well are actually atmospheric rather than juvenile. In any case it is not possible to make reliable quantitative estimates of juvenile emanation in hot springs, or even to assess how important hot springs really are as channels for juvenile emanation. It is important to note, however, that in hot springs the $^3He/^4He$ ratio is usually substantially higher than the air ratio (Polak *et al.*, 1975; Nagao *et al.*, 1981), which indicates that much of He at least is indeed juvenile.

As noted in Section 7.6, Jenkins *et al.* (1978) have observed a remarkable correlation between ^3He concentration and temperature in submarine hydrothermal vents at the Galapagos spreading center (Fig. 7.5). The inferred correlation, which corresponds to 7.6×10^{-8} cal/atom (of ^3He), suggests a common process for primordial ^3He emanation and heat flow, not unreasonable since both He and heat are presumably both transported principally by magma ascending beneath the oceanic ridge. If the proportionality between ^3He and heat is general rather than a unique local value, the Craig *et al.* (1975) average oceanic ^3He flux of

4 atoms cm^{-2} sec^{-1} corresponds to a total heat flow of 6.5×10^{12} W. This is of an order of magnitude similar to the recent estimate of a global heat flow of 4.2×10^{13} W by Sclater *et al.* (1980).

Qualitatively at least, there is also a relationship between He emanation and heat flow on land as well as at ocean ridges. Polak *et al.* (1975) found high $^3He/^4He$ ratios ($\gtrsim 10^{-5}$) in continental volcanic gases and hydrothermal waters in young tectonically-active volcanic and rift zones characterized by high heat flow, and low ($\lesssim 10^{-7}$) $^3He/^4He$ ratios, essentially in the radiogenic range, in older Precambrian platforms with low heat flow.

10.3   Natural gases

Use of the term 'natural gas' is here restricted to natural gaseous emanation which is not (at least not obviously) associated with volcanic activity. Usually, but not always, natural gas conforms to the common nontechnical definition as a gas mixture dominated by methane. The term also connotes gas accumulated in a crustal reservoir after migration over perhaps considerable distances until trapped in a suitable geological structure, e.g. at the top of an anticline in which a permeable stratum such as sandstone (the reservoir) is overlain by an imperme- able stratum such as shale. Again usually but not always, the source of the gas is organic sedimentary material, and typically natural gases are associated with petroleum deposits. The gas is under pressure and while leakage occurs, e.g. along cracks or fault planes in the trapping structure, the term natural gas usually suggests a well tapping the reservoir with often considerable wellhead pressures (up to hundreds of atmospheres).

Noble gases in natural gases range from low trace levels to surprisingly high concentrations. The most abundant noble gas is He, and in a few wells He occurs at the per cent level, up to nearly 10%. Such wells are rare, but are the principal source for commercial He production and as such are important natural resources. A major and well-known region of commercial-grade He wells is centered near the Texas Panhandle area (USA). The most extensive survey of noble gases in natural gas wells is that of Zartman *et al.* (1961).

The noble gases in natural gas wells are predominantly radiogenic, i.e. typic- ally greatly enriched, relative to air, in ^4He and ^{40}Ar. Ne composition data for gas wells are limited but available data indicate prominent radiogenic Ne as well (cf. Emerson *et al.*, 1966; Phinney *et al.*, 1978). In some cases Rn is detectable but there is little or no correlation between He and Rn content. In the Zartman *et al.* (1961) survey He concentrations ranged from 37 ppm to 6.2%; the ratio of ^4He to radiogenic ^{40}Ar ranged from 1.6 to 130, with most values in the rela- tively narrow range 6 to 25. Nonradiogenic noble gases in natural gases are usually and plausibly thought to be atmospheric rather than juvenile, but the mechanisms for their introduction – possibly air infiltration, solution in groundwater, trapped gases in sediments – remain unknown.

The relatively narrow range of most of the radiogenic $^4He/^{40}Ar$ compositions observed by Zartman *et al.* merits further remark. The range is largely consistent with current or integrated past production for K/U ratios characteristic of common crustal rock (cf. Section 11.5). Zartman *et al.* considered the observed values of the $^4He/^{40}Ar$ ratio and factors which might account for its variations, such as the greater tendency of ^4He to be released from rock, different solubilities in water, elemental fractionation during migration, etc. They concluded that there was no reason to invoke source rocks with K/U ratios other than those directly observed in ordinary crustal rocks. This point is significant because it might otherwise have been supposed that the noble gas sources, particularly for those wells extremely rich in He, would be local deposits of U/Th minerals or ores. Such deposits would have extremely low K/U ratios and would produce correspondingly high $^4He/^{40}Ar$ ratios. Such high ratios are not observed, even in the wells richest in He, so highly radioactive U/Th-rich rocks are not the major sources of radiogenic well gases.

The general problem of noble gas emanation in natural gases is different from and more complex than that of volcanic emanation. In volcanic emanation the gases are presumably mobilized from crystalline lattices during partial melting and transported over long distances by flow of the magma. In nonvolcanic emanation the mobilization from the source rock is evidently more complex and may involve multiple processes, e.g. diffusion over short distances, chemical reorganization, mechanical disturbance, etc. It is very difficult to be quantitative but it seems that such processes would be slow, and so the gases would accumulate over geologically long times. The transport processes are also mysterious and probably also slow. It is thus not possible to identify the source regions. Presumably natural gas reservoirs accumulate noble gases from extensive regions, an assumption supported by the analogy with petroleum reservoirs and by the probable averaging over a variety of rock compositions suggested by the $^4He/^{40}Ar$ ratios discussed above, but this cannot be shown conclusively. Indeed, Pereira & Adams (1982) argue that in many if not most ordinary wells the ^4He can be accounted for primarily by emanation from the reservoir sediments themselves, without the need for concentration of gas from other areas. It is usually considered that the noble gases (as well as other volatiles) in natural gases are crustal rather than mantle-derived, but it is difficult to advance definitive arguments for even this generalization, and it appears not to be completely true in at least some instances (see below).

In some wells the principal gas is not CH_4 but is instead CO_2. This is of major interest in noble gas geochemistry because of the special results obtained from wells in the Bueyeros Field, Harding County, New Mexico. The gas is very nearly pure ($\approx$99.9%) CO_2, and noble gases are present only at very low levels (17–45 ppm ^{4}He, less than 0.001 ppb ^{130}Xe). The ^{4}He/^{40}Ar ratio is low, about 1.6 in the wells sampled by Zartman *et al.* (1961), and even lower, 0.5, in the sample analyzed by Phinney *et al.* (1978). The ^{21}Ne is substantially radiogenic (Phinney *et al.*, 1978). The Xe contains a significant fission component, about 12% of the ^{136}Xe, whose composition matches that of ^{238}U spontaneous fission (Phinney *et al.*, 1978). The most remarkable feature, however, is an enrichment of ^{129}Xe, which is elevated about 10% above air abundance. This feature was initially observed by Butler *et al.* (1963), which was the first known case of excess ^{129}Xe in any terrestrial sample. Excess ^{129}Xe has been found in repeated observation in these samples (Wasserburg & Mazor, 1965; Boulos & Manuel, 1971; Hennecke & Manuel, 1975c; Phinney *et al.*, 1978; Smith & Reynolds, 1981) and can be considered conclusively confirmed (see Section 6.9).

There is no known plausible source for this excess ^{129}Xe other than the obvious one: decay of ^{129}I which was part of the original material which accreted to form the earth. Since it is rather surprising that excess ^{129}Xe in terrestrial materials is not accompanied by ^{244}Pu-fission Xe (Sections 6.9 and 13.8), it is worth re-emphasizing that the most precisely determined fission Xe composition for the Harding County well gas, that of Phinney *et al.* (1978), is quite compatible with ^{238}U fission; no admixture of ^{244}Pu-fission Xe is suggested, and Phinney *et al.* place an upper limit of 20% on the possible ^{244}Pu contribution.

A crustal source for the excess ^{129}Xe seems impossible. Even if the crust formed while ^{129}I was still extant, it is hardly plausible that it would have retained ^{129}Xe during subsequent reworking. A mantle source is thus indicated, a relation supported by observations of excess ^{129}Xe in other mantle-derived materials (Chapters 9 and 11). Mantle emanation is also suggested by the relatively high ^{3}He/^{4}He ratio (4.4×10^{-6}) which Phinney *et al.* (1978) observed in this gas. Boulos & Manuel (1971) suggested that the excess ^{129}Xe was carried in a magma whose heat decomposed crustal carbonates to produce CO_2, but Hennecke & Manuel (1975a) found no evidence for excess ^{129}Xe in local igneous rocks. It thus remains uncertain whether the mantle-derived gases (excess ^{129}Xe and ^{3}He, possibly others) originated in a magma or by nonmagmatic gaseous emanation from the mantle.

The source of the mantle-derived ^{129}Xe and ^{3}He is not necessarily the source of the other components, and the radiogenic ^{4}He, ^{40}Ar, ^{21}Ne, and fission Xe may be crustal. The origin of the major gas, CO_2, is also a mystery. Decomposition of crustal sediments, by whatever mechanism, is the obvious and simple supposition. A mantle origin for the CO_2 is also an attractive hypothesis, however. There are widespread occurrences of CO_2 fluid inclusions in mantle peridotite xenoliths, and a relationship between mantle CO_2 and ^{129}Xe is suggested by observation of excess ^{129}Xe in CO_2 inclusions in Hawaiian xenoliths (Hennecke & Manuel, 1975b). The presence of CO_2-rich fluid in the mantle has also been emphasized in explaining certain characteristics of the low velocity zone (cf. Green, 1972) and of basaltic magma genesis (cf. Mysen & Boeftchen, 1975). Carbonatites (igneous carbonates) have low ^{87}Sr/^{86}Sr ratios (cf. Powell *et al.*, 1962; Hamilton & Dears, 1963) and are generally attributed to mantle sources, and so might be a suitable vehicle. Direct nonmagmatic gaseous emanation of CO_2 and other volatiles remains a possibility, however.

10.4 Radon emanation

Rn has no stable isotopes; it occurs in nature, however, because it has three isotopes, ^{219}Rn (half-life 3.9 sec), ^{220}Rn (56 sec), and ^{222}Rn (3.825 days), which are members of the decay chains of the long-lived radioisotopes ^{235}U, ^{232}Th, and ^{238}U, respectively (Fig. 3.1f). Each Rn isotope is produced by α decay of an isotope of Ra and produces, again by α decay, an isotope of Po. Naturally occurring quantities of Rn are quite low in absolute terms, but they can be measured with high sensitivity by observation of the α decays. Quantities of the order of 10 pCi (in one liter of gas) can be measured routinely; this activity is 0.37 dps = 22 dpm, equivalent to 1.8×10^5 atoms = 6.6×10^{-15} cm^{3}STP of ^{222}Rn, and proportionately smaller amounts of the shorter-lived isotopes. For comparison, the specific activities of the parents (and thus, in secular equilibrium, the activities of all the radioactive daughters, including Rn isotopes) are: ^{235}U = 15.4 nCi and ^{238}U = 334 nCi per gram of U, and ^{232}Th = 110 nCi per gram.

The geochemical relevance of Rn is due primarily to the phenomenon that it escapes, as a gaseous emanation, from the rocks in which it is produced. The efficiency of escape is surprisingly high; emanation loss to production ratios for rocks in the laboratory are often in the range one to several per cent, and sometimes approach unity. In general, Rn emanation at observed rates is far too high to be explained simply by volume diffusion out of the mineral grains in which it is produced. It is generally believed that the escape is promoted by recoil (about 100 keV for ^{226}Ra decay to ^{222}Rn), in that recoiling atoms stopped in microfractures can migrate substantial distances as gas. Also, radiation damage caused by U/Th system decays may lead to substantially faster local diffusion and there is evidence that this too is important in Rn emanation (cf. Baretto, 1975). The emanated Rn is useful as a geochemical tracer in a variety of geological applications, and the disequilibrium in U/Th decay series caused by Rn mobility has a number of chronological applications. Once released from its source minerals, the Rn can migrate over macroscopic distances, sometimes surprisingly great distances, either by gas diffusion in pore spaces or by transport in groundwater or seawater. In general, the rapidly decaying ^{219}Rn and ^{220}Rn do not survive long enough to be observed, so in most geochemical usage 'radon' means ^{222}Rn, the ^{238}U-series member.

The continental average flux of ^{222}Rn into the air is of the order of 0.75 atom cm^{-2} sec^{-1}, corresponding to a 'standing crop' column density of 3.6×10^5 atoms cm^{-2} (20 pCi/cm^2) or volume density 460 atoms/l(STP) (0.026 pCi/l). For comparison, the global average production rate of ^{222}Rn is about 2×10^5 atoms/cm^2. The abundance of ^{222}Rn in seawater is generally in equilibrium with its immediate parent ^{226}Ra (half-life 1622 yr) at activity around 33 dpm (15 pCi)/100 kg (and thus an average column density close to that of air); local enhancements by emanation are observed, e.g. excess activity of 80–229 dpm/100 kg in the ^{3}He-enriched hydrothermal plumes in the Galapagos Rift observed by Lupton *et al.* (1977b) (cf. Section 7.6).

As noted in Section 1.2, the factors controlling the natural distribution of Rn are typically rather different from those for the other noble gases, and our scope does not include Rn except as it relates to other noble gases. A number of more general discussions of Rn geochemistry are available, e.g. Turekian *et al.* (1977); Tanner (1978); and Austin & Droullard (1978).

10.5   Earthquake prediction

Several years prior to an earthquake ($M = 5.3$) in Tashkent (USSR), Rn levels in deep wells in the epicentral region increased significantly, to about double normal value. The high levels were maintained until the earthquake, at which time or soon after they returned to normal (Ulomov & Mavashev, 1971). A remarkable example of a similar kind is illustrated in Fig. 10.1. As reported by Wakita *et al.* (1980), Rn levels in spring water monitored about 25 km from the epicenter showed an abrupt increase shortly before the Izu-Oshima-Kinkai earthquake ($M = 7.0$), with a more gradual subsequent return to normal levels.

It is still vividly memorable that, partly on the basis of anomalous increases in Rn levels in several wells in Yunnan Province, the Chinese Seismological Agency issued a warning for an earthquake which indeed occurred ($M = 7.2$) shortly after the warning (see Wakaita, 1978). The warning, which was highly publicized, saved many lives and averted much property damage. Similar successful earthquake warnings based on anomalous Rn (and/or He) emanation have also been made in the Soviet Union.

These examples illustrate a number of cases that clearly demonstrate that earthquakes are frequently associated with (and preceded by!) anomalous noble gas emanation. The Soviet Union, China, and Japan, and to a lesser extent the United States, now have earthquake observation and prediction research programs in which extensive routine monitoring of Rn emanation plays an important role.

Nevertheless, of course, reliable earthquake prediction remains a goal rather than a reality: noble gas emanation (like all other potential earthquake precursory signals) cannot be depended on to predict earthquakes, and there have been many earthquakes which did not produce observed noble gas emanation effects as well as many emanation effects not apparently related to earthquakes. In spite of much enthusiasm and great effort, the problem remains that there is insufficient knowledge of how tectonics affects emanation, and how such effects may be distinguished from those due to factors unrelated to tectonic phenomena.

A good example of this problem is the apparent long-range nature of the relationship between earthquake and emanation. There are a few reports in which precursory anomalous emanation was observed at distances of hundreds of kilometers from the epicenter. It seems unlikely that gases originating in the hypocentral region could have migrated over such distances at all, and extremely unlikely that they could have done so in times short enough to be identified with the earthquake or to avert essentially complete decay of the Rn. Presumably then the source of the gas was in the locality of the observing station, or the local effect is only spuriously associated with the remote earthquake.

Adequate understanding and application of noble gas emanation phenomena precursory to an earthquake requires address of a number of questions. One class of questions relates to the source of the gas, whether in the immediate neighbourhood of the observing station, in the focal region, or in some intermediate reservoir tapped by tectonic factors and transported to the observation site. The answers may differ in different specific cases. Another class concerns mechanism, whether tectonic influences actually enhance liberation of gases from their source rocks (see Section 10.6) or whether they only facilitate release of gases from previously existing reservoirs. There must also be a better understanding of the normal conditions and the factors which control the level and variability of emanation in the absence of unusual tectonic influences.

The existence of a relationship between Rn emission and earthquakes was first clearly demonstrated by Okabe (1956), who discovered a correlation between daily variation in atmospheric Rn content near the ground surface and local seismicity in western Japan. Subsequently, a number of reports have been made in support of the same conclusion. A direct relationship between Rn emission and stress was also demonstrated by the Chinese Group of Hydro-Chemistry (GHCSBHP, 1975), who studied the effect of artificial explosions (equivalent magnitudes from 3.1 to 4.4). They monitored Rn in six springs located from 1.6 to 14 km from the explosion site. The responses in Rn level varied in amplitude to more than 10 times the amplitude of normal variations, with durations from a few hours to several days (Fig. 10.2). Note that both increases and decreases were observed. A problem in associating emanation effects with earthquakes, however, is that the effects might be due to other factors. Rainfall affects Rn emanation, for example, and if an earthquake occurs during heavy rain it is not always clear whether an observed Rn effect should be attributed to the earthquake or the rain.

As one example, King (1978) studied Rn emanation in 20 shallow dry holes along a 60 km segment of the San Andreas and Calaveras faults in central California over a three year (1975–1977) observation period. During this period two moderate earthquakes (M = 4.3 and 4.0) occurred. Rn emanation gradually increased to peaks at the times of the earthquakes and then gradually decreased. Along with the Rn measurements King also monitored rainfall and atmospheric temperature. He found no correlation between Rn and the (sparse) rainfall but an apparently close correlation between Rn and atmospheric temperature. However, since Rn emanations in climatically similar areas showed rather different variations with temperature, King concluded that the correlation between Rn and temperature in his study was fortuitous and that the Rn emanation was indeed correlated with local seismic activity. A contrasting example is the study of Yamanouchi & Shimo (1980), who monitored Rn activity in a tunnel over a four year period. They concluded that rainfall was primarily responsible for Rn variations, arguing that water permeation enhanced Rn emanation, and noted a clear seasonal variation in Rn emission. These examples are contrasting but not necessarily contradictory, and it seems likely that the differences arise simply from differences in local geological conditions where the observations were made. It seems evident, or at least plausible, that at some sites variations are primarily sensitive to tectonic disturbances while at other sites variations are more sensitive to other factors such as rainfall, atmospheric temperature, or atmospheric pressure. This emphasizes that if Rn emanation is to be useful in earthquake prediction an important consideration will be selection of appropriate observational sites.

Despite some uncertainty about effects at very long range (hundreds of kilometers) it is clear that tectonic disturbances affect Rn emanation at shorter ranges (tens of kilometers), but even in such cases there are problems about the source and transportation of the Rn. Intuitively it seems unlikely that all the observations can be accounted for in terms of Rn actually produced in the immediate neighbourhood (a few meters) of the observing stations, since this would require that Rn release be extremely sensitive indeed to very minor changes in stress (cf. Section 10.6). Whether local emanation anomalies represent gas produced in the actual focal region which fails in the earthquake, or the release of gas stored in some intermediate reservoir, it seems that Rn

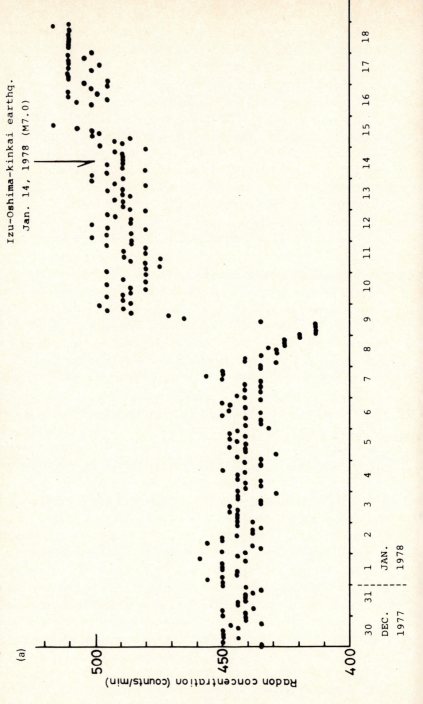

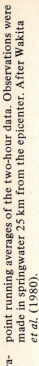

Fig. 10.1. Short-term and long-term variations in Rn concentration in groundwater before and after the Izu–Oshima–Kinkai earthquake. (a) Data for two-hour measurements. (b) Nine-point running averages of the two-hour data. Observations were made in springwater 25 km from the epicenter. After Wakita et al. (1980).

Izu-Oshima-kinkai earthq.
Jan. 14, 1978 (M7.0)

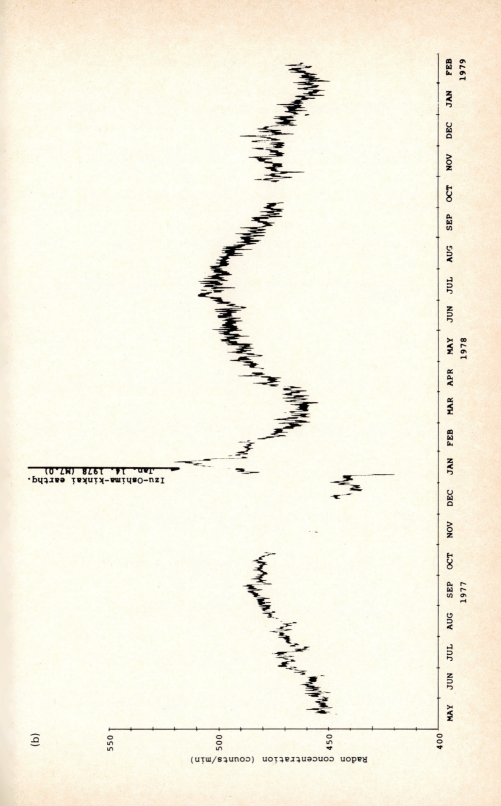

(b)

Radon concentration (counts/min)

Izu-Oshima-kinkai earthq.
Jan. 14, 1978 (M7.0)

migration is required over some considerable distance, hundreds of meters to kilometers. Migration over such distances is not quantitatively understood. Diffusion through soil is apparently far too slow, since a characteristic diffusion length in the 4 day half-life of ^{222}Rn is only about a meter (cf. Tanner, 1964). Transport in moving groundwater is also too slow, since its characteristic speed is only a few centimeters per day.

Mogro-Campero & Fleischer (1977) have proposed a fluid convection model for Rn transport. In a fluid-filled porous medium convection can occur when $kgh^2\rho\alpha(\mathrm{d}T/\mathrm{d}Z)_{\text{net}} > 4\pi^2 D\eta$. (Here k is hydraulic permeability, g is gravitational acceleration, h is a characteristic vertical dimension, D is thermal diffusivity of

Fig. 10.2. Response of groundwater Rn concentration to nearby artificial explosions. Δ is distance between explosion and observation site, M is equivalent magnitude of the explosion. Data were obtained by the Chinese Group of Hydro-Chemistry (GHSCBHP 1975). Reproduced from Teng (1980).

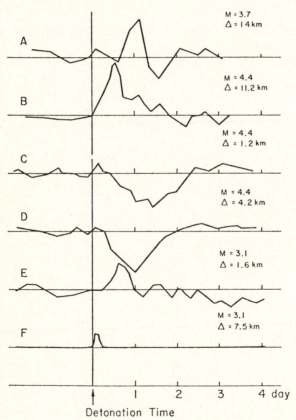

the fluid-filled medium, and η, ρ, and α are viscosity, density, and thermal expansion coefficient of the fluid. ($(\mathrm{d}T/\mathrm{d}Z)_{net}$ is the excess of the thermal gradient, $\mathrm{d}T/\mathrm{d}Z$ over the adiabatic gradient, which is negligible for water and $3-10°/\mathrm{km}$ for air, depending on its humidity.) For $h = 300\,\mathrm{m}$ and $(\mathrm{d}T/\mathrm{d}Z)_{net} = 30°/\mathrm{km}$, convection can occur, for air or water, for permeability $k > 3 \times 10^{-8}\,\mathrm{cm}^2$, a reasonable value for sand, soil, or high permeability sandstone. The convective circulation time is approximately $(D/kgh)(\mathrm{d}T/\mathrm{d}Z)$. In optimum conditions – convecting cells extending to the approximately 5 km depth at which fractures tend to be closed by overburden pressures, and a medium with permeability $k \approx 2 \times 10^{-6}\,\mathrm{cm}^2$, corresponding to loose sand – the transport time for moving Rn to the surface from a depth of about 100 m is about 20 days. Mogro-Campero & Fleischer argued that in regions of higher than usual permeability such thermal convection might be an important transport mechanism for Rn, and further suggested that control of thermal convection by surface temperature provides a logical explanation of natural seasonal variations in Rn emanation. At present this model is the only one available for a quantitative description of Rn migration over substantial distances, but in view of the favorable parameter choices necessary it seems unlikely to apply to the general case; indeed, it nicely illustrates the difficulties involved in a quantitative understanding of Rn transport.

The other noble gas whose emanation is sensitive to tectonic influences and thus has potential for earthquake prediction is He (cf. Reimer, 1980). Since He is stable, He data do not provide the time dimension inherent in Rn observations, but the greater abundance and higher diffusivity of He suggest a possibly greater sensitivity to tectonic disturbances. Also, He data allow the possibility of better constraints in source regions by means of isotopic signature, information which could be very important in understanding the extent of a tectonically disturbed region.

Wakita *et al.* (1978) analyzed He in soil gases along the fault zone formed by the Matsushiro earthquake swarm (maximum $M = 5.3$), in which seismic activity persisted for two years (1965–1967). Their measurements were made ten years afterwards. Away from the fault zone the soil gas He was not much different from atmospheric. Along the fault zone, however, they found anomalously high He concentrations, up to 70 times atmospheric levels. Furthermore, in the fault zone the $^3\mathrm{He}/^4\mathrm{He}$ ratios were significantly higher than atmospheric, up to 8.9×10^{-6}, i.e. mantle He rather than crustal (radiogenic) He. Wakita *et al.* suggested that both the high He emanation and the seismic activity could be attributed to an upper-mantle origin, speculating that the seismic activity was caused by intrusion of a diapir beneath the earthquake region and that juvenile He emanation from the magma reached the surface only along the fault zone.

As with Rn, however, He emanation varies in response to other factors besides tectonic disturbances. Reimer (1980), for example, found that He concentration in near-surface soil gases varied (within about 4% of normal air concentration) in a diurnal cycle which he concluded was controlled by soil moisture, wind speed, and temperature-induced atmospheric pumping. At depths below about one meter, however, he found no'corresponding variations, with He concentration constant within about 10 ppb (0.2% of air concentration).

Since most emanating He is radiogenic, there is usually an ambiguity whether anomalous concentrations reflect tectonic influences or extraneous influences such as meteorological conditions. To circumvent such ambiguity, Sugisaki (1978) suggested the use of the He/Ar ratio as an earthquake precursor. The thesis is that normal background variations in emanation level would not much affect the He/Ar ratio, but tectonic disturbances would tap different reservoirs with He/Ar ratios different from the background composition, and so be more easily distinguished from background. Sugisaki gave an example suggested to show correlation between He/Ar ratio and seismicity, but the association appears inconclusive.

10.6   Stress emanation

The release of noble gases from rocks as a response to stress is clearly relevant to the general problem of nonvolcanic noble gas emanation, but it has a particular immediacy to the problem of monitoring tectonic stress and thus potential applicability in earthquake prediction (Section 10.5). It seems intuitively evident that one response to stress might be enhanced gas release, and this is indeed observed to be so, but there is little understanding of the mechanism involved. Present knowledge of this problem is thus primarily empirical and closely tied to laboratory studies of gas release from stressed rocks. Such experiments are still few, but nevertheless suggest interesting implications.

The mechanical parameter most commonly linked to gas release is dilatancy, an inelastic volume strain (increase in volume) under nonhydrostatic stress. Dilatancy reflects microfracturing, and typically begins (in the laboratory) at uniaxial stress about half the macroscopic fracture strength. It is noteworthy that dilatancy is often suggested to be a key precursor to earthquakes, and prominent among observable effects pursued as potential earthquake warnings are those which may reflect dilatancy: changes in elastic wave velocities and electrical resistivity, water table variations, and gas emanation.

Holub & Brady (1981) studied the effect of uniaxial stress on Rn emanation from granitic rock. They found that during initial stress loading Rn emanation decreased, an effect they interpreted as due to closure of previously existing cracks. At a higher stress, about half the compressive strength, they observed a temporary 50% increase in Rn emanation, interpreted as an enhancement of emanation efficiency by accelerating microfracturing.

Honda *et al.* (1982) studied the effect of uniaxial stress on He and Ar release from a variety of rocks. Their results, some of which are illustrated in Fig. 10.3, indicate the possibility of major loss of these radiogenic gases as a result of inelastic strain. As might be expected, He is characteristically liberated more readily than Ar, and the extent of gas release increases with dilatant strain.

The effects of water on the release of gases from stressed rocks is unclear. It has been suggested that the presence of water in microcracks may promote Rn emanation by decelerating the recoiling atoms so that they do not become implanted in the crystalline lattices or by masking potential adsorption sites (cf. Holub & Brady, 1981), but empirical results on the effects of water on Rn emanation are inconclusive. The Honda *et al.* (1982) experiments allowed comparison of gas release in rocks stressed dry (room air) and wet (immersed in water). Significant release of Ar from the Inada granite occurred during the wet but not the dry compression (Fig. 10.3). They suggested that this reflects a two-step nature of the gas release process, first liberation of gas into microfracture channels, which should depend primarily on the amount and distribution of microfractures, followed by escape from these channels, which might depend more sensitively on the identity of the gas. For the Inada granite, escape of the Ar was thus apparently less efficient than that of the He except when the movement of Ar could have been promoted by transport in water. For He in general and Ar in other samples they found no significant difference between dry and wet conditions.

On the basis of even the limited empirical data available, it is readily possible to imagine that a buildup of stress which increases dilatancy can result in enhanced release of noble gases, Rn but also He and Ar. Once released these gases can reach the surface through cracks and pores, transported either as gases or in groundwater. Such a scenario is a plausible explanation for changes in noble gas emanation in the focal region of an impending earthquake, where significant changes in dilatancy presumably occur. It hardly seems adequate for changes observed at greater distances, up to hundreds of kilometers, away from the focal regions, since migration of gases from the focal region to such distances in the relevant times appears quite implausible. Nevertheless, such effects are noted, and if they are indeed the result of changes in the focal region then some mechanism must exist. Presumably the role of stress in producing changes in gas emanation at remote sites is limited to facilitating passage of preaccumulated local reservoirs to the surface rather than actually increasing release from the interiors of rocks. Still, the possibility of geological structures which produce local concentrations of stress, or of significant changes in emanation due to rather modest changes in

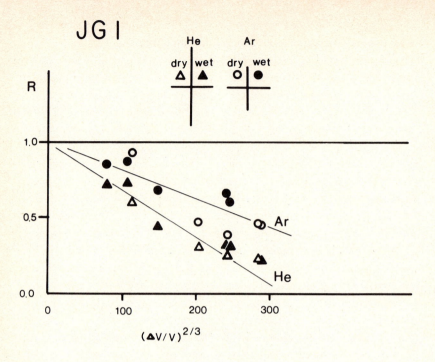

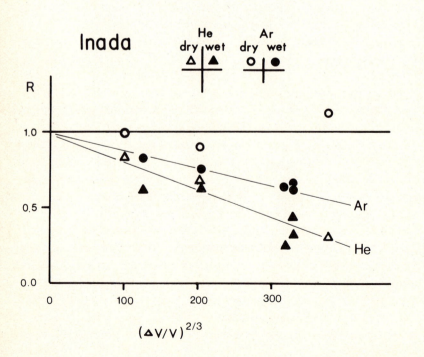

elastic stress (cf. Holub & Brady, 1981), is difficult to dismiss conclusively. In any case it is clear that emanation of noble gases cannot be used for dependable earthquake prediction without a much better understanding of the mechanisms by which stress changes influence their release.

Fig. 10.3. Fraction of original He and Ar remaining (ordinate) in granitic rocks after uniaxial compression to indicated inelastic dilatant strain (abscissa), measured in units of $\Delta V / V \times 10^{-6}$. Use of the 2/3 power of dilatant strain reflects an expectation that dilatancy emanation should be proportional to surface area created. Reproduced from Honda *et al.* (1982).

11 The mantle

11.1 Introduction

In this chapter, we will consider what may be inferred about the present noble gas status of the mantle. This is not easy to do, and unqualified statements about mantle noble gases are few, but the importance of such an evaluation is great also. The mantle constitutes the bulk of the earth and is the source of its atmosphere (Section 12.2). While it is frequently considered that the mantle is nearly completely degassed, this may not be true and even if it is there is much to be learned about the history of degassing by comparison of mantle and atmospheric gases. The mantle is not completely degassed: juvenile radiogenic and primordial noble gases are observed trapped in mantle-derived rocks and in the process of being added to the atmosphere. It is important to consider what fraction of the total terrestrial noble gas inventory is still in the mantle (Chapter 12), and it is important to compare atmospheric and mantle ratios of radiogenic to primordial gases (Chapter 13).

The basic data available are those for igneous rocks (Chapter 9) and gaseous emanation (Chapter 10). Identification of some mantle gases – excess radiogenic gases and primordial He – is easy when they are isotopically distinct from atmospheric gases, but quantitative evaluation is more difficult. In cases without isotopic distinctions, even the identification is difficult. The basic problem is relating the observations to the state of the mantle. Elemental fractionations will arise in the partitioning of gases from parent solid into magma, and the resultant rocks may not even reflect faithfully the gases in the magma because of loss directly *to* the atmosphere and contamination *by* the atmosphere.

We will not explicitly undertake a general description of primordial gas abundances and abundance patterns here, as this has been considered in Chapter 9, particularly Section 9.2. Few generalizations are possible. Absolute abundances in igneous rocks vary widely, and while it is thus not possible to specifically cite any representative concentrations for the mantle, it appears that mantle gases are low in abundance in the sense that there is presently considerably less gas in the mantle (or part of the mantle – see Section 9.2) than in the atmosphere. Nor is it possible to specify even a representative relative abundance pattern for the mantle.

Some samples have abundance patterns resembling a putative primitive – i.e. planetary – composition, but the relationship such samples bear to each other or the source regions is unclear. Rapidly-quenched marine basalt glasses frequently (if not typically) exhibit a characteristic abundance pattern illustrated in Fig. 9.11, arguably representing the composition of the mantle source of normal seafloor basalts, but the origin of this pattern, and thus its significance, have yet to be satisfactorily explained by a quantitatively defensible model.

Even if these problems could be resolved and the nature of mantle gases elucidated, there is still the problem of provenance and of trying to form a picture of the mantle as a whole. Regional and global scale heterogeneities in the mantle may complicate such a picture: conversely, however, the noble gases may themselves be useful in helping clarify the problem of mantle heterogeneity.

11.2   Noble gases and mantle structure

It is a truism, almost a cliché, that the mantle is not homogeneous. Describing the nature and distribution of the heterogeneities is a difficult task, however, as much or more so for the noble gases as for other kinds of observations. The data surveyed in Chapters 9 and 10 indicate a nonuniformity of mantle noble gases, and characterization of this nonuniformity in terms of the kinds of noble gases to be found in different mantle regimes is one of the major problems of noble gas geochemistry.

A simple model which is often advanced is that there are basically two kinds of mantle, or two prominent poles in a spectrum of mantle characteristics. One kind is 'depleted' mantle, material which has undergone at least one episode, probably more, of partial melting and extraction of magma with the consequent dimunition in the residue of a number of incompatible accessory and trace elements, those which are preferentially partitioned into the melt. In this model, partial melting of the depleted mantle produces the magma responsible for normal ocean floor volcanic rocks, erupting predominantly but not exclusively at oceanic ridges. The second kind of mantle, described by the terms 'undepleted', 'pristine', and others, has higher concentrations of these incompatible elements, presumably because it has undergone less or no such differentiation. The inference is that undepleted mantle is closer to an original composition, retained since the formation of the earth, while the depleted mantle is the residue of original mantle whose initial complement of incompatible elements is now mostly concentrated in the crust. The undepleted mantle is invoked as the source, by partial melting, of other kinds of magmas, in particular the alkalic magmatism frequent in oceanic islands such as Hawaii or otherwise associated with 'hot spots' (Morgan, 1972). This model emerges principally from geochemical observations, including both trace element abundance and radiogenic isotope systematics (e.g. Schilling, 1973; Sun *et al.*, 1975), and sometimes seismic data are interpreted in its support as well. The model is undoubtedly oversimplified, but may retain its essential validity nonetheless. We will not attempt a review of the general problem here, except to note that interpretation of the evidence is not unanimous and that other models, fundamentally different, can also be advocated (e.g. Tatsumoto, 1978; Anderson, 1981).

Depleted and undepleted mantle are also frequently referred to as upper and lower mantle, respectively. This is essentially a separate additional hypothesis, and the basic geochemical distinction between depleted and undepleted mantle may be valid even if their spatial assignments are not. It is possible, for example, that undepleted mantle is dispersed as relatively small pods in depleted mantle, or vice versa. The basic structural distinctions might also be lateral rather than vertical, e.g. if undepleted mantle is undepleted because it is mostly stagnant in the interior of global scale convection cells, or if depleted/undepleted reservoirs mysteriously remain associated with ocean/continent structures in spite of continental drift. In general, it is difficult to specify the physical location of a magma source. In the common view of depleted upper and undepleted lower mantle, hot spot magmatism is considered to be caused by upwelling of a deep convective plume from the lower mantle (Morgan, 1972).

Noble gases, and the other volatiles which constitute the atmosphere, are also incompatible elements and so should have different characteristics in different kinds of mantle. There are a number of noble gas observations which seemingly reflect mantle heterogeneity, although considerable differences of opinion have been voiced as to whether they actually do. Interpretations in terms of heterogeneity are usually made within the model of the depleted/undepleted dichotomy or cited as supporting evidence for the model and sometimes used to infer constraints otherwise unobtainable. Such interpretations are often also ambiguous or controversial, however. An important generalization is that if atmospheric gases have been extracted predominantly from, and nearly completely from, depleted mantle, then in many respects noble gases in undepleted mantle should resemble atmospheric noble gases more than either variety resembles noble gases remaining in depleted mantle. When observations conform to such expectations the devil's advocate may always suggest that the gases in question actually are atmospheric noble gases (Section 9.5) and that a great deal of interpretation is only misinterpretation of experimental artifact. It is not clear how great this danger is, but it should not be forgotten. Fortunately there are a variety of kinds of observations, chiefly involving isotopic variations, which are not subject to this ambiguity and which well illustrate the potential of noble gas data for providing real and important constraints on models of mantle structure.

11.3   Argon

As developed in Chapter 13, the mantle $^{40}Ar/^{36}Ar$ ratio is a key parameter in models for the degassing of the earth, high values (many times the air ratio) corresponding to rapid and nearly complete degassing and low values (not much larger than the air ratio) corresponding to slow and incomplete degassing. In recent years, considerable effort has been devoted to identifying a representative $^{40}Ar/^{36}Ar$ ratio for the mantle on the basis of gases trapped in mantle-derived materials such as ultramafic xenoliths, diamonds, and volcanic rocks, especially the glassy margins of submarine basalts. Observed Ar compositions in such materials are illustrated in Fig. 11.1. The range is wide, from essentially the air value to very high ratios greater than 10^4.

Fig. 11.1 Histogram of trapped $^{40}Ar/^{36}Ar$ ratios in various mantle-derived materials. There are numerous observations of ratios close to the air value; these are excluded for the sake of clarity. For data references see Fig. 9.1.

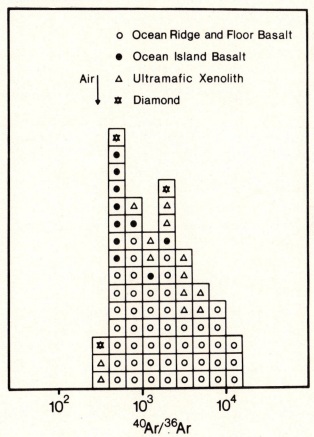

At face value, assignment of a representative value on the basis of these observations seem hopeless. At least some of the scatter, however, can presumably be attributed to variable air contamination. As pointed out by Takaoka & Nagao (1980), this hypothesis can be examined by an ordinate-intercept analysis. If measured Ar is a mixture of a mantle component (designated by subscript m) and an atmospheric component (subscript a), then the measured composition (unsubscripted) will be

$$(^{40}Ar/^{36}Ar) = (^{40}Ar/^{36}Ar)_a + S/(^{36}Ar) \qquad (11.1)$$

where

$$S = (^{36}Ar)_m [(^{40}Ar/^{36}Ar)_m - (^{40}Ar/^{36}Ar)_a] \qquad (11.2)$$

If, in a given suite of samples, there is a single mantle composition present in constant concentration, i.e. both $(^{40}Ar/^{36}Ar)_m$ and $(^{36}Ar)_m$ are constant, admixture of air Ar in variable concentration will produce a linear correlation between $(^{40}Ar/^{36}Ar)$ and $(^{36}Ar)^{-1}$. The value of $(^{40}Ar/^{36}Ar)_m$ must be larger than the largest observed $(^{40}Ar/^{36}Ar)$ ratio.

Fig. 11.2. Ordinate-intercept plot for Ar in modern glassy submarine basalts; see text and eq. (11.1) and (11.2). The line is the locus for mixing variable amounts of atmospheric Ar with a constant concentration (3.2×10^{-6} cm^3 STP/g) of highly radiogenic Ar. Modified from Hart *et al.* (1979) (figure supplied by R. Hart).

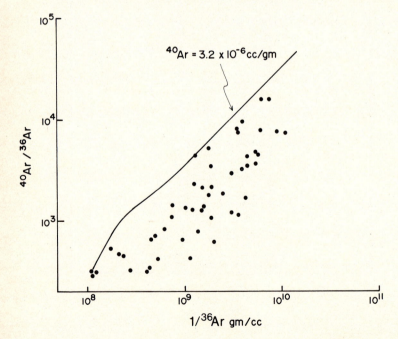

Figs. 11.2 and 11.3 are ordinate-intercept plots by which this hypothesis may be examined. Fig. 11.2 shows a clear trend of positive correlation, i.e. the more ^{36}Ar the lower the $^{40}Ar/^{36}Ar$ ratio, qualitatively supporting the hypothesis. There is considerable scatter, however, reflecting failure of the assumptions: either a variable mantle composition or a variable concentration (which could be intrinsic to the magmas involved or partial loss of magmatic gas on eruption). It would be expected that for smaller sets of more closely related samples, the assumptions of constant mantle Ar composition and concentration would be more nearly satisfied, at least for samples of the same magma. As seen in Fig. 11.3, this indeed appears to be the case.

On the basis of data illustrated in Figs. 11.1–11.3, we can infer the existence of a mantle Ar component which is highly radiogenic, $(^{40}Ar/^{36}Ar)_m \gg (^{40}Ar/^{36}Ar)_a$. For such highly radiogenic Ar the slope S is essentially the $(^{40}Ar)_m$ concentration. This analysis is thus not useful for determination of the actual value $(^{40}Ar/^{36}Ar)_m$, but this is unimportant since the relevant qualitative consideration is that it is large.

The slopes in Fig. 11.3 correspond to mantle component $(^{40}Ar)_m$ concentrations of 1.0×10^{-6} cm^3 STP/g in the drilled Cretaceous ocean floor basalts of Takaoka & Nagao (1980) and 0.7×10^{-6} cm^3 STP/g in the recent Hawaiian submarine basalts of Dalrymple & Moore (1968). The line drawn by Hart *et al.* (1979) in Fig. 11.2 indicates an upper limit concentration of 3.2×10^{-6} cm^3 STP/g; they interpret the points well below this line as reflecting loss of magmatic Ar. There is a noteworthy similarity in the ^{40}Ar contents of the magma sources of these diverse rocks.

A representative value of ^{40}Ar concentration of 10^{-6} cm^3 STP/g in these magmas is low in an absolute sense. Partitioning of noble gases between solid and melt presumably favors the melt, so that ^{40}Ar is evidently present at no more than 10^{-6} cm^3 STP/g in the mantle source. In the entire mass of the mantle, 4×10^{27} g, the corresponding inventory of ^{40}Ar is 4×10^{21} cm^3 STP, about 10% of the amount in air. By this reasoning we would conclude, in consonance with the implications of the high $^{40}Ar/^{36}Ar$ ratio, that the mantle is extensively degassed of radiogenic ^{40}Ar and for all practical purposes fully degassed of primordial ^{36}Ar.

It is clear that there is a mantle Ar component with high $^{40}Ar/^{36}Ar$ ratio and low ^{40}Ar concentration that corresponds to extensive degassing. It is not clear, however, whether this applies to the entire mantle or just part of the mantle, the source of the normal oceanic basalts. There are numerous reports of low trapped $^{40}Ar/^{36}Ar$ ratios (cf. Fig. 11.6) which are interpreted as real mantle Ar rather than as air contamination. Low ratios are usually considered to correspond to only minor degassing and such interpretations are thus generally applied to samples thought to tap the putative undepleted lower mantle, such as continental ultramafics or ocean island (e.g. Hawaii) basalts (cf. Saito *et al.*, 1978;

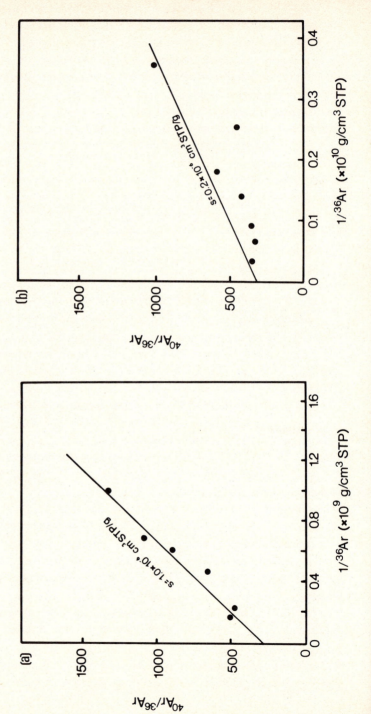

Fig. 11.3. Ordinate-intercept plot (a) for Ar in Cretaceous-age ocean floor basalts (data from Takaoka & Nagao, 1978), and (b) for recent Hawaiian submarine basalts (data from Dalrymple & Moore, 1968); cf. Fig. 11.2. The Cretaceous data have been corrected for *in situ* ^{40}Ar production in an assumed age of 10^8 y1

The greater scatter in the Hawaiian data is perhaps attributable to eruption in relatively shallow (low pressure) water, with consequently greater opportunity for vesiculation and loss of magmatic Ar.

Thompson *et al.*, 1978; Kaneoka *et al.*, 1978; Hart *et al.*, 1979; Kaneoka
& Takaoka, 1980). There are even reports of trapped $^{40}Ar/^{36}Ar$ ratios apparently
lower than the current air value: Saito *et al.* (1978) report a measured value
which, when corrected for *in situ* ^{40}Ar in an assumed geological age, becomes
less than the air ratio; Melton & Giardini (1980) report a *measured* value less
than the air ratio, but their technique is different from most and their datum
should be confirmed. Association of low $^{40}Ar/^{36}Ar$ ratios with high $^3H/^4He$
(Fig. 11.6) and $^{129}Xe/^{130}Xe$ ratios is also cited as evidence that they represent
a source different from the upper mantle ridge basalt source.

There are difficulties with this view, however, and caution is in order. The
basic problem is that, given that atmospheric Ar contamination clearly accounts
for many cases of low $^{40}Ar/^{36}Ar$ (Figs. 11.2 and 11.3), it is difficult to make an
unambiguous assessment of whether or not it accounts for all of them. Also,
samples from an 'undepleted' reservoir would be expected to have higher concentrations of trapped ^{40}Ar and other gases as well, and it is unclear whether any
such systematic difference between high and low $^{40}Ar/^{36}Ar$ samples exists.
Furthermore, the same geographical areas and sample classes that produce low
$^{40}Ar/^{36}Ar$ ratios also produce high ratios: the Hawaiian volcanics illustrated in
Fig. 11.3 have high $^{40}Ar/^{36}Ar$, for example, and while some diamonds analyzed
by Ozima *et al.* (1982) had $^{40}Ar/^{36}Ar$ only slightly larger than atmospheric,
others had much higher values.

In summary, our present understanding is insufficient to assign a representative
$^{40}Ar/^{36}Ar$ ratio for the mantle. Clearly there is a major mantle reservoir, the
source of most marine basalts, with 'low' ^{40}Ar concentration and high $^{40}Ar/^{36}Ar$,
samples of which are sometimes extensively modified by admixture of an air
component. The major problem is whether this description fits the mantle as a
whole or whether there is another, more 'primitive' major mantle reservoir with
much lower, nearly atmospheric $^{40}Ar/^{36}Ar$ ratios.

11.4   Helium

The $^3He/^4He$ ratio in the mantle, like the $^{40}Ar/^{36}Ar$ ratio, is an important parameter in models for mantle evolution and atmospheric degassing, and
considerable attention has been directed to its identification. Also like $^{40}Ar/^{36}Ar$,
the $^3He/^4He$ ratio varies over a wide range. Fig. 11.4 illustrates the distribution
of observed He compositions and Fig. 11.5 shows the compositional ranges associated with various tectonic environments; these figures include both igneous rock
and gaseous emanation data. The distribution of observed compositions (Fig. 11.4)
is distinctly bimodal. It is noteworthy that the atmospheric $^3He/^4He$ ratio lies in
the trough between the modes; as noted in Sections 6.6 and 12.5, the atmospheric
composition is not closely related to geological source compositions.

In contrast to $^{40}Ar/^{36}Ar$, however, there is general agreement about the nature of the major variations. The low $^{3}He/^{4}He$ ratios are those associated with continental crust and represent radiogenic He (cf. Section 6.2). The higher values in the mode at around 10^{-5} are mantle He. (In the mantle He the ^{3}He is essentially all primordial but the ^{4}He is mostly radiogenic: cf. Section 6.6). In gases from the Kamchatka peninsula, for example, Polak *et al.* (1975) note a sharp decrease in $^{3}He/^{4}He$ from the east coast, a region of relatively thin crust adjacent to an oceanic trench, to the west coast, underlain by thick continental crust.

Fig. 11.4. Histogram showing distribution of measured $^{3}He/^{4}He$ ratios in (1) rocks and (2) natural gases; samples are from the Soviet Union and Iceland. Reproduced from Polak *et al.* (1975).

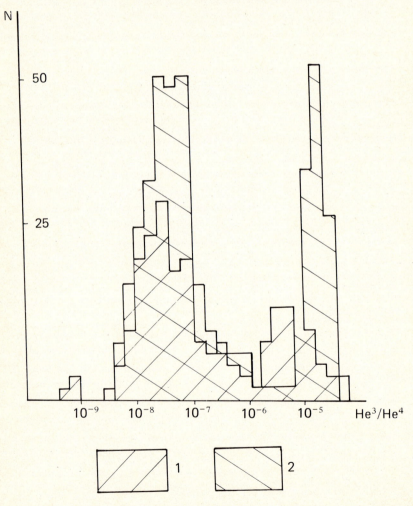

Mantle He composition, as measured in oceanic samples, varies over only a relatively modest range, from the mantle (oceanic) value to much lower values of ^{3}He/^{4}He (Fig. 11.5). These lower values are usually interpreted as indicating crustal (i.e. radiogenic) contamination. With this interpretation the ^{3}He/^{4}He ratio is a sensitive indicator of the source of the He, a feature of importance in a variety of studies, e.g. of earthquakes (cf. Section 10.5).

In normal seafloor/ridge basalts trapped He composition is nearly uniform. Craig & Lupton (1976) found ^{3}He/^{4}He to vary only narrowly about 1.4×10^{-5} in a variety of both Atlantic and Pacific basalts, and advocated the thesis that this value represented a single composition for the global-scale mantle reservoir from which normal marine basalts are derived. Significantly lower values, down to 1.1×10^{-5}, were subsequently found (Jenkins *et al.*, 1978; Lupton, 1979), however, in the juvenile He in hydrothermal plumes (cf. Fig. 7.4), and since this He is presumably from the local basalts or from the same source, He in the mantle reservoir parent to normal basalts is apparently not quite uniform. The near

Fig. 11.5. Ranges of observed ^{3}He/^{4}He ratios in rocks and gases in various tectonic environments. Data sources are: (1) Polak *et al.* (1975); (2) Nagao (1979); (3) Craig *et al.* (1978a); (4) Craig *et al.* (1978b); (5) Craig *et al.* (1975); (6) Craig & Lupton (1976); (7) Lupton (1979); (8) Jenkins *et al.* (1978); (9) Lupton *et al.* (1977a,b); (10) Lupton & Craig (1975); (11) Kaneoka & Takaoka (1978); (12) Kaneoka & Takaoka (1980).

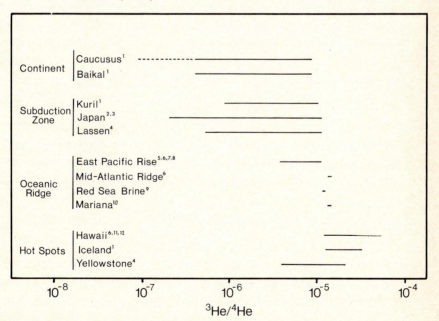

constancy of trapped ^{3}He/^{4}He in basalts in which trapped ^{40}Ar/^{36}Ar varies so widely (Fig. 11.1) is striking but understandable, since contamination with atmospheric gas is so much less important for He than for Ar (Section 9.5). It may even be that mantle He is even more strikingly uniform, the range (1.1–1.4) $\times$ 10^{-5} representing either experimental error or an interlaboratory bias.

It should be noted explicitly that the 'mantle' He component, with ^{3}He/^{4}He nominally around 1.2×10^{-5} or in the range $(1.1-1.4) \times 10^{-5}$, is defined primarily by those samples associated with depleted mantle, the source of normal oceanic volcanics. Since the ^{3}He is primarily primordial and the ^{4}He primarily radiogenic, the uniformity in their ratio indicates an impressive degree of global-scale mixing in this reservoir. It is also worth reemphasizing here the point noted in Section 9.6: that depleted (upper?) mantle is not the most logical place where one would expect to find primordial ^{3}He.

There are scattered observations of substantially higher ^{3}He/^{4}He ratios, e.g. up to 3.2×10^{-5} in Iceland (Polak *et al.*, 1975), 5.2×10^{-5} in Hawaii (Kaneoka & Takaoka, 1980), and the remarkably high value of 3.2×10^{-4} in a South African diamond (Ozima & Zashu, 1983a) (cf. Section 9.8). High ^{3}He/^{4}He, i.e. a lower ratio of radiogenic to primordial gas, is a feature which would be expected in the putative undepleted mantle, and high ratios are indeed frequently interpreted to represent samples of this undepleted mantle (e.g. Hart *et al.*, 1979; Kaneoka & Takaoka, 1980). This interpretation is strengthened by observation of high ratios in tectonic regimes already suggested to represent undepleted mantle, i.e. 'hot spots' at Hawaii and Iceland, for example, and further strengthened by association with low ^{40}Ar/^{36}Ar ratios (Fig. 11.6).

Not all ^{3}He/^{4}He ratios at 'hot spots' are high, of course. Kaneoka & Takaoka (1980), for example, observed normal ('mantle' He, highly radiogenic Ar) results as well in Hawaiian samples (Fig. 11.6). Their highest ^{3}He/^{4}He ratios (and lowest ^{40}Ar/^{36}Ar ratios) were observed in phenocrysts in alkalic lavas (cf. Table 9.1). Their interpretation is that these phenocrysts, which presumably formed at substantial depth prior to eruption, are least subject to loss or contamination at eruption and so most likely to preserve original magmatic gases. Xenoliths in these same lavas, and other lavas at Hawaii, have the lower ^{3}He/^{4}He and higher ^{40}Ar/^{30}Ar ratios characteristic of normal oceanic basalts and are accordingly inferred to originate in the upper (depleted) mantle.

Agreement with such interpretations is not universal, however. Some caution is in order because the relevant data are still relatively sparse and it is not clear how well they correlate with the corresponding geochemical and tectonic associations. Also, given apparent fractionations in mantle-derived Ne (Section 11.6), the possibility that high ^{3}He/^{4}He ratios arise by fractionation, while seemingly remote, cannot be entirely dismissed. Altogether, however, these interpretations of noble gas structures in terms of the depleted/undepleted mantle model must nevertheless be considered quite attractive.

11.5   The ^{4}He/^{40}Ar ratio

Essentially all the ^{4}He and ^{40}Ar now observable on the earth are radiogenic. Their ratio in any reservoir is governed by the relative abundances of U, Th and K and by the epoch and duration of their accumulation. The ^{4}He/^{40}Ar ratio in important reservoirs, such as the mantle, is of interest because of the constraints it places on these parameters and because of its potential use as a geochemical tracer, e.g. in identifying the source rocks of radiogenic noble gases in natural gas wells (Section 10.3).

The time dependence of the ^{4}He/^{40}Ar ratio is illustrated in Fig. 11.7. For representative values such as Th/U = 3.3 and K/U = 10^4, the current production rate is ^{4}He/^{40}Ar = 5.5 and integrated production from 4.5 Ga ago to the present is ^{4}He/^{40}Ar = 2.1 (cf. Section 6.2). As seen in Fig. 11.7, the instantaneous production ratio has not varied much over the last few times 10^8 yr, and the integrated production ratio over such times varies even less from the current instantaneous rate. The ratio for radiogenic gas accumulated since 4.5 Ga ago has changed only

Fig. 11.6. He and Ar compositions in Hawaiian samples. The tholeiite field (circled) is defined by 'mantle' ^{3}He/^{4}He ($\approx 1.2 \times 10^{-5}$) and high ^{40}Ar/^{36}Ar. Other samples exhibit higher ^{3}He/^{4}He ratios in conjunction with lower ^{40}Ar/^{36}Ar, both features suggested to represent a more primitive, undepleted mantle. Reproduced from Kaneoka & Takaoka (1980).

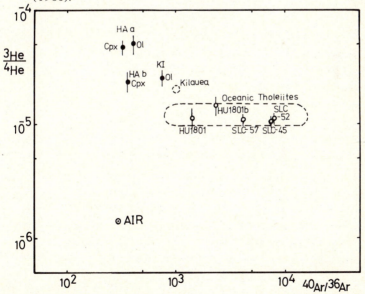

relatively little over the entire history of the earth and hardly at all in the last 10^9 yr or so.

The value of ^{4}He/^{40}Ar in the mantle has been the subject of considerable interest because of its relevance to the history of mantle degassing. Quantitative predictions for the history of mantle ^{4}He/^{40}Ar can be developed in terms of formal degassing models such as those described in Section 13.7, but it is interesting to consider some semiquantitative relationships here. Thus, assuming that the chemical compositions used in Fig. 11.7 are appropriate, it can be seen that

Fig. 11.7. Variations in ^{4}He/^{40}Ar production as a function of age τ. Curve 1 is the instantaneous production ratio, curve 2 is the integrated production ratio for accumulation from 4.55 Ga until τ, and curve 3 is the integrated production ratio for accumulation from τ until the present. Input parameters are K/U $= 10^4$ and Th/U $= 3.3$, with decay parameters in Table 3.1; different choices for K/U correspond to scale changes on the ^{4}He/^{40}Ar axis. The format is that of Zartman et al. (1961).

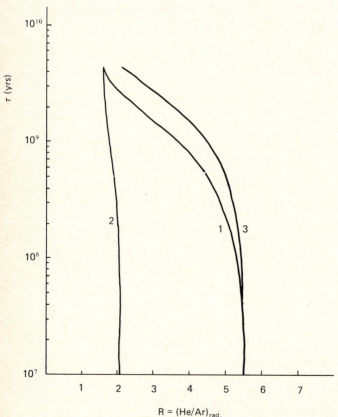

if the mantle is degassed reasonably efficiently on a short ($<10^9$ yr) timescale, it will retain only recent production and will have a relatively high ratio, $^4He/^{40}Ar \approx$ 5–6. Lower ratios, in the range $^4He/^{40}Ar \approx 2$ could be provided if both He and Ar were rapidly degassed from the mantle but He a few times faster than Ar. Since such a disparity seems unlikely for mantle-scale degassing, in which the limiting process is presumably convective and/or magmatic transport (Section 13.3), a more plausible interpretation for low $^4He/^{40}Ar$ is retention of accumulated production since >4 Ga ago, i.e. over nearly the entire history of the earth. If it is further supposed that the mantle is extensively degassed, as follows from a high $^{40}Ar/^{36}Ar$ ratio for example, it must then be concluded that the degassing must have been initiated – and (almost completely) terminated – very early in the history of the earth.

It is of course implicitly presumed that there exists a representative $^4He/^{40}Ar$ value for the mantle. This assumption is questionable, but at least for the (depleted?) mantle which is the source of normal oceanic basalts, it seems plausible in view of the rather uniform $^3He/^4He$ ratio (Section 11.4). The chief difficulty is experimental determination of this composition, which is presumably sampled in the excess (trapped) radiogenic gases characteristic of young marine volcanics. A prominent early work is that of Gramlich & Naughton (1972), who inferred, from analyses of Hawaiian volcanic xenoliths, a very low ratio for the mantle, $^4He/^{40}Ar \simeq 1.2$. Dymond & Hogan (1973) and Fisher (1975), however, observed trapped $^4He/^{40}Ar$ ratios in submarine glasses which ranged widely, approximately 1–20, and other investigations have likewise indicated variations in this range with no particular preferred value (Fig. 11.8). Fisher assumed that the highest measured values set a lower limit to the mantle ratio, arguing that the spread resulted from preferential loss of He from magma, and thus that mantle $^4He/^{40}Ar$ was at least 15–20. Such high values are incompatible with even the current production rate (Fig. 11.7) except for lower K/U than usually assumed, and Fisher concluded that the mantle was characterized by K/U $\approx 1.5 \times 10^3$. Schwartzman (1973b), however, argued that the predominant elemental fractionation in sampling was in the opposite sense; he argued that the mantle ratio was low, about 2, and that observed values higher than this reflected preferential extraction of He from source rocks into magma. Fisher (1979) later supported this argument on the basis of his study of the ratio of ^4He to excess ^{136}Xe (from spontaneous fission of ^{238}U) in submarine glasses.

The situation remains unresolved and there is no clear determination of $^4He/^{40}Ar$ in even the upper mantle. Observed values in volcanics range widely and there are plausible mechanisms to produce elemental fractionation in either direction. Much of the attention has thus shifted to comparable considerations involving different isotopes of the same element, as described in Chapter 13.

Fig. 11.8. (a) Trapped $^4He/^{40}Ar$ and $^{40}Ar/^{36}Ar$ ratios in various mantle-derived samples. Reproduced from Hamano & Ozima (1978). (b) Histogram of $^4He/^{40}Ar$ ratios in various mantle-derived materials.

(a)

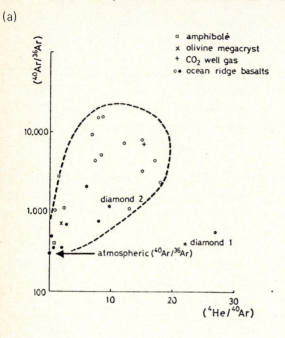

(b)

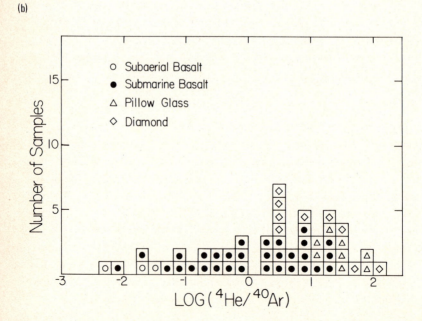

11.6   Neon

Ne compositions observed in a variety of mantle-derived samples are illustrated in Fig. 11.9. In most cases both the $^{21}Ne/^{22}Ne$ and/or the $^{20}Ne/^{22}Ne$ ratios are higher than the air values. The enhanced ^{21}Ne abundance is plausibly attributed to a contribution of radiogenic Ne (Section 6.2); the values are, in general, plausible quantitatively and often a correlation between ^{21}Ne enhancement and radiogenic ^{4}He (cf. Tolstikhin, 1978; Kyser & Rison, 1982) can be observed, a relationship which supports this attribution. The variation in $^{20}Ne/^{22}Ne$ ratio cannot be attributed to a radiogenic component and is usually assumed to reflect isotopic fractionation. It is perhaps significant that nearly all the $^{20}Ne/^{22}Ne$ values are within about 5% of the air ratio, a range consistent with simple fractionation mechanisms such as single stage diffusion (Section 4.7).

If the mantle, or part of the mantle, was incompletely degassed in the process of forming the atmosphere, it is not unreasonable to suppose that the residue of Ne, still being sampled now by igneous rocks, might be fractionated. The expected sense of the fractionation would be depletion of ^{20}Ne, however, opposite to the bulk of the observations (Fig. 11.9). Loss of magmatic gas during ascent or erup-

Fig. 11.9. Ne compositions in various mantle-derived samples. The range is believed to represent a combination of two effects, a variable contribution of radiogenic Ne manifested primarily in elevated ^{21}Ne abundances and isotopic fractionation which accounts for variation in $^{20}Ne/^{22}Ne$.

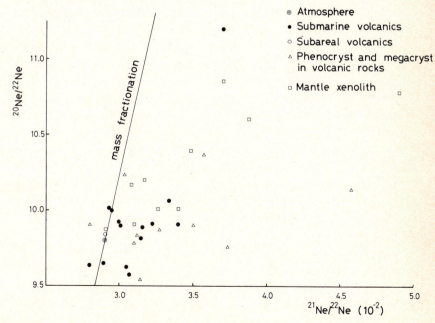

tion would also be expected to deplete ^{20}Ne if it had any effect at all. On the other hand, diffusive fractionation enhancing ^{20}Ne might arise if incomplete diffusion of Ne from solids to magma in the mantle source region is important (cf. Ozima & Alexander, 1976). Thus, potential fractionation effects in either sense are plausible, but essentially impossible to evaluate quantitatively.

In the paragraph above it is implicitly assumed that the earth's original Ne had air composition (specifically, ^{20}Ne/^{22}Ne ratio) and that this composition is probably preserved in residual juvenile Ne in the mantle, so that the fractionation in question arises in the sampling. Alternative views are certainly possible; these are usually based on the premise that air Ne is the residue of major exospheric escape of atmospheric Ne, so that air Ne is fractionated – in the sense of depressed ^{20}Ne/^{22}Ne – with respect to the earth's original Ne and to the mantle regions which still preserve its composition. Craig & Lupton (1976), for example, suggest an original solar Ne composition; as described elsewhere (Sections 6.5, 9.2 and 9.3), we consider such a hypothesis implausible – or at least unnecessarily complicated – and lacking in supporting evidence in the abundances and compositions of the other gases.

11.7   Xenon and krypton

The isotopic structures of Xe and Kr are particularly complicated in terms of the variety of components available in the solar system, both nuclear components generated within the solar system and primordial components never fully homogenized with each other (Section 5.6). Terrestrial (atmospheric) Xe and Kr, especially Xe, are ill-understood mixtures of these components, with an apparent severe and global but equally ill-understood isotopic fractionation superimposed. In this light it is perhaps a nontrivial observation that no known sample of terrestrial Xe preserves traces of possible initial heterogeneity: all measured Xe and Kr isotopic compositions are either identical to air compositions or are modifications of air composition by local isotopic fractionation or addition of specific nuclear components, both kinds of modification having occurred in the earth. The fractionations are plausible but not understood in detail. Most of the nuclear components (see Chapter 3, Table 3.3) are quite uncommon and reflect unusual local conditions, but are not especially mysterious. Two components, radiogenic ^{129}Xe and fission Xe, merit further discussion.

A number of mantle-derived samples exhibit excesses of ^{129}Xe which are difficult to attribute to any source other than decay of ^{129}I (Section 6.9). This is a rather remarkable result, since the short half-life (17 Ma) of ^{129}I requires that this Xe in the mantle has retained its integrity – i.e. has not been homogenized with the atmosphere or the reservoirs which were degassed to form the atmosphere – since the first 10^8 yr of solar system history. Indeed, if the earth had not even accreted in this 10^8 yr (cf. Section 12.7), the suggestion is that the reservoir containing the excess ^{129}Xe was never mixed with the rest of the earth during its entire evolution as a planet. In any case the observations of excess ^{129}Xe require rather severe constraints on any detailed models for global geochemical evolution in that they prohibit any thorough homogenization of earth materials any time later than about 4.4 Ga ago.

It has proved rather difficult to establish the provenance of the ^{129}Xe excesses. Most of the positive observations of excesses have been in various kinds of Hawaiian samples (cf. Hennecke & Manuel, 1975b; Kaneoka & Takaoka, 1978, 1980; Rison, 1980a), and the obvious suggestion is that ^{129}Xe excesses are to be found (only?) in 'undepleted' mantle. Excesses have also been reported in Skaergaard samples (Smith, 1978). The first-known and best-documented occurrence of excess ^{129}Xe, in a CO_2-dominated natural gas field (Section 10.3), remains particularly mysterious. Despite careful attention to the possibility, excesses of ^{129}Xe have not been found in many other samples of presumably mantle-derived Xe. The absence of excess ^{129}Xe in normal ocean floor basalt glasses has been particularly conspicuous. Recently, however, Staudacher & Allègre (1982) have reported excesses of ^{129}Xe in several normal (tholeiitic) basalts from Atlantic, Pacific, and Indian Ocean sites; these observations underscore the puzzle of ^{129}Xe provenance.

There are a number of observations of heavy isotope excesses which are interpreted in terms of fission Xe contributions. Many of these are for continental crustal rocks and are not particularly remarkable because they are readily attributed to *in situ* spontaneous fission of ^{238}U during crustal residence ages. In most cases the heavy isotope enrichments are not large enough to permit determination of the fission Xe composition with precision enough to identify the parent. In some cases, however, sufficient precision is possible; except for bizarre cases like Oklo (Chapter 3) the composition is always that of ^{238}U. No fission Xe excesses relative to air have yet been found to have ^{244}Pu composition, a circumstance worth particular note because of the expectation that there should be large excesses in the long-isolated samples which show ^{129}Xe excesses (Section 13.8).

Nevertheless, there have been suggestions (cf. Staudacher & Allègre, 1982) that in mantle-derived Xe there is a correlation between ^{129}Xe and fission Xe. Assessment of the existence of such a correlation is rather subjective, and if the correlation is real it could have a number of explanations, including experimental artifact. The suggested interpretation of a real correlation is obviously the existence of a radiogenic Xe component – ^{129}Xe plus fission Xe – which is more or less characteristic of the mantle in general. The maximum enrichment so far known is still that of the CO_2 well gas (Section 10.3), about an 8% enrichment of ^{129}Xe. It may be that such a value represents a real end-member mantle Xe component, and that less radiogenic compositions represent dilution with atmospheric Xe. It could equally well be that the observed compositional range represents only a limited sampling of mixtures of more highly radiogenic Xe with Xe of atmospheric composition, perhaps juvenile Xe of atmospheric composition or perhaps actual atmospheric Xe. Irrespective of the origin of the atmospheric-like component, if there really is a mantle radiogenic Xe component then its composition – the ^{136}Xe/^{129}Xe ratio – has an important significance in degassing/evolution models (Section 13.8).

12 Inventories

12.1 The atmospheric inventory

It has been recognized for a long time that air and the oceans undergo large scale and complex interactions with each other and with sedimentary rocks, and that appreciable amounts of material can be transferred between these three reservoirs. Accordingly, it is common to include parts of the ocean and sedimentary rock inventories along with the air inventory in what is designated the atmosphere. Indeed, under circumstances not drastically different from those now prevailing, many chemical species would move from the ocean or sedimentary rocks into the air, and the earth's air would be much more dense and dominated by H_2O and CO_2. Throughout this book we have consistently used the term 'atmosphere' in the generalized sense which designates the surface of the earth, including 'air' (= the gaseous phase of the atmosphere), the ocean and other surface waters, ice, sedimentary rocks, soil, the biosphere, etc., and excluding the core, mantle, and igneous rock crust.

The elements comprising these surface reservoirs fall into two groups. The first group includes those (e.g. Si, Mg, Ca) which, by geochemical balance studies, can be inferred to have been supplied by weathering of primary igneous rocks. Rubey (1951) was the first to clearly describe the second group, those which are present in excess of the igneous rock supply. These elements (or their compounds) share the common feature of relative volatility and are those generally considered to constitute the atmosphere (Table 12.1), and so are usually designated 'excess volatiles'. For the volatiles listed in Table 12.1, the igneous rock supply is generally quite small in comparison with the total inventory. There are a few other species which also appear to be excess volatiles, and which should thus be considered part of the atmosphere, but for which the igneous rock supply may not be so small. Rubey (1951) and Horn & Adams (1966) identify B, F, Sb, Se, Br, and I in this category.

It seems a matter of taste, unlikely to cause confusion, whether a generalized 'atmosphere' should include the total inventory or just the excess part. We will consider 'atmosphere' to include the total (Table 12.1) surface inventory of those elements which are excess volatiles.

Table 12.1. Inventory of atmospheric volatiles[a]

	Air[b]	Ocean	Sedi-ments[n]	Total	Concentration[c]
Major volatiles (amounts $\times 10^{18}$ moles)					
H	–	$165\,000^d$	$16\,500^e$	$181\,500$	31 ppm
C[f,g]	–	–	$6\,100$	$6\,100$	12 ppm
Cl[f]	–	824	590	$1\,414$	8 ppm
S[f]	–	42	491	533	3 ppm
N[e]	276	2	100	378	885 ppb
O[h]	74	–	–	$(1\,860)^i$	–
Nonradiogenic noble gases (amounts $\times 10^{10}$ moles)					
^{3}He[j]	0.1	–	–	$(9\,140)^j$	$(3.43 \times 10^{-10}$ cm^3 STP/g)[j]
^{20}Ne	$290\,000$	–	–	$290\,000$	1.09×10^{-8} cm^3 STP/g
^{36}Ar	$560\,000$	$7\,500$	–	$567\,500$	2.11×10^{-8} cm^3 STP/g
^{84}Kr	$11\,500$	300 [k]		$11\,800$	4.42×10^{-10} cm^3 STP/g
^{130}Xe	63	3 [k]		65	2.45×10^{-12} cm^3 STP/g
Radiogenic[l] noble gases (amounts $\times 10^{16}$ moles)					
^{4}He	0.1	–	–	$(361)^m$	$(1.35 \times 10^{-5}$ cm^3 STP/g)[m]
^{40}Ar	164	2	–	167	6.25×10^{-6} cm^3 STP/g

[a] Blank entries are cases where reservoir accounts for small fraction ($<$1%) of inventory in other reservoirs. Other excess volatiles properly considered part of the atmosphere are B, F, Sb, Se, Br, and I (Rubey, 1951; Horn & Adams, 1966).
[b] All air data from Tables 2.1 and 2.2.
[c] Total atmospheric inventory divided by mass of earth (5.976×10^{27} g); major volatile concentrations are ppm (parts in 10^6) or ppb (parts in 10^9) by weight.
[d] For 1.5×10^{24} g H_2O, including fresh water and glacial ice.
[e] Walker (1977).
[f] Garrels & Mackenzie (1972).
[g] 51×10^{20} moles of inorganic C plus 10×10^{20} moles of organic C.
[h] Uncombined O (as O_2) only.
[i] Li (1972) estimates this value for the total uncombined O (as O_2) ever produced; only a small fraction now exists in air, the majority having oxidized primordial volatiles (e.g. H_2S, HCl) or Fe.
[j] Estimate for nonradiogenic ^{3}He released into atmosphere *assuming* planetary ratio ^{3}He/^{20}Ne $= 0.0315$ (Tables 5.3, 5.4); present atmospheric abundance (Tables 2.2 and 2.3) is much lower because of escape.
[k] It has been proposed that atmospheric Xe in sediments occurs in quantity much greater than that in air (Section 5.3). If atmospheric Xe actually is present in planetary proportions (Table 5.4), the total atmospheric inventory would be a factor of 23 higher, and the concentration would be 5.64×10^{-11} cm^3 STP/g, as computed from atmospheric Kr; if this much Xe resides in sediments, it is likely that sediments could also make an appreciable contribution to atmospheric Kr also (cf. Chapter 8), however.
[l] Also see Section 6.2 and Table 6.1.

The major constituent of the terrestrial atmosphere is seen to be H, present as H_2O. The next most abundant constituent is C, present primarily as carbonate. The atmosphere accounts for a fraction of the total earth equal to about 8×10^{-4} by atom or about 5×10^{-5} weight. Fig. 12.1 illustrates the absolute abundances of atmospheric constituents, and Fig. 12.2 illustrates the extent of their depletion in the earth relative to their cosmic abundances.

The dominant source of atmospheric volatiles is generally agreed to be the solid earth, and our present atmosphere is accordingly designated as secondary (Section 12.2). Atmospheric elements, in preference to the others, are transferred from the solid earth to surface reservoirs because they are volatile, and the process is accordingly designated 'degassing', in contrast to weathering. Possible sinks, channels by which atmospheric elements are removed from the atmosphere, are discussed in Section 12.4.

Atmospheric O is an interesting and important case which merits special attention here. Free O, not combined with other elements, is not expected to have been present in significant abundance in the early solar nebula and is not found in any extraterrestrial samples. Similarly, the O fugacity in the interior of the earth is very low and the abundance of uncombined O can be considered vanishingly small. Uncombined O is thus not expected to be significant component of a planetary atmosphere, nor is it found to be so elsewhere.

Generation of uncombined O in geologically important quantities is attributed to either of two processes, photodissociation of H_2O in the upper atmosphere and biological photosynthesis. In both cases, the source of the O is H_2O, and the essential step is removal of the H so that it cannot recombine with the O. In the photodissociation mechanism, the H is removed by escape from the earth's gravitational field. In photosynthesis, the H is added to CO_2, thereby converting the $\bar{C}$ from 'inorganic' C to 'organic' C. Whether the photodissociation mechanism has in fact been geologically significant in the past is somewhere uncertain. It is generally agreed, however, that for present production, and for production integrated over geological time, the photosynthetic process is by far the dominant one.

Notes to Table 12.1 (*cont.*)

[m] Estimate for radiogenic ^{4}He released into atmosphere, taken as 2.2 times atmospheric ^{40}Ar; the ratio corresponds to integrated production in 4.55×10^9 yr with $K/U = 10^4$ (these figures correspond to minimum $K = 78$ ppm and $U = 8$ ppb; actual abundances are believed to be higher; cf. Section 12.7). The present atmospheric abundance of ^{4}He (Table 2.3) is much lower because of escape.

[n] Sedimentary inventories are rather uncertain; other estimates of C and N abundances are severalfold higher.

Photosynthetic free O is rapidly (on a geological timescale) recycled back into combined O by the processes which undo photosynthesis, i.e. respiration and decay. O_2 has accumulated in air only because some of the organic C (with the H that balances the O_2) is sequestered in sedimentary rocks and not reoxidized. The organic C in sedimentary rocks thus provides a measure of the net amount of photosynthetic O_2 which has been added to the atmosphere. The O_2 in modern air is much less than this (Table 12.1); most of the O_2 which would otherwise be in air has been consumed in the oxidation of Fe^{2+} to Fe^{3+} in igneous rock weathering and in the oxidation of 'reduced' juvenile volatiles such as HCl, H_2S, CO, etc.

Fig. 12.1. The abundances of the principal elements in the earth's atmosphere (cf. Table 12.1). Mole fractions (of the earth) calculated assuming average atomic weight 27. The He and Ar compositions are dominated by radiogenic ^4He and ^{40}Ar; all other elements are dominantly primordial. He and O are shown as present air abundances (solid) and as abundances (dashed line extensions) which would be present except for losses from the atmosphere.

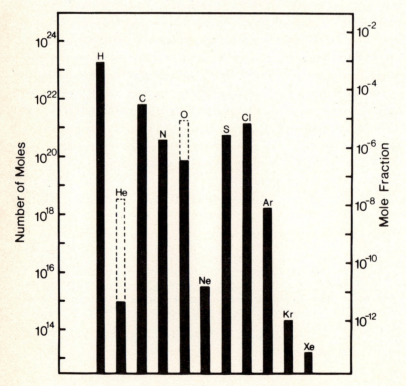

12.2   A secondary atmosphere

It is nearly universally believed that the earth's present atmosphere is secondary, an assumption which is implicit throughout this book. The contrast is with a hypothetical primary atmosphere, one which is acquired, as gas, during the formation of the planet, i.e. by gravitational capture of ambient solar nebula gas. Jupiter and Saturn, for example, have primary atmospheres, and are indeed largely constituted of atmosphere in this sense. The earth does not have a primary atmosphere, nor do any of the other terrestrial planets.

The basic reasoning, first clearly described by Brown (1952), is simple: the composition is wrong. A primary atmosphere should have cosmic composition (that of the sun), and the earth's atmosphere does not. More specifically, not counting He (which escapes) the most depleted element in the earth's atmosphere, relative to cosmic composition, is Ne (Fig. 12.2). Even if it is supposed that a primary atmosphere accounts for all the Ne, other elements would be present only in cosmic proportion to Ne; their abundances would plot only as high as Ne in Fig. 12.2, and the primary atmosphere could only account for a small to negligible fraction of any other present atmospheric species. (Actually, arguments based on isotopic structures (Sections 5.5 and 6.5) indicate that little if any of even the Ne is solar, but the main argument here is independent of such subtleties.)

More generally, it can also be inferred that the present atmosphere is not the remnant of a primary atmosphere mostly now lost, at least not if the loss involves escape, as gas, from the earth's gravitational field. Depletion by gravitational escape would be regular in mass of element or plausible compound, substantially uniform for hydrodynamic flow or nearly all-or-none for atom-by-atom Jeans escape (i.e. if some element, e.g. Ne, is partially depleted, all lighter species, e.g. H_2, H_2O, CH_4, should be much more depleted and heavier species, e.g. Ar, CO, CO_2, HCl, hardly depleted at all). Similarly, any discrimination in the original capture of nebular gas should also display a regular mass dependence of depletion factor.

Uniform or regularly mass-dependent depletion is a conspicuously inadequate description of atmospheric composition (Fig. 12.2). It is thus concluded that most of the present atmosphere did not enter the earth's gravitational field as nebular gas, i.e. that the earth's atmosphere is not primary. If the earth ever did have a primary atmosphere (Section 12.9) it was lost, totally or nearly totally, and the present atmosphere is secondary, generated by other sources.

12.3   Sources

It is generally agreed that the source of the present atmosphere must be the solid earth, i.e. that atmospheric volatiles were originally part of the solids which accreted to form the earth and were subsequently degassed to form the atmosphere. It is well to consider alternatives, however, which we do in this section. It is also well to recognize that there must be indeed some source, since the present atmosphere cannot be an original feature of the earth, one which formed at the same time and in the same process as the earth as a whole (Section 12.2).

An obvious potential source is external: accumulation of solar wind. Satellite measurements and direct collection of the solar wind during Apollo missions indicate that solar wind flux and composition are variable on a short timescale. For a rough calculation we will adopt a representative ^{4}He flux of 10^7 atoms cm^{-2} sec^{-1}, based on direct solar wind collection on foils exposed during several Apollo missions (Geiss *et al.*, 1972). For a cosmic ratio ^{4}He/^{20}Ne =

Fig. 12.2. Ratios of observed abundances of atmospheric elements (Fig. 12.1 and Table 12.1) to abundances expected for cosmic composition (Table 5.1). For He and O the comparison is not relevant and these elements are excluded from the figure; for Ar the comparison is for (nonradiogenic) ^{36}Ar.

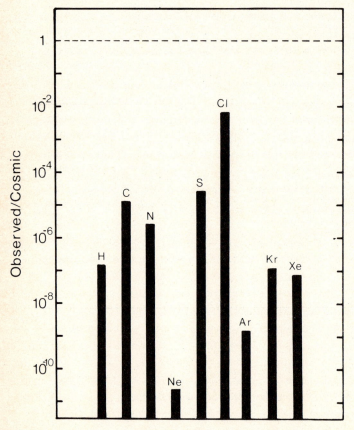

694 (Table 5.3), at this flux in 4.5 Ga the cross-section of the earth will intercept about 10×10^{19} cm^3 STP of ^{20}Ne, slightly more than the present air inventory (about 7×10^{19} cm^3 STP). While this agreement is suggestive, this is an inadequate source: the earth's magnetic field prohibits most solar wind ions from reaching the atmosphere, and only a very small fraction of the intercepted flux could actually leak through the field and accumulate. The estimated auroral precipitation mechanism for He (Section 12.5), for example, amounts to only about 0.5% of the intercepted flux. At face value these calculations lead to the conclusion that the solar wind is quantitatively inadequate to account for the earth's present atmosphere.

It is of course possible to argue against this conclusion by postulating persistently higher solar wind flux and/or lower geomagnetic field intensity in the past. It should be noted, however, that this representative calculation was done for Ne, the element (after He) which is most depleted in the atmosphere relative to cosmic composition (Fig. 12.2). The corresponding shortfall for any other element would be correspondingly greater.

This last point is more general and should be noted explicitly (cf. Section 12.2). No source of cosmic (solar) composition can simply accumulate to form the atmosphere. Even if a source can be found which is quantitatively adequate to supply the Ne, it can account for only a small fraction of any other atmospheric species, all of which are considerably less depleted, relative to cosmic composition, than Ne (Fig. 12.2).

Another potential external source is the continuing infall of extraterrestrial material, which will contribute its trapped gases to the atmosphere. This is very close in spirit to what is actually believed, of course, i.e. that the earth's volatiles were contained in accreting solids, and is particularly close to the variant in which it is postulated that most of the volatiles were in a late-accreting veneer. The question here is thus whether or not this can be understood as a continuing process.

For an order of magnitude calculation we will adopt a present infall of 10^{10} g/yr (cf. Hughes, 1978) of extraterrestrial matter, but it should be noted that such estimates are difficult to make and rather uncertain. In the age of the earth this rate of infall accumulates 5×10^{19} g, i.e. 10^{-8} of the earth's mass. In order for this source to be significant, the volatile content of the infalling material must thus be 10^8 times 'air abundances' (Table 2.3). This seems highly unlikely. Again considering Ne, the 'air abundance' of ^{20}Ne is 10^{-8} cm^3 STP/g, so the infall would have to contain about 1 cm^3 STP/g. The only data available for infalling material indicates trapped ^{20}Ne at about 10^{-4} cm^3 STP/g (Hudson *et al.*, 1981), and even lunar soils range only to about 10^{-2} cm^3 STP/g (Table 5.2). We thus conclude that infall of extraterrestrial material could supply only a very minor fraction of atmospheric abundance, even of Ne. Also, both the values cited above are for Ne in a solar pattern, so there would again be greater shortfall for other elements. If infalling material had planetary trapped gases, the composition would be right but no known meteoritic material, even the separated phases in which planetary gases are concentrated (Section 5.2), contains planetary gases enriched relative to air abundance by more than a factor of about 10^3 (Table 5.2), so that this source is insufficient by a factor of about 10^{-5}. It can, of course, be postulated that the infall was much greater in the past, but this is hardly different from the major accretion of the earth.

By default, the source of the earth's atmosphere must be the original solids which accreted to form the earth. This is certainly quite plausible, since both the absolute abundances and basic compositional features of the earth's atmosphere have quite adequate analogs in known extraterrestrial materials. This is especially true for the noble gases, which occur in a regular (planetary) pattern in meteorites and the same pattern in air (Sections 5.2 and 5.3), except for Xe (Sections 6.7 and 12.6).

It is also clear that the solid earth still contains juvenile volatiles and that these are being added to the atmosphere at present. It is rather more difficult to make quantitative comparisons of compositions of the atmosphere, the juvenile flux, and the residual inventory, however. Still, the present atmospheric composition is intermediate among estimated juvenile compositions from various

Fig. 12.3. Estimated major element compositions of various sources of juvenile volatiles. Atmospheric composition ('excess volatiles') is intermediate, qualitatively illustrating the principle that the atmosphere can be derived from processes similar to those still operating at present. After Rubey (1951).

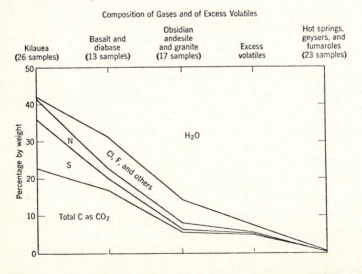

Composition of Gases and of Excess Volatiles

sources (Fig. 12.3), and it can be seen that, qualitatively at least, the present atmosphere can be accounted for in terms of degassing processes which are still ongoing.

It would be of particular interest to quantify juvenile noble gas compositions, but this is unfortunately not possible at present. As discussed in Chapters 9–11, there are no reliable objective criteria by which juvenile compositions can be identified in igneous rocks or in the mantle as a whole; indeed, the shoe is on the other foot, in that observed gases are suggested to be primordial (juvenile) if their compositions resemble a supposed juvenile (typically planetary) composition (cf. Section 9.2). The only volatile species which can actually be identified quantitatively in transit from the solid earth to the atmosphere is ^{3}He (cf. Sections 7.6 and 10.2), and this is principally because the He abundance in air is so low. The same is not true of the other noble gases, and it has not yet been possible to identify the ratio of any other primordial noble gas to ^{3}He in the process of being added to the atmosphere.

12.4   Sinks?

It is usually tacitly assumed that the atmosphere integrates juvenile volatile fluxes, at least for the noble gases, i.e. that once gas enters the atmosphere it remains there. In quantitative evolution models, however, it is important to consider at least the possibility that there are sinks as well as sources for atmospheric gases (cf. Section 13.4).

The obvious possibility for an atmospheric sink is, of course, exospheric loss. Geologically rapid exospheric loss of He is well known (Section 12.5). At present, however, and presumably throughout at least most of the earth's history, exospheric loss of any atmospheric species heavier than He is unimportant to its inventory, and this sink can be ignored.

It is possible that atmospheric loss might have been important very early in earth history, and indeed such a scenario arises in many contexts. At one extreme there may have been bulk hydrodynamic loss. Such loss would not be selective according to chemical identity and it is difficult to imagine that it would be partial: it should be either very minor or total. In the latter case the present atmosphere would have to be generated subsequent to the loss and the problem of atmospheric inventory and evolution would deal only with this post-loss period. At the other extreme there may have been significant loss atom by atom. In this case the loss would probably be very selective, and it is difficult to imagine conditions which would result in partial loss of one species, (e.g. Ar) without total loss of some species (e.g. Ne) and trivial loss of others (e.g. Kr, Xe). On the basis of a number of regularities described in Chapters 5 and 6, there is no evidence that this has happened. The only plausible candidate for selective loss is Xe (Section 5.3), which is at the wrong extreme in atomic weight and for which exospheric loss is unlikely to be important except in considerably more complicated scenarios (cf. Section 6.7). We thus conclude that exospheric loss has not been important in determining the absolute abundances of atmospheric species except possibly in the sense of total loss of an early atmosphere (cf. Section 12.9), nor in determining present atmospheric composition, for any noble gas other than He and probably not for any other species either.

The only other possible sink for atmospheric species is downwards, back into the solid earth. The extent to which such recycling occurs, and in what contexts it might be important, has been a perennial problem in geochemistry. Among atmospheric species for which recycling might be important, attention has usually been focused on water and carbon, but the possibility of noble gas recycling should not be dismissed out of hand.

The principal mechanism for recycling is presumably tectonic subduction at plate margins. Subduction of juvenile gases, both primordial and radiogenic, in igneous rocks would not constitute recycling, since these gases never entered the atmosphere. If water is subducted, so too, presumably, will be its dissolved atmospheric gases, but considering the small fraction of atmospheric gases dissolved in even the entire ocean (Table 7.8), it seems unlikely that this could be an important atmospheric sink. (This does not preclude the possibility that recycled gases are an important *source* for volcanic rocks, however; cf. Section 9.2). Subduction of marine sediments, which might be recycled with greater efficiency than water, is probably more likely to be significant. Noble gases in marine sediments extant at any given time are probably only a small fraction of the atmospheric inventory (Chapter 8), but this is much less evident for sediments than for water, and if subduction is geologically rapid then subductive recycling over the age of the earth might be a significant atmospheric sink. This possibility must be further qualified, however, since it is uncertain whether sediments along with their volatiles are significantly subducted at all or, if they are, whether the volatiles might be promptly returned to the atmosphere in arc/trench volcanism rather than being carried into the deep mantle for long times.

It is not at all clear that recycling is important for any atmospheric noble gas, but if it is, the best candidate is Xe. In both water and sediments, but especially in sediments, the heavier gases are enriched relative to air composition (Figs. 7.1, 8.2-8.4 and 9.9), the greatest relative enrichment being of Xe. This feature might be invoked to account for the apparent Xe deficiency in air (Section 5.3), but aside from the qualitative suggestion, this question has received very little detailed or quantitative attention. Ozima & Podosek (1980), however, have pointed out that recycled sedimentary gases seem an unlikely source for volcanic gases. As illustrated in Fig. 12.4, the abundance of Xe in volcanics might be accounted for by an *ad hoc* postulate of a small (one part in 10^2 to 10^4) admixture

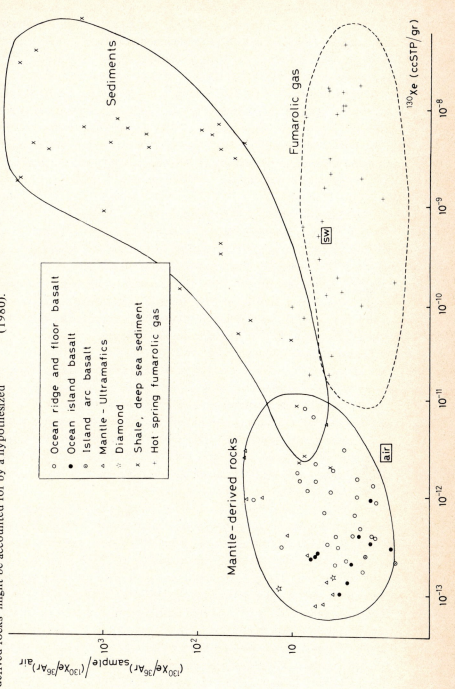

Fig. 12.4. Relationships between concentrations and compositions of noble gases in various terrestrial samples. 'SW' is seawater, 'air' is the quantity in air divided by the mass of the earth. The figure suggests that while gas quantities in 'mantle-derived rocks' might be accounted for by a hypothesized inclusion of minor amounts of seawater or sedimentary rocks, the sedimentary rocks would produce too high an Xe/Ar ratio and so are probably not a significant source for the noble gases in 'mantle-derived rocks'. Reproduced from Ozima & Podosek (1980).

of sediments (or seawater), along with their noble gases, but the compositional mismatch is much greater for sediments than for water. If, as the available evidence suggests (Fig. 12.4), the Xe/Ar ratio in sediments is characteristically about two orders of magnitude higher than in volcanics, recycled sedimentary gases can account for no more than a very small fraction of the volcanic gases.

In summary, it is difficult to exclude categorically the possibility that the inventory of atmospheric gases has been significantly affected by loss to sinks, either upwards or downwards. Except for He, however, there is no positive or presumptive evidence that this is the case. Occam's razor thus suggests that pending such evidence the atmosphere be considered conservative, with no sinks for gases other than He.

12.5   The atmospheric helium budget

He enjoys the distinction of being the only species, at least at present, whose atmospheric abundance is significantly influenced by its escape through the top of the atmosphere. The problems involved in assessing its global inventory are therefore different from those for species which accumulate in the atmosphere. The discussion below is summarized in Table 12.2. Budgetary considerations are usually conducted in terms of fluxes averaged over the surface of the earth, but with no implication that the fluxes are actually uniform.

The major source term in the He budget is emanation from the solid earth. Such emanation can proceed in a number of ways: direct discharge in volcanic vents and fumaroles, extraction by ground water, diffuse gaseous emanation, rock erosion, etc. Direct observation in a fashion amenable to a global census for these items is essentially impossible, so recourse to indirect methods must be made. The simplest expedient is to relate the atmospheric flux to the current production of radiogenic ^{4}He from U and Th. Global concentrations of these elements are, in turn, usually estimated by assuming that their heat production matches current heat flow.

One simple model can be generated by assuming radioactive element concentrations equal to chondritic meteorite concentrations: $U = 12$ pbb, $Th/U = 3.3$, $K/U = 7.4 \times 10^4$. These concentrations produce 59 erg cm^{-2} sec^{-1} average heat flow and 2.6×10^6 atoms cm^{-2} sec^{-1} of radiogenic ^{4}He. These values also correspond to most radiogenic ^{40}Ar being still retained inside the earth, so the degassing rate must be low (Chapter 13) and the flux of ^{4}He must be lower than its production, of the order of 0.5×10^6 atoms cm^{-2} sec^{-1}. Many models currently considered stipulate much lower K/U ratios, e.g. $U = 32$ ppb, $Th/U = 3.3$, $K/U = 10^4$. These figures correspond to average heat flow of 83 erg cm^{-2} sec^{-1} and ^{4}He production 6.9×10^6 cm^{-2} sec^{-1}. This K value also corresponds to rapid degassing, so the ^{4}He flux would equal its production rate.

The predicted ^{4}He fluxes range from 0.5×10^6 to about 7×10^6 atoms cm^{-2} sec^{-1} We have arbitrarily inserted a mean 2.3×10^6 atoms cm^{-2} sec^{-1} in Table 12.2a, but these calculations indicate the range of uncertainty which must be attached to this figure. These extremes cover the range of likely models, but it should be noted that they are only models, and the ^{4}He fluxes are not based on observations. Of the total flux, the only part which is based on direct observation is the apparently minor 0.3×10^6 atoms cm^{-2} sec^{-1} (Craig *et al.*, 1975) corresponding to excess ^{4}He in seawater (Chapter 7), evidently injected primarily at accretional plate boundaries.

The major nonoceanic ^{4}He flux, as described above, is radiogenic, and should be accompanied by ^{3}He (Section 6.2) in the ratio ^{3}He/^{4}He $= 10^{-7}$ (Morrison & Pine, 1955); this is a minor item in the ^{3}He budget. The oceanic He flux is rich in primordial ^{3}He (Sections 6.6, 7.6), contributing about 4 atoms cm^{-2} sec^{-1} (Craig *et al.*, 1975), a major, perhaps *the* major source of ^{3}He. Additional primordial ^{3}He is vented to the atmosphere via volcanoes, fumaroles, etc. (Chapter 10) but it is not known whether the quantities are important to the global budget.

Aside from geological sources, there are other sources of the flux injected directly into the atmosphere. Johnson & Axford (1969) have reviewed these, and we cite their evaluations. Production of ^{3}He from galactic cosmic-ray interactions in the atmosphere, about one-third of it channelled through tritium, is about 0.6 atoms cm^{-2} sec^{-1}, and is the most widely as well as the most reliably known component. Other sources considered are direct accumulation of galactic cosmic rays, solar flares and solar wind, solar cosmic-ray interactions, and accretion in extraterrestrial solids (meteoroids, cosmic dust), all evidently unimportant ($\leqslant 0.1$ atoms cm^{-2} sec^{-1}). Johnson & Axford (1969) noted that a hitherto overlooked source, precipitation of auroral primary ions (presumably of solar wind origin) into the atmosphere, could be important, estimating a ^{3}He flux of 4 atoms cm^{-2} sec^{-1}. All these sources involve relatively high ^{3}He/^{4}He ratios, and will be unimportant for ^{4}He; the largest will be the auroral precipitation, 10^4 atoms cm^{-2} sec^{-1} for 'solar' composition (Table 5.4).

He is lost from the atmosphere at high altitudes (>500 km), and evaluation of the loss rates involves a different category of considerations, that of upper atmosphere physics. A review of aspects relevant to the He budget is given by Kockarts (1973).

The most obvious loss is thermal (Jeans) escape from the high-temperature exosphere. For 1951–1961 Kockarts & Nicolet (1962) calculated average escape fluxes of 6×10^4 atoms cm^{-2} sec^{-1} for ^{4}He and 3.5 atoms cm^{-2} sec^{-1} for ^{3}He. For 1947–1968, Johnson & Axford (1969) calculated 5.9 atoms cm^{-2} sec^{-1} for ^{3}He. The calculations are quite sensitive to a number of variables (cf. Kockarts,

1973). The escape flux is strongly peaked at periods of high solar activity and can be quite different from one cycle to the next (the difference between the two ^{3}He estimates above is mostly due to the strong solar maximum of 1947). The difference between ^{3}He and ^{4}He escape rates is seen to be quite large: the ^{4}He lifetime against Jeans escape is 70 times greater than the ^{3}He lifetime.

The dominant loss mechanism for ^{4}He must be nonthermal, and appears to be escape of photoionized He${}^+$ (along 'open' magnetic field lines) in the 'polar wind' (Axford, 1968; Banks & Holzer, 1968), which can evidently accommodate a ^{4}He flux of $(2-4) \times 10^6$ atoms cm^{-2} sec^{-1} (Kockarts, 1973). This mechanism should not discriminate severely between ^{3}He and ^{4}He, and since isotopic fractionation is apparently small even in the exosphere, also leads to ^{3}He escape at 3–6 atoms cm^{-2} sec^{-1}.

For as long as the only known sources of He were the radiogenic flux and cosmic-ray production, and Jeans escape the only losses, the He budget presented

Table 12.2a. *Major fluxes in atmospheric He (atoms cm^{-2} sec^{-1})*

Flux term	^{3}He	^{4}He ($\times 10^6$)
Escape[a]		
Thermal (Jeans)	5.9	0.06
Nonthermal (polar wind ?)	4.2	3.0
Total	10.1	3.1
Sources[a]		
From world ocean	4.0	0.3
Solid earth other than ocean	0.2	2.0
Cosmic-ray spallation	0.6	–
Auroral precipitation	4.0	0.01
Other	⩽0.1	–
Total	8.8	2.3

[a] See text for references.

Table 12.2b. *Atmospheric He budget summary*

Item	^{3}He	^{4}He	^{3}He/^{4}He
Inventory[a] (atoms cm^{-2})	1.5×10^{14}	1.09×10^{20}	1.4×10^{-6}
Flux[b] (atoms cm^{-2} sec^{-1})	9.5	2.7×10^6	3.5×10^{-6}
Mean residence time (yr)	0.50×10^6	1.3×10^6	–

[a] From Tables 2.1 and 2.2.
[b] Averages from Table 12.2a.

a problem: how to maintain an atmospheric $^3He/^4He$ ratio higher than the total production ratio in spite of losses which must surely remove ^3He at least as fast as ^4He. Specifically, the problems were to find a loss mechanism for ^4He which could match the assumed geological influx, and to find a source for ^3He which could match the Jeans escape. With the more or less simultaneous introduction of a nonthermal loss mechanism for ^4He, an atmospheric source for ^3He, and a comparable geological source for ^3He, these problems are apparently resolved. As seen in Table 12.2, the He budget seems to be in balance, at least as well as can be expected given that none of the major fluxes can be considered known within a factor or two, although there is certainly room for modification of some of the terms or introduction of new ones.

Actually, there is no real reason to require the He budget to be in balance. On the 10^6 yr scale of the mean residence time, variations in fluxes will be smoothed out, and the present epoch may not be typical. The geological and cosmic-ray sources are probably pretty steady on this timescale, but the thermal loss is very sensitive to solar activity, the nonthermal loss is sensitive to the geomagnetic field, and auroral precipitation sensitive to both, and both are variable on 10^6 yr time scales. At an extreme, the fluxes may be highly irregular; Sheldon & Kern (1972), for example, have suggested that major loss of ^4He and gain of ^3He occurs during geomagnetic field reversals.

Whether or not the He budget fluxes are episodic on a timescale longer than the solar cycle, variations in atmospheric He on a timescale longer than the 10^6 yr lifetimes would not be surprising. Secular trends in average solar activity and geomagnetic field are possible and, on this timescale, variations in the geological sources are likely as well, e.g. in overall tectonic activity, rock erosion, etc. Variations, to whatever extent they occur, would probably not be monotonic. On the basis of the same seawater data which show small additions of primordial ^3He we can conclude that there have been no large changes in atmospheric $^3He/^4He$ over the last 10^3 yr or so (a representative mixing time for the oceans), but beyond that there are no observational constraints. In any case it is evident that the present composition and abundance of He are not closely related to any major geological reservoir, and so are only local and probably temporary features.

12.6   The xenon inventory

As described in more detail in Chapters 5 and 6, the match between atmospheric noble gases and planetary gases in meteorites is sufficiently good that the analogy between the two is difficult to ignore. There are significant differences or uncertainties, but most are at least plausibly interpretable in relatively simple terms. The biggest exception is Xe, for which there are two major effects, each difficult to understand and of uncertain relationship to each other. One effect is isotopic: overall, a severe isotopic fractionation, with further complications involving heavy isotope components (Sections 5.6 and 6.7). The second is elemental: a deficiency of Xe abundance in air by more than an order of magnitude (nominally a factor of 23) from what would be expected by the analogy (Section 5.3). Also, the most straightforward models for the isotopic fractionation characteristically predict that the elemental abundance of Xe should be even higher than planetary proportion (Section 6.7), which would correspond to even greater deficiency of Xe in air. We summarize here considerations relevant to this Xe deficiency, an interesting problem whose resolution, if it could be achieved, might be important in a variety of contexts.

If the meteorite analogy holds and most (more than 90%) of the expected Xe is missing from air, there are three possibilities: (i) The Xe is still in the solid earth, i.e. Xe has been degassed much less efficiently than the lighter gases and most of the Xe never entered the atmosphere. This would impose strong constraints on models for the mechanism by which gases enter the atmosphere (cf. Section 13.4). We would also expect to find very high Xe contents in igneous rocks. Xe is indeed typically relatively (to air proportions) abundant in igneous rocks (Section 9.2 and Figs. 9.1 and 9.6) but, on available evidence, not rich enough. (ii) The Xe entered the atmosphere but was lost from it. Selective exospheric loss of Xe seems highly improbable. Selective loss of Xe by subductive recycling is qualitatively plausible but again it is difficult to justify this scenario quantitatively (Section 12.4 and Fig. 12.4), and we would again expect to find evidence of very high Xe contents in igneous rocks. (iii) The Xe entered the atmosphere and is still there, but is hidden because most of it is in some reservoir other than air, e.g. sedimentary rocks. Yet again, this hypothesis is qualitatively plausible but cannot be supported quantitatively (Section 8.5 and Figs. 8.1 and 8.7).

All three of these possibilities are qualitatively plausible in that if the relevant effect occurs at all it is reasonable that it should be selective for Xe. All three are quantitatively inadequate to account for the magnitude of the effect, however, at least on the basis of presently available data, and without the meteorite analogy it is doubtful that any special Xe inventory effect would be predicted on the basis of observation. It is certainly not possible to rule out the basic premise that a large quantity of Xe is hidden somewhere, but if this is the case then the relevant reservoir has not yet been found. It is also interesting to note that if most of the Xe inventory is hidden, there is no assurance that its isotopic composition is that of air Xe, and it is this hidden component, not air, which should be evaluated in terms of meteoritic isotopic structures (cf. Sections 5.6 and 6.7).

As long as the postulated hidden Xe remains hidden, it is prudent to retain an alternative working hypothesis that there is no hidden Xe at all, and that the earth never had Xe in planetary proportions. This partial abandonment of the meteorite analogy thus stipulates that the origin of terrestrial Xe is fundamentally different from the origin of the other gases, a view already suggested by the isotopic peculiarities of Xe. Indeed, an independent argument on the basis of the inventory of fission ^{136}Xe (Section 12.7) suggests that the earth cannot have this much Xe. In this case, an explanation for terrestrial Xe must be sought in the early solar system, prior to the earth's accretion, and not in the earth's subsequent planetary evolution (cf. Ozima & Nakazawa, 1980).

Rather similar considerations arise for Xe inventories in the atmospheres of Venus and Mars (Section 12.10). Albeit within rather broad experimental limits, Xe appears underabundant in comparison with other noble gases in both Venus and Mars. Such, observations lend support to the hypothesis that the origin of Xe is qualitatively different in the major terrestrial planets than in meteorites (cf. Bernatowicz & Podosek, 1978), either because of size or position within the solar system.

12.7 Inventory of radiogenic gases

Direct and indirect production of noble gases in radioactive decay processes make significant contributions to several specific isotopes (Section 6.2). It is interesting to consider the atmospheric inventory of such radiogenic species separately from the primordial gases because the abundances of radiogenic gases can be related to the abundances of the relevant parent elements and because the radiogenic gases were produced at various times and so may have experienced different degassing histories from those of the primordial gases.

For comparison with atmospheric inventories the relevant parent element abundances are those for the whole earth. Such values are, in general, difficult to determine and estimates are often rather uncertain. Most of the abundances in Table 12.3 are from the model calculations of Morgan & Anders (1980). The U and K abundances are based on an assumed geochemical coherence (i.e. constant ratios in magmatic differentiation processes) of K (and Th) with U, with the overall level of these trace elements set to match average heat flow. The I abundance, which is very difficult to constrain observationally, is based on the assumptions of coherence between Tl and U and a cosmic I/Tl ratio. The Pu abundance assumes coherence between Pu and U.

Table 12.3 includes the noble gas abundances corresponding to these parent abundances as well as the identifiable atmospheric inventories. We will not discuss here two other possible significant radiogenic species, ^{21}Ne and ^{86}Kr, since it is quite difficult to relate them to parent abundances or even to identify the radiogenic inventories (Sections 6.2, 6.3 and 6.5). Also, while ^{3}He is often a prominent radiogenic isotope in rocks (Section 6.2), its atmospheric abundance is evidently dominated by nonradiogenic contributions (Sections 6.6 and 12.5).

Actually, Table 12.3 lists air inventories rather than atmospheric inventories. This neglects the amounts stored in other atmospheric reservoirs (Chapters 7 and 8); it is presently impossible to quantify these contributions adequately, but on the basis of available evidence it does not appear that these contributions are large compared to the air inventory. Xe is a possible exception (Section 12.6) and is treated as a special case below.

The abundances of radiogenic gases remaining in the interior of the earth are also unknown. On the basis of available data it seems that this contribution is also small compared to the air inventory, but this is a highly qualified conclusion. Indeed, there is considerable interest in approaching the problem from the opposite direction, that of considering radiogenic inventories as a means of gaining insight into the amounts of noble gases remaining in the interior of the earth. This is the perspective adopted in the discussion below.

The atmospheric inventory of ^{4}He is approximately four orders of magnitude lower than expected radiogenic production (Table 12.3). It is thus concluded that ^{4}He does not accumulate in the atmosphere, and the relevant inventory considerations are those of fluxes (Section 12.5) rather than accumulation.

The case of ^{40}Ar is quite interesting and important. The air inventory is 57% of the nominally expected radiogenic production (Table 12.3); alternatively stated, the minimum K content of the earth dictated by radiogenic ^{40}Ar abundance is 77 ppm, which is 57% of the Morgan & Anders (1980) model value of 135 ppm. At face value, this indicates that the solid earth has lost about half its integrated production of radiogenic ^{40}Ar. Since this much ^{40}Ar was not even

Table 12.3. *Inventories of radiogenic noble gases*

Daughter[a]	Parent[a]	Parent abundance[b,c]	Nominal radiogenic production (cm^3 STP/g)[a,c]	Air inventory (cm^3 STP/g)[d]
^{4}He	U	14 ppb	24×10^{-6}	35×10^{-10}
^{40}Ar	K	135 ppm	10.8×10^{-6}	6.2×10^{-6}
^{129}Xe	I	14 ppb	240×10^{-12}	1.04×10^{-12}
^{136}Xe	Pu	0.20 ppb	129×10^{-14}	24×10^{-14}

[a] Cf. Tables 3.1–3.3, 6.1 and 12.1; radiogenic production assumes accumulation for 4.5 Ga.
[b] Model abundances from Morgan & Anders (1980), except for Pu.
[c] Initial ^{129}I and ^{244}Pu abundances as in Table 3.1.
[d] Cf. Table 2.3 and Sections 5.6 and 6.7.

produced until well after the first Ga of earth history, the inference is that the earth transports gases from the interior to the atmosphere fairly efficiently, and any primordial gas or radiogenic gas produced with a half-life shorter than that of ^{40}K should be at least this efficiently degassed, and probably more so. This would be an important conclusion if it were reliable, but unfortunately it is not. The assumptions and data base leading to this K abundance are not so firm that a significantly lower value could be excluded; in such case the degassing of ^{40}Ar, and thus presumably all primordial volatiles, might be very nearly complete. Such a conclusion is indeed suggested by high trapped $^{40}Ar/^{36}Ar$ ratios in many igneous rocks (Section 11.3). On the other hand, the uncertainties would also permit a significantly higher K abundance. It has been suggested, for example, that most of the earth's K is in the lower mantle or core (e.g. Hall & Murthy, 1971; Goettel, 1976), and that the crustal K/U ratio is unrepresentative of the earth as a whole. As a likely limit the earth's K content might be as high as the chondritic value, 880 ppm. This would correspond to only 7% efficiency of ^{40}Ar degassing, qualitatively a quite different conclusion.

The amount of radiogenic ^{129}Xe in air is less than 0.5% of the expected production (Table 12.3). Even allowing for substantial uncertainty in the earth's I content, the inferred conclusion about degassing efficiency for ^{129}Xe is quite different from that for ^{40}Ar, particularly since the parent ^{129}I is much shorter-lived than ^{40}K. It is in this sense that the remarkable aspect of excess ^{129}Xe in air is not that it exists at all but there is so little of it. One possible resolution of this apparent discrepancy is that air Xe is only a very small fraction of atmospheric Xe (Section 12.6), but this explanation remains hypothetical and conflicts with the analogous considerations for ^{136}Xe (see below). Alternatively, it might be supposed that most of the earth's I has been sequestered in very deep reservoirs since the first few Ma of earth history, and thus that its daughter ^{129}Xe has not been degassed with nearly as high an efficiency as other radiogenic or primordial gases. Such an interpretation is indeed suggested by the presence of excess ^{129}Xe without accompanying excess fission Xe.

The amount of radiogenic ^{136}Xe in air is too high to attribute to spontaneous fission of ^{238}U but is quite compatible with production by fission of ^{244}Pu (Table 12.3), nominally corresponding to about 20% degassing efficiency. The major uncertainty here is probably in determining the actual fraction of radiogenic (fission) ^{136}Xe in air (cf. Section 5.6). That the inferred degree of degassing of ^{136}Xe is much larger than that of ^{129}Xe, in spite of the longer half-life of its parent, is suggestive of a major and very early separation of (perhaps chalcophile?) I from (lithophile) Pu.

Inventory considerations for ^{136}Xe allow an interesting constraint on the Xe inventory problem (Section 12.6). If it is assumed that the initial Pu abundance was 0.47 ppb (corresponding to 14 ppb U and the Podosek (1972) initial ratio $^{244}Pu/^{238}U = 0.016$), and that radiogenic ^{136}Xe is adequately identified (Section 5.6), there should be no more than 13 times as much fission ^{136}Xe in the earth as a whole than there is in air, i.e. no more than 13 times as much Xe as there is in air. This would be somewhat low but still within the bounds of planetary Xe abundance. If, however, the initial ^{244}Pu abundance was lower, as now appears to be the case, the limit is correspondingly stronger. The Hudson *et al.* (1983) value for $^{244}Pu/^{238}U$ (Tables 3.1 and 12.3) corresponds to 20% of the earth's fission ^{136}Xe in the atmosphere and thus no more than four times the identifiably atmospheric inventory anywhere else in the earth. If these figures and assumptions are correct, the hypothesis of planetary Xe in the earth must be abandoned.

For both ^{129}Xe and ^{136}Xe the parents are sufficiently short-lived that an alternative view is feasible: the 'shortage' of radiogenic daughters may be due not to incomplete degassing but to underabundance of the parents because of delayed 'formation' of the earth, so that the earth's 'initial' supply of ^{129}I and ^{244}Pu was lower than the undifferentiated meteorite values used for Table 12.3. If all the radiogenic Xe is in air, the deficiency of radiogenic ^{129}Xe is a factor of 230, corresponding to 133 Ma of ^{129}I decay; the corresponding figure for ^{136}Xe and ^{244}Pu is 227 Ma. An alternative calculation is a 'Xe–Xe' age for the earth (Wetherill, 1975): the ratio of radiogenic ^{129}Xe to radiogenic ^{136}Xe can be expressed in terms of an I/U ratio and a decay time, and it need not be assumed that all radiogenic Xe is in air, only that equal fractions of each isotope are in air. For the parameters given in Table 12.3 the Xe–Xe age of the earth is about 121 Ma. The exact meaning of such Xe retention ages is somewhat unclear, however, but there is no need to assume that formation in the sense of Xe retention is synchronous with formation defined by other chronometers. If it is assumed that no Xe is lost after accretion, these calculations indicate a minimum formation interval of approximately 100 Ma, the time between formation (in the sense of Xe closure) of undifferentiated meteorites and the accretion of the earth. Accretion timescales of this same order also result from theoretical mechanical considerations (cf. Wetherill, 1981).

12.8 Is the earth degassed?

While there is general agreement that the present atmosphere accumulated by degassing of volatiles from the interior of the earth, there is less agreement about whether the process is nearly complete, with nearly all the earth's original volatiles already in the atmosphere, or whether it is only partial, with much of the volatiles remaining in the interior. The question is analogous to that of whether or not the major chemical differentiation of the earth is nearly complete, and in this sense the atmosphere is often treated as part of the crust. The problem can be addressed by a variety of approaches, and noble gases often play a prominent role.

The considerations are different for primordial and radiogenic gases, since the radiogenic gases are produced at different times and locations and their abundances can be related to those of their parents. The possibility of comparing radiogenic and primordial gases is one of the principal approaches to the problem of atmospheric evolution. Formal quantitative degassing models are described in detail in Chapter 13. Here we address the question by simple inventory considerations.

Radiogenic gas inventories (Section 12.7) do not clearly suggest either a high or low degree of degassing. The ^{4}He inventory is essentially irrelevant to this problem. The data for ^{40}Ar do suggest a relatively high degree of degassing, but unfortunately uncertainty about the abundance of its parent K is such that it is possible that the degree of ^{40}Ar degassing is quite high indeed but also that it is possible that it is rather low. Both of the principal radiogenic Xe isotopes indicate a rather low degree of degassing ($<10\%$), but here the principal complicating uncertainties are the possibilities of low parent abundances because of delayed 'formation' of the earth (Section 12.7) or of major under-representation of atmospheric Xe by air Xe (Section 12.6); either of these effects could change the sense of this assessment.

As noted in Section 5.3, the atmospheric abundances of the primordial noble gases are quite consistent with the levels of trapped gases in meteorites. This alone cannot be construed as an argument that the earth is fully degassed, since a total inventory higher by an order of magnitude or more would also be quite compatible to the meteorite comparison. It has been argued, however, that the compositional match between atmospheric and meteoritic noble gases (with an excusable exception of Xe) does indicate nearly complete degassing, on the premise that minor partial degassing would most likely result in severe elemental fractionation. This is not necessarily a compelling argument either, since compositional fidelity would also follow from nearly complete degassing of a small part of the earth (cf. Bernatowicz & Podosek, 1978). Similar views are often advanced for the extraction of nonvolatile incompatible elements in the crust.

The observations of primordial ^{3}He emanation into the atmosphere (Chapters 7 and 10) and of juvenile primordial ^{3}He in igneous rocks (Chapters 9 and 11) clearly demonstrate that the earth is not completely degassed of its original volatile inventory. Unfortunately, it is not possible to use these observations to establish strong constraints on the total quantities of primordial noble gases remaining in the solid earth. Nevertheless, crude semiquantitative evaluations are useful for illustration of the orders of magnitude involved. The present flux of juvenile ^{3}He into the atmosphere is evidently at least 4 atoms cm^{-2} sec^{-1}

(Section 7.6); this is the estimated flux into the world ocean (Craig *et al.*, 1975), and ignores any emanation directly into air. If we make the *ad hoc* and presently untested assumption that this ^{3}He flux is accompanied by other primordial gases in nominal planetary proportions (Table 5.3), accumulation of this flux for 4.5 Ga will produce 5% of the present air abundance of ^{20}Ne (10% of the ^{36}Ar, 6% of the ^{84}Kr, 136% of the ^{130}Xe). Again with the exception of Xe, these values are relatively low, although certainly not trivial. The inference is that the rate of degassing must have been higher in the past, that juvenile emanation is slowing down, and, within the framework of uniformitarian models such as first-order degassing (Chapter 13), that the degassing process is thus nearly complete. Once again, however, the results of this calculation are near the borderline range such that the uncertainties in the data and assumptions permit either qualitative conclusion, i.e. that the degassing process could be either substantially complete or relatively incomplete.

12.9   A primary atmosphere?

The earth's present atmosphere is clearly secondary rather than primary; more specifically, a residual primary atmosphere can account for no more than a very small fraction of the total atmospheric inventory (Section 12.2). There is, moreover, no evidence in the geological record that the earth ever had a primary atmosphere. We cannot thereby conclude that the earth actually never had a primary atmosphere, however. The question remains of interest because it can be argued, as by Hayashi *et al.* (1979), that the earth could or even should have had a large primary atmosphere.

The basic reasoning is that the earth is large enough to have gravitationally captured a primary atmosphere from the solar nebula. Assuming spherical symmetry, hydrostatic equilibrium, and a constant net energy flow through the atmosphere, Hayashi *et al.* (1979) estimate that the earth should have acquired a massive atmosphere, with total mass about 10^{26} g and surface pressure about 10^3 atm. There are at least two major qualifications to this conclusion, however. One is the question of whether hydrostatic equilibrium could actually have been achieved. The second is whether or not the solar nebula had been dispersed by the time the earth accreted to a size large enough (about 10% of its present mass) to initiate a significant gravitational concentration of any ambient gas.

If the earth actually had such a dense atmosphere, the effect on early evolution would have been profound. For example, the atmosphere would have been optically thick and greatly impeded loss of accretional energy; Hayashi *et al.* estimated a surface temperature about 4000 K. Once such an atmosphere exists, getting rid of it is also a substantial problem. Simple hydrodynamic flow could account for part of the loss, but appeal to some external agency, such as solar ultraviolet radiation or solar wind at intensities much greater than at present, is also required; Sekiya *et al.* (1980) consider that the necessary conditions could have been achieved during a T Tauri stage in the sun's early evolution.

Mizuno *et al.* (1980) noted that absorption of noble gases from so dense a primary atmosphere would not be trivial. At 10^3 atm and solar composition (Table 5.1), the pressure of ^{20}Ne would be about 0.2 atm. If the Kirsten (1968) molten enstatite solubility (Table 4.3) applies, the level of dissolved ^{20}Ne in equilibrium would be $1.4 \times 10^{-5} \, cm^3 \, STP/g$. This cannot be the origin of all the earth's noble gases since such solution would produce the wrong kind of elemental fractionation relative to solar composition (cf. Section 5.7), but the level is quite impressive. It is three to six orders of magnitude higher than typical ^{20}Ne concentrations in igneous rocks (Fig. 9.3), and this concentration in just 0.1% of the earth's mass would more than account for the total present atmospheric inventory of Ne. Mizuno *et al.* estimated that dissolved Ne from the primary atmosphere would have been 10^1–10^2 times the present atmospheric inventory.

Clearly, present noble gas observations do not support such a scenario: present abundances are too low, and isotopic compositions are wrong as well (Ne in a primary atmosphere would presumably have solar composition). If the outlines of this history are correct, the dissolved Ne must have been very efficiently exsolved and dispersed along with the rest of primary atmosphere. In such a scenario it seems questionable that any gases at all would remain in the interior for the subsequent development of the secondary atmosphere.

12.10 Venus and Mars

The recent Viking missions to Mars and Pioneer and Venera missions to Venus have yielded a wealth of previously unobtainable data concerning the state and constitution of the atmospheres of these two planets, the members of the solar system most nearly like the earth. The availability of these data (Table 12.4) has stimulated substantial interest not only in these planetary atmospheres themselves but also in the bearing they have on theories for the origin and evolution of the major terrestrial planets. The opportunity for comparative planetology in this area has also broadened perspectives in theories of the evolution of the earth. In this section we will examine some of the volatile inventory considerations for Venus and Mars that were made for the earth earlier in this chapter. Noble gases play a major role in such considerations.

As on the earth, atmospheric volatiles are not necessarily all in the gas phase. The surface temperature of Venus is so high (about 460 °C) that it is doubtful that relatively unreactive species such as noble gases and N are anywhere other than in the gas phase, but species such as S and O (from dissociation of H_2O)

may have reacted extensively with near-surface rock. For Mars, however, the major atmospheric gas, CO_2, is condensable, and most of the H_2O in the generalized atmosphere is in polar caps or permafrost rather than in the air; there is also a possibility that much of its atmospheric Xe is adsorbed on the regolith (Fanale *et al.*, 1978).

In spite of substantial uncertainties in the measurements and ambiguities about the location of atmospheric species, it is immediately obvious that the atmospheres of Venus and Mars are compositionally much more similar to the earth's atmosphere than to cosmic abundances. By the same reasoning described in Section 12.2, it can be inferred that Venus and Mars, like the earth, have secondary atmospheres rather than primary atmospheres.

The major volatiles C and N are present in comparable abundances in the atmospheres of Venus and Earth; assuming that the degrees of degassing of primordial volatiles are not grossly different, it is thus suggested that the initial endowments of these species in Venus and Earth were comparable as well. By extension, it is further suggested that their initial inventories of the other major volatiles H, S, and Cl were comparable as well. For S and Cl it is not possible to evaluate the Venus atmospheric inventory since, as for Earth, these species are expected to reside elsewhere than the gas phase, e.g. in cloud droplets, aerosols, or in the crust. This is not true for H, however, and it is clear that H is very underabundant in the atmosphere of Venus, approximately four orders of magnitude lower than in the earth's atmosphere. This can be interpreted as reflecting exospheric loss (cf. Donahue & Pollack, 1983) of H dissociated from H_2O. The principal contrast with the earth would then be that on the earth H_2O is primarily liquid while in the early Venusian atmosphere, much hotter than that of Earth, the H_2O would have been gaseous (and thus the dominant constituent) and there susceptible to dissociation and exospheric loss. There is, however, no direct evidence that this happened, and Venus might simply always have been poor in H_2O.

By similar considerations it appears that Mars' volatile inventory was considerably lower than that of Venus and the earth, but it is difficult to make such an argument rigorously. The present observable atmospheric quantities of the major volatiles H, C, and N are all quite low in comparison with Venus and Earth, but for all three of these it is possible to argue that substantially greater quantities presently reside in the crust (and are thus not observable) and/or were once much more abundant in the gas phase but suffered substantial exospheric loss (cf. McElroy *et al.*, 1978; Anders & Owen, 1977). Further uncertainty arises in the possibility that Mars may have degassed a significantly different – smaller – fraction of its total volatiles than did the earth (or Venus). The noble gases are important in volatile inventory considerations since they are typically too heavy to escape (except for He and possibly, on Mars, Ne) and too inert to reside in the crust (again for Mars, possibly excepting Xe).

As in the case of the earth (Section 12.7), radiogenic gas inventories provide important constraints on the degree of global-scale degassing (cf. Pollack & Black, 1983). The ^{40}Ar inventory in the atmosphere of Venus is a factor of four lower than that of the earth. It seems unlikely that this can be due to a lower global K inventory: the K/U ratio in surface rocks on Venus is rather similar to the nominal terrestrial value of 10^4, and it would be surprising if the global abundance of a refractory element such as U were significantly lower for Venus than for Earth. It thus seems more likely that ^{40}Ar is less efficiently degassed on Venus than on Earth. This is a rather unexpected inference considering the surface thermal regime on Venus, but it has been suggested that this indicates fundamentally different tectonic style (cf. Section 13.3) on Venus. This inference cannot, however, be extended to primordial volatiles, which might have been degassed catastrophically and early (Section 13.2), before the generation of ^{40}Ar.

The same arguments apply to Mars as well, for which the atmospheric concentration of ^{40}Ar is substantially lower than the earth's. In part, but only in part, this is apparently attributable to a lower K content for Mars, since its K/U ratio is evidently about a factor of three lower than the earth's. If U contents in Earth and Mars are comparable, this then corresponds to an ^{40}Ar degassing efficiency for Mars which is nearly an order of magnitude less than the earth's. This is not implausible, since the size of Mars suggests, and the antiquity of its surface indicates, that it has considerably less volcanic activity than the earth does.

On Venus the atmospheric concentration of ^{4}He and the atmospheric ^{4}He/^{40}Ar ratio are both considerably higher than they are on Earth. The ^{4}He/^{40}Ar ratio is still at least an order of magnitude lower than any plausible production ratio, so that it can be inferred that Venus has lost most of the ^{4}He ever liberated into its atmosphere, but the lifetime of ^{4}He in the atmosphere of Venus is equally clearly seen to be much longer than in the earth's atmosphere (cf. Section 12.5).

The primordial noble gases are more abundant in the atmosphere of Venus than in the earth's atmosphere by a factor between one and two orders of magnitude, a rather surprising observation in view of prior expectations and the more so because the major volatiles C and N are *not* comparably overabundant. This must reflect a comparably large enrichment factor for the global inventory of noble gases in Venus as compared to the earth, rather than different degassing efficiencies, since it is not really plausible to maintain that only a few per cent of the earth's volatiles are in the atmosphere (Sections 12.7 and 12.8). In contrast, the primordial noble gases in the atmosphere of Mars are about two orders of magnitude less abundant than in the earth's atmosphere. Some of this deficiency, about an order of magnitude as suggested by the above considerations for ^{40}Ar,

Table 12.4. Atmospheric volatile inventories for Venus, Earth and Mars

Quantity	Units	Venus[a]	Earth	Mars[a]	Relative to Earth Venus	Relative to Earth Mars
Planetary characteristics						
Surface pressure	bar	91	1.01	0.007	91	0.007
Surface gravity	cm/sec^2	888	978	373	0.91	0.38
Radius	km	6 097	6 378	3 380	0.96	0.53
Atmospheric mass	10^{20} g	4 750	51	0.27	94	0.005
Planetary mass	10^{27} g	4.87	5.98	0.64	0.815	0.108
Air composition[b]						
N_2	%	4	78	2.7	—	—
O_2	%	0.002	21	0.13	—	—
CO_2	%	96	0.0315	95.3	—	—
H_2O	%	—	~0.3	~0.03	—	—
^{4}He	ppm	12	5.24	—	—	—
^{20}Ne	ppm	4	16	2	—	—
^{36}Ar	ppm	30	31	5	—	—
^{84}Kr[c]	ppm	0.03–0.6	0.65	0.2	—	—
^{130}Xe	ppb	<2	3.54	2	—	—
^{40}Ar/^{36}Ar	—	1.0	296	3 000	—	—

Atmospheric concentrations[a]

	Units					
N[e]	ppm	2	0.9	>0.0006	—	—
C	ppm	26	12	>0.01	—	—
H	ppm	0.07	31	>0.5	—	—
^{4}He	10^{-8} cm^3 STP/g	65	0.35	—	186	
^{20}Ne	10^{-8} cm^3 STP/g	23	1.09	0.0056	21	0.005
^{36}Ar	10^{-8} cm^3 STP/g	150	2.11	0.013	74	0.006
^{84}Kr[c]	10^{-10} cm^3 STP/g	15–300	4.42	0.042	3–70	0.009
^{130}Xe	10^{-12} cm^3 STP/g	<100	2.45	0.071	<40	0.03
^{40}Ar	10^{-6} cm^3 STP/g	1.6	6.25	0.30	0.26	0.05

[a] Trace constituent abundances for Venus and Mars have relatively large experimental uncertainties, often of the order of a factor of two.

[b] Venus data from von Zahn et al. (1980), Oyama et al. (1980), Hoffman et al. (1980), Istomin et al. (1980), and Donahue et al. (1981); Mars data from Owen et al. (1977); terrestrial data from Tables 2.1 and 2.2.

[c] Lower value (for Venus) from Donahue et al. (1981); higher value from Istomin et al. (1980).

[d] Venus and Mars values are those adopted by Donahue & Pollack (1983) and Pollack & Black (1983), except for N; terrestrial values from Table 12.1. Concentrations are atmospheric inventories divided by planetary mass.

[e] Mars concentration is value corresponding to present quantity in air; the enrichment of ^{15}N (^{14}N/^{15}N = 277 for Earth, 165 for Mars) has been interpreted to indicate substantial exospheric loss (McElroy et al., 1978).

is presumably due to a lower degassing efficiency for Mars, but some, again about an order of magnitude, evidently reflects a deficiency of global inventory of noble gases in Mars as compared to the earth. The same conclusions also follow from consideration of atmospheric $^{40}Ar/^{36}Ar$ ratios, which reflect both planetary evolution and global $K/^{36}Ar$ ratio (Sections 13.5 and 13.6): compared to the terrestrial value atmospheric $^{40}Ar/^{36}Ar$ is quite low on Venus and quite high on Mars. Much of this variation probably reflects global $K/^{36}Ar$ ratio variation, and presumably this ratio varies more as a function of ^{36}Ar abundance differences than K abundance differences. A comparable result is also obtained from Mars' Xe composition: for $^{129}Xe/^{132}Xe \approx 2.5$ (Owen *et al.*, 1977), the ratio of radiogenic ^{129}Xe to primordial Xe is somewhat more than an order of magnitude higher than it is in the terrestrial atmosphere, a factor comparable to the relative enrichment of radiogenic ^{40}Ar to primordial ^{36}Ar.

This unexpected pattern of primordial noble gas abundances, substantially higher on Venus and lower on Mars, both relative to the earth, has stimulated interest in the question of whether or not the origin of noble gases in the major terrestrial planets is significantly different from the origin of trapped noble gases in meteorites (see Section 5.7). It is thus particularly interesting to evaluate noble gas *compositions* in the atmospheres of Venus and Mars. Except for the radiogenic/primordial ratios $^{40}Ar/^{36}Ar$ and $^{129}Xe/^{132}Xe$, however, isotopic composition data are not sufficiently precise to be useful in this regard. To first order, noble gas *elemental* compositions in the atmospheres of Venus and Mars, as in the earth's atmosphere, are clearly planetary (Fig. 5.5), at least in the sense that they resemble planetary composition much more closely than they resemble solar composition (Section 5.3). At a finer level of consideration, however, it remains possible that elemental abundance patterns can shed some light on potentially significant differences between meteorite and planetary atmosphere compositions. Unfortunately, experimental uncertainties in the Venus and Mars compositions are sufficiently large that it is difficult to make such arguments conclusive.

For Mars, the atmospheric noble gas composition (Fig. 5.5) is essentially identical to nominal planetary composition with the probable exception of Xe. The reported Xe/Kr ratio is between the meteoritic and terrestrial atmospheric values, closer to the terrestrial values. Within estimated uncertainties, it could be identical to the terrestrial value; with somewhat greater stretching of limits, however, it could also be rather close to the nominal meteoritic value. Fanale *et al.* (1978) suggest that atmospheric Xe/Kr on Mars may actually be planetary but appear to be lower by essentially the same mechanism that Fanale & Cannon (1971a) propose to account for the same effect on Earth (Section 12.6): sequestering of Xe in the crust. Alternatively, however, the apparent underabundance of Xe in both Earth and Mars, compared to meteorites, suggests that there may be a qualitative difference in the origin of Xe between meteorites and large planets (cf. Bernatowicz & Podosek, 1978).

The Venus atmospheric data have attracted particular attention, and a rather wide spectrum of opinions, because the primordial noble gas abundances, at least for the lighter gases Ne and Ar, are so high, at the high end of the range observed in bulk meteorites. Such high abundances are somewhat of a strain on the view that volatiles in the major terrestrial planets are essentially the same as volatiles in meteorites, particularly heterogeneous accretion models which postulate that most volatiles originate in a late-accreting 'veneer'. Wetherill (1981) suggests that the high ^{36}Ar abundance indicates that the primordial Ar is solar rather than planetary, specifically that it originates in a solar wind irradiation (experienced only to a much smaller degree by Earth or Mars materials) of dispersed material before accretion to form Venus. The ^{20}Ne/^{36}Ar ratio for Venus is about 0.15, however, in the normal planetary range and two orders of magnitude lower than the solar or cosmic ratio (Table 5.3); if the ^{36}Ar is solar, a secondary loss of solar Ne must be postulated, coincidentally producing a planetary Ne/Ar ratio. Owen *et al.* (1983), in contrast, suggest that Venus noble gases are fundamentally planetary rather than solar in origin, albeit a special and unusual variety of planetary gas termed the 'subsolar' component (Crabb & Anders, 1981), found in enstatite chondrites and differing from more typical planetary composition chiefly by relatively high ^{36}Ar abundance, especially a high Ar/Kr ratio.

Heavy noble gas abundances are particularly important for assessing models for Venus, but unfortunately the experimental situation is presently unsettled. Venera and Pioneer analyses yield inconsistent results for ^{84}Kr atmospheric abundance differing by a factor of 20, a high value of about 600 ppb (by volume) for Venera (Istomin *et al.*, 1980), and a low value, about 30 ppb, for Pioneer (Donahue *et al.*, 1981). The high value corresponds to ^{36}Ar/^{84}Kr = 50, identical to the terrestrial value; if this is valid, then Venus, like Earth and Mars, would have relative abundances of Ne, Ar, and Kr well within the meteoritic planetary range and special explanations in terms of a solar component or an unusual planetary component would be unwarranted. The low value corresponds to ^{36}Ar/^{84}Kr = 1000, substantially higher than planetary ratios and rather more similar to a solar ratio (Table 5.3); if valid, this result indeed would favor explanation in terms of a real solar (solar wind) or 'subsolar' planetary component. For Xe in the Venus atmosphere, only an upper limit abundance is available (Donahue *et al.*, 1981): a ^{130}Xe volume ratio of <2 ppb (assuming normal Xe composition). The high value of ^{84}Kr then corresponds to ^{84}Kr/^{130}Xe>300, somewhat higher than the terrestrial value of 180; if valid, the suggested generalization is planetary proportions of Ne, Ar, and Kr but progressively lower proportions of Xe in the sequence Mars, Earth, Venus, i.e. sunward of the probable location of undifferentiated meteorites. The low value of ^{84}Kr corresponds to ^{84}Kr/^{130}Xe>15, higher than the planetary value (nominally about 8), perhaps as high or higher than the terrestrial value, but considering experimental uncertainties consistent with either.

13 Atmospheric evolution

13.1 Introduction

For as long as it has been evident that the atmosphere is not just
a remnant of the earth's accretion process, but instead has a secondary origin by
degassing of the solid earth (Section 12.2) and thus has evolved, attempts have
been made to exploit noble gases in the study of that evolution. One advantage
of noble gases in this regard is that they are very low-abundance trace elements
which will respond to but not influence geochemical processes. More important,
their chemical behavior is very much simpler than that of the major volatiles.
Early studies focused primarily on the radiogenic gas inventories, particularly
^{40}Ar. More recently, the importance of the nonradiogenic gases has also been
recognized; as described in Chapter 5, in extraterrestrial materials the noble gases
have a number of interesting features in both elemental abundance patterns and
isotopic structures which offer the opportunity to gain information about the
evolution of the atmosphere as well as the origin of planet earth. A number of
simple inventory considerations are presented in Chapter 12. In this chapter we
consider formal quantitative models which treat not only inventories but also
the evolution of the atmosphere in time.

The principal feature which makes noble gases so important in the study of
atmospheric evolution is that they are the only atmospheric species which have
not only primordial components but large and easily distinguishable radiogenic
components as well. The primordial components were present when the earth
was formed, but some or most of the radiogenic components were not, and so
have different responses to the earth's geochemical evolution. Thus, isotopic
ratios as well as elemental abundances have evolved differently in different parts
of the earth, and the ratio of a radiogenic to a stable isotope carries chronological
information. Also, there are obviously much brighter prospects for characterization
of isotopic ratios than there are for absolute elemental abundances. The traditional
radiogenic isotopes exploited in degassing models are ^{40}Ar and ^{4}He; more recently,
models have included ^{129}Xe and ^{136}Xe as well.

Even the most abundant noble gas is only a very minor constituent of any
geochemical system, including the atmosphere. Conclusions reached for the

evolution of noble gases can thus be transformed to the atmosphere as a whole, or any specific constituent of it, only with the assumption that volatiles in general have evolved similarly. This seems a reasonable assumption, and there is no evidence to contradict it, but it should not be forgotten that it is an assumption. It does appear to be a good assumption in that volatile transport is seemingly governed more by the overall evolution of the earth than by individual chemical identity (cf. Sections 13.2 and 13.3).

In the same sense, noble gas evolution models provide information on the overall thermal and chemical evolution of the earth. Models based on the accumulation of radiogenic noble gases in the atmosphere have many parallels with models for the accumulation of nonvolatile radiogenic isotopes (^{87}Sr, ^{143}Nd, ^{206}Pb, ^{207}Pb, ^{208}Pb) in the crust. This is not only in the sense of providing complementary information about global evolution but also in the methodological sense of having achieved comparable levels of sophistication and suffering from comparable levels of ambiguity and incomplete information.

13.2   Catastrophic degassing?

In many discussions of atmospheric evolution the concept of catastrophic degassing plays an important role. The term 'catastrophic' is used in the geological sense as a description of a singular event, in this case a unique event, in which conditions are markedly different from those before or after. Catastrophic degassing is thus the sudden release of volatiles from the earth's interior to the atmosphere in a single event. In principle, the event is instantaneous; more generally, the sense of a catastrophic event would be preserved if the event had a finite duration which was small compared with the age of the earth, e.g. of the order of 10^8 yr. In practice, an era of finite duration would be considered a catastrophic event if the duration is too small to be resolved by the applicable geochronometer. At the present state of knowledge, any atmospheric evolution model based on ^{4}He or ^{40}Ar would be unable to distinguish events on times finer than the order of 10^8 yr, so that anything that happened on this timescale would be considered an event; finer resolutions are potentially possible for models based on radiogenic ^{129}Xe or ^{136}Xe because of the much shorter parent lifetimes. It is, however, important to preserve the sense of event, specifically that conditions were different after the event was over. This point is significant in atmospheric evolution models, in that catastrophic degassing would affect whatever volatiles were present at the time, including all the primordial volatiles, but that the degassing regime was different for subsequently produced radiogenic species; if this is not true, then the degassing regime in question is not a catastrophic event.

Typically, catastrophic degassing is thought of as early catastrophic degassing, an event that happened very early in earth history. Often, the connotation of 'early catastrophic degassing' is that the event is essentially coincident with the formation (accretion) of the earth, but a short degassing event delayed by a time of the order of 10^8 yr after formation would also qualify as 'early', at least in the sense that it would be difficult to distinguish delay on this scale. Catastrophic degassing is not necessarily early, however, and some degassing models treat the time of catastrophic degassing as a free parameter whose value is potentially determinable from observation.

Catastrophic degassing is also usually thought of as extensive, nearly complete. Really complete degassing cannot have occurred, of course, since the earth's interior assuredly contains at least some primordial gas (Chapters 9 and 11), but catastrophic degassing could have been nearly complete. This is not required, however, and indeed some models also treat the extent of catastrophic degassing – the fraction of the total inventory released to the atmosphere – as a free parameter. In another sense, however, catastrophic degassing must be considered extensive in that it accounts for a major fraction of the present atmosphere (which is not necessarily a major fraction of the total inventory); while it is possible in principle to consider a single event which accounts for a small fraction of the atmosphere, in practice there would be little hope of distinguishing such an event.

The antithesis of catastrophic degassing is continuous degassing. Ideally this designates mathematically continuous degassing, but the effects would be similar if degassing were episodic as long as there were many episodes over long periods of time (times not small compared to the age of the earth or, for radiogenic species, not small compared to the half-life of the parent). Catastrophic and continuous degassing are not mutually exclusive: it is easily possible to imagine continuous degassing with a catastrophic event superimposed, and many models are so constructed. Continuous degassing sometimes carries the connotation of slow or incomplete degassing, but it need not. It could be supposed, for example, that all volatiles are rapidly flushed from the earth's interior; this would amount to producing most of the earth's atmosphere, and depleting the interior of volatiles, very early in its history, but if such conditions continued, and subsequently produced radiogenic gases were also rapidly flushed into the atmosphere, this would constitute continuous rather than catastrophic degassing.

Continuous degassing certainly occurs. Thus, the observation that some volcanic rocks have 'excess' ^{4}He and ^{40}Ar (Section 9.4), while others from the same magma source do not, indicates that gases from the latter are released into the atmosphere; this is continuous degassing of radiogenic ^{4}He and ^{40}Ar from the mantle. Similarly, radiogenic gases in crustal rocks are released into the atmosphere in weathering; indeed, crustal rock gas-retention ages typically lower than nongaseous daughter (Sr, Nd, Pb) ages, and direct observation of radiogenic gas emanation (Chapter 10), indicate degassing independently of weathering. Current degassing is not limited to radiogenic species: excess ^{3}He in seawater (Chapter 7) is a manifestation of presently ongoing degassing of a primordial volatile species. The uniformitarian precept that the present epoch is not exceptional, at least in comparison with a considerable geological time, leads to the conclusion that continuous degassing has persisted throughout much, at least, of the earth's history.

What is less clear is how important continuous degassing has been in the evolution of the atmosphere. Alternatively stated, the question is whether or not there has been a significant catastrophic event responsible for generation of most of the atmosphere. This question is certainly addressable by means of degassing models, and we will follow this approach in this chapter. It should be noted, however, that many theoretical discussions incorporate a prejudice, explicit or implicit, that catastrophic degassing should have occurred, indeed must have occurred, and that catastrophic degassing is the default assumption to be considered valid until shown otherwise (cf. Fanale, 1971). Often it is assumed that catastrophic degassing not only accounts for most of the present atmosphere but also a nearly complete degassing of the total terrestrial volatile inventory. Such a conclusion often emerges from consideration of degassing models, but not always, and it is important to distinguish conclusion from assumption.

A bias in favor of catastrophic degassing has not always been fashionable (cf. Rubey, 1951). Its present ascendency seems related as much to the broader context of general models of planetary formation and evolution as to studies of the volatiles themselves. Many models, for example, stipulate that the early earth would be very hot because of accretional energy; this heat source is presumed to have produced a 'magma ocean' on the moon and would be even more effective on the earth. A rapid and efficient release of volatiles would also result, even for an initially cold outer earth, in heterogeneous accretion models which stipulate that most of the volatile inventory was carried by a late-accreting veneer of low-temperature condensate, perhaps even cometary material bearing adsorbed or clathrated noble gases (cf. Section 5.7) along with other volatiles. In both these cases, the degassing would be early as well as catastrophic. Other possibilities include major reorganization of structure or tectonic style. Core formation, for example, would liberate enough energy to produce a very hot earth (unless, as in heterogeneous accretion models, the core accreted first and the mantle around it). Another example is tectonic change, as in a change from some prior regime to the present plate tectonic mode, which could also provide opportunity for major gas release. In both of these cases there might be catastrophic degassing which is not necessarily early. In these and other possible scenarios, catastrophic degassing, particularly early catastrophic degassing, is

certainly plausible but not always guaranteed. A good parallel is the likely
existence of undepleted mantle (cf. Section 11.2); if, whatever the history, there
remains a chemically primitive mantle, it may be primitive in terms of volatiles
as well. In any case we will approach catastrophic degassing as a hypothesis to
be tested rather than a boundary condition to be postulated.

13.3   Continuous degassing mechanisms

As described in the previous section, 'continuous degassing' is the
release of gases continuously over long periods – time not short compared to
the age of the earth or, for radioactive species, compared to the parent half-life.
The process may be episodic in detail, but as long as the individual episodes are
frequent on the same timescale and their average cumulative effect varies but
slowly, treatment of this process as mathematically continuous is justified.

In essentially all models proposed to describe global-scale evolution, the
problem of continuous degassing is framed in terms of two or more discrete
reservoirs (Section 13.4), each of which is uniform. Each reservoir is imagined
to contain a quantity Q of some species of interest, and the transport of this
species to another reservoir usually described in terms of a first-order rate law
in which the loss rate of Q is proportional to Q:

$$dQ/dt = -\alpha Q \qquad (13.1)$$

According to context the proportionality constant α is termed the degassing
coefficient or the transport coefficient. The transport can be considered fast or
slow according to whether $\alpha T \gg 1$ or $\alpha T \ll 1$, respectively, where T is the age of
the earth.

Use of first-order transport, eq. (13.1), has the obvious virtues of simplicity
and intuitive appeal, as well as appropriateness for a reservoir which is presumably 'stirred' in some fashion in order to maintain the assumed uniformity.
It is also possible to provide a physical basis for this phenomenological
assumption, however.

Thus, the (only) obvious alternative is to imagine a reservoir in which
material transport is effected by diffusion. This is the customary approach for
degassing of laboratory specimens (cf. Section 4.8), for example. On a global
scale, however, diffusion is ineffective. For illustration, we can adopt
$D \approx 10^{-3}$ cm^2/sec as an optimistic upper limit for a He diffusion coefficient
(cf. Fig. 4.6); values for He below about 1500 °C and for any other gas are
likely to be substantially lower. In the age of the earth ($T \approx 10^{17}$ sec), the
characteristic distance scale for diffusion, $(DT)^{1/2}$, for even this D is only about
100 km, too low to account for global-scale transport.

The problem with diffusion as the principal agent for global-scale material transport is comparable to the problem with lattice conduction as the principle agent for global-scale heat transport: the earth is too big. The resolution is presumably the same in both cases: bulk mass transport, i.e convection. The basic approach in degassing models is thus to imagine a series of discrete reservoirs, each uniform, with transport from one to another mediated by convection. The hope is that a tolerable approximation to reality can be achieved within a manageable level of complexity (in practice, no more than about two solid earth reservoirs). Given the limitations on plausible diffusion distances, the principal mechanism for transport of gases from the mantle to the atmosphere would seem to be generation of magma within the mantle and eruption or emplacement of this magma on or near the surface.

Degassing models generally treat transport coefficients such as that in eq. (13.1) as free parameters whose value is to be inferred from observations. It is possible, however, to make an approximate analysis which permits estimation of a plausible order of magnitude. If a solid with initial gas concentration C undergoes partial melting to a fractional degree f, the concentration C_L in the melt will be

$$C_L = \frac{C}{f + K_D(1 - f)} \tag{13.2}$$

where $K_D = C_s/C_L$ is the partition coefficient. (C_s is concentration in the solid residue.) If the rate of eruption of magma to the surface is p, and the magmatic gas is efficiently lost (to the atmosphere), the rate of loss is

$$dQ/dt = C_L p \tag{13.3}$$

Since the reservoir inventory is $Q = Cm$, where m is the reservoir mass, eq. (13.3) can be rewritten as first-order loss, eq. (13.1), with

$$\alpha = \frac{p/m}{f + K_D(1 - f)} \tag{13.4}$$

The major uncertainty in evaluation of α by eq. (13.4) is in estimating p, specifically the rate of volcanic eruption close enough to the surface to cause efficient loss of magmatic gases. Loss by vesiculation is probably negligible, and diffusion lengths following surface eruption are probably small. Most loss probably occurs by grain boundary diffusion and thence in cracks and fissures (cf. Section 9.5). For an order of magnitude guess, an appropriate estimate of p may be the crustal generation rate of the oceanic ridge system, some 3×10^{16} g/yr (10 km^3/yr). If the relevant reservoir is the upper mantle, with $m \approx 10^{27}$ g, then $p/m \approx 3 \times 10^{-11}$ yr^{-1}. Presumably K_D is less than unity, but perhaps not much less (Section 4.4), in which case the denominator in eq. (13.4) is near unity. If $K_D \ll 1$ the denominator approaches f; for representative $f \approx 20\%$, α will be about a factor of 5 higher than p/m. An order of magnitude estimate thus gives $\alpha \approx 10^{-10}$ yr^{-1}, in the midrange between fast and slow degassing; considering the uncertainties involved, both fast and slow degassing constants are plausible, and rates inferred from degassing models are indeed of this general order of magnitude. Volcanism associated with the postulated undepleted mantle (cf. Section 11.2) is probably at least an order of magnitude less than ridge volcanism, so that the same calculation suggests a degassing coefficient in the slow degassing range.

If volcanism is indeed the dominant mechanism of degassing from major interior reservoirs, and the depth from which efficient loss occurs is similar for different volatile species, the major control on elemental fractionation in degassing may be the K_D term in eq. (13.4). It seems unlikely that order of magnitude fractionation would result from this bias, and if $K_D \ll 1$ for all relevant species, fractionation would be minimal. Control of volatile degassing by bulk mass transport thus provides some justification for the view that degassing is governed by the same laws with similar transport coefficients for different elements, i.e. that volatiles are degassed together, without sensitive dependence on chemical identify. Eq. (13.4) contains no dependence on isotopic mass, so isotopic fractionation is likely to be minor, perhaps arising primarily in diffusion of gas into magma at depth or out of magma at the surface, but in any case the usual assumption that different isotopes of the same element are degassed at equal rates seems justified.

If the degassing coefficient is proportional to the rate of magma generation, it has probably been reasonably constant over recent geological time. The problem of variations over periods comparable to the age of the earth is obviously important, but does not appear close to resolution. In some cases the rate of magma generation, and thus transport coefficients, are assumed to scale with the rate of heat production, and thus exhibit substantial decline over the age of the earth; O'Nions *et al.* (1979), for example, assume transport coefficients decrease exponentially in time. Subsolidus convection models and geochronological studies of the rate of crustal formation are ambiguous on this point, however, and it might be argued that volcanic activity has been substantially uniform over most of the earth's history. An important systematic uncertainty is the possibility that the present plate tectonic regime may not characterize most of the earth's history and thus that in some prior style of heat transport the rate of volcanism, and probably the rate of continuous degassing, may have been substantially different from what it is now.

13.4   Model structures

As noted in Section 13.3, degassing models are generally constructed in terms of a framework of a small number of discrete reservoirs with material transfer between them governed by a first-order rate law. The models differ in number of reservoirs, the nature and direction of flow between them, and whether or not catastrophic degassing is superposed on continuous degassing. The choices reflect the generalized 'boundary conditions' of a model for the earth which establishes the structure of a degassing model.

One of the reservoirs is always the atmosphere. Typically, it is assumed that the atmosphere is conservative, i.e. nothing is lost from the atmosphere. This is not true for He (Chapter 12), of course, and has at least been challenged for Ne. For Xe there is the added problem that atmospheric Xe may not be adequately represented by air (Section 12.6), not in terms of inventory and perhaps not even in terms of composition.

The simplest possible model is one which employs only two reservoirs, the atmosphere and the solid earth (mantle). Many authors have explored such models, e.g. Damon & Kulp (1958b), Turekain (1959, 1964). Ozima & Kudo (1972), Ozima (1975), Tolstikhin (1975), Fanale (1976), Bernatowicz & Podosek (1978). The general features of a two-reservoir model are described in the following section.

In view of the major structural division of the solid earth into core, mantle and crust, a two-reservoir model is clearly inadequate for the geochemical evolution of the earth as a whole and must be presumed inadequate for volatiles as well. Still, the extent of this inadequacy is not yet determined, and it remains important to compare the results of two-reservoir models with those of more sophisticated models. An important problem, as emphasized by Bernatowicz & Podosek (1978), is that there are not enough reliable observational constraints to overconstrain even the simplest models. More sophisticated models are more realistic but also require assumption of more parameters, and are thus liable to the common danger of indicating parameter solutions which match constraints well but are physically implausible nonetheless.

Degassing models generally ignore the earth's core. Inventory considerations aside, the relevant feature is the stipulation that the core is neither source nor sink for mantle gases. This presumably reflects the intuitive assessment that molten Fe–Ni is unlikely to contain significant amounts of noble gases. This certainly seems plausible. There are, however, no experimental data for gas partitioning between metals and silicates. The metal phases of iron meteorites are essentially devoid of noble gases, but graphite and sulfides (as well as silicate inclusions) in iron meteorites do contain noble gases, so the possibility of non-trivial noble gases in the core is not necessarily absurd. It is also possible that ignoring the core is inappropriate if it contains significant K, as hypothesized by Goettel (1976), for example; if so, the core could be a source of ^{40}Ar and assessments based on comparison of atmospheric ^{40}Ar with mantle plus crustal K might be erroneous.

Three-reservoir models – mantle, crust, and atmosphere – for Ar evolution have been discussed by Schwartzman (1973a,b), Ozima (1975), and Hamano & Ozima (1978), among others. Details are presented in Section 13.6. The principal qualitative difference between two- and three-reservoir models is that in the latter class not only the gases but also other elements are transported as well: for Ar, for example, one must consider not only degassing of Ar from mantle and crust but also the transport of K between mantle and crust. In the models cited above the transport of K is assumed to be unidirectional: from mantle to crust.

The question just raised, that of whether material transport between mantle and crust is just one way or whether there is a significant return flow from the crust back into the mantle, is controversial. Arguments for one-way flow, based on Sr, Pb, and Nd isotopic systematics, have been presented by Hurley *et al.* (1962), Moorbath (1975), and Carlson *et al.* (1978), among others. Contrary arguments that return flow is or should be important have been presented by Armstrong (1968), Russell (1972), Russell & Birnie (1974), O'Nions *et al.* (1979), and Veizer & Jansen (1979), again among others.

If return flow does occur the mechanism is presumably tectonic subduction at convergent plate boundaries. Subduction of oceanic crust – the volcanic rocks of immediate mantle derivation – would not constitute return flow in the sense used here, since it would contribute no net flux. Return flow requires subduction of sediments and/or water (or continental crust), and the sediments may not be subducted, instead being scraped up into accretional wedges on the over-riding plate. Evidence is inconclusive. Some major subduction zones (e.g. the Japan, Chile, and Aleutian trenches) lack major accretional wedges, suggesting sediment subduction, but it can be argued by Pb compositions that sediments are an insignificant contribution to the island-arc volcanics (Meijer, 1976; Sinha & Hart, 1972). Conversely, high ^{87}Sr/^{86}Sr values in arcs such as the Scotia, Banda, and Chile systems (Hawksworth *et al.*, 1977; Magaritz *et al.*, 1978; Francis *et al.*, 1977) do suggest sedimentary contribution to arc volcanism, but it could also be argued that this reflects only crustal contamination of ascending magma, not the magma source. Ambiguity about the importance of tectonic recycling is also augmented by uncertainty about how long plate tectonics has been dominant.

The question of tectonic return flow to the mantle must be addressed separately for the noble gases themselves and for the parents of radiogenic species. Since the principal reservoir for the gases is air, it is generally considered that return flow is negligible for gases (see Section 12.4) even if not for other elements, but even this question cannot be considered definitely resolved.

Return flow to the mantle should be kept in mind as a possible major shortcoming in models for noble gas evolution. This feature has not played a prominent role in degassing models so far. Aside from the ambiguity about whether it occurs at all, it seems very difficult to guess at a quantitative evaluation of the rate of return flow or to find constraints by which it could be inferred from the model.

Probably the most serious deficiency in quantitative degassing models, and one which must always be expressed as a qualification and reservation, is the treatment of the mantle as a single reservoir. As described in Section 11.2, it is frequently asserted that the mantle has two major structural components with quite different geochemical characteristics, a primitive undepleted mantle and a depleted mantle. It is correspondingly argued that the noble gas state of the two types of mantle is quite different as well, particularly in terms of Ar and He isotopic compositions (Sections 11.3 and 11.4). If the dichotomy is valid, and the qualitative noble gas characterizations valid as well, application of simple degassing models (Section 13.5) to each type separately leads to quite different conclusions: a rapidly and extensively degassed depleted mantle, a slowly degassing and still only partially degassed primitive mantle. This in turn inspires the view that part of the mantle was indeed rapidly and thoroughly degassed, and has been primarily responsible for the generation of the atmosphere, but that a substantial or major fraction of the earth's total volatile inventory remains in undepleted mantle (cf. Bernatowicz & Podosek, 1978; Hart *et al.*, 1979), leaking out slowly even today. Quite similar views are expressed on the basis of other isotopic systems for crustal accumulation, of course.

Formally, the mathematics of three-reservoir degassing can be applied to a two-mantle model as well as a mantle–crust model. In practice, there has been little development along these lines. The noble gas characteristics of undepleted mantle are still a matter of contention, and it is not clear whether there are any data which do not include major depleted mantle contributions. Similarly, it would be very difficult to make independent quantitative estimates of the transport coefficients which were not equivalent to assuming a solution to the problem. The undepleted/depleted mantle model thus remains a probably very important factor in noble gas geochemistry which has yet to receive adequate quantitative treatment.

13.5   Two-reservoir degassing

In this section we describe the simplest possible degassing model, employing only two reservoirs: the atmosphere, which is assumed to be conservative (no losses) and the solid earth (the mantle). Degassing from the mantle to the atmosphere is assumed to be first order. While such a model is clearly inadequate, it possesses many of the basic qualitative features of more elaborate models and illustrates them well. Aside from simplicity, the principal advantage of this model is the minimal number of parameters and assumptions needed. The discussion below follows that of Bernatowicz & Podosek (1978).

The model can be framed in terms of a generalized stable primordial isotope S and radiogenic daughter isotope D. The daughter D is produced by decay of a radioactive parent P with decay rate λ and yield y (e.g. for $D = {}^{40}Ar$ and $P = {}^{40}K$, the yield $y = 0.105$). Subscripts a and m designate the atmosphere and mantle reservoirs, respectively. Initial abundances S_0, D_0 and P_0 are assumed to reside exclusively in the mantle, i.e. at time $t = 0$ there is no atmosphere.

The governing equations are

$$\frac{d}{dt} S_a = \alpha S_m \qquad \frac{d}{dt} S_m = -\alpha S_m \qquad (13.5)$$

$$\frac{d}{dt} D_a = \alpha D_m \qquad \frac{d}{dt} D_m = -\alpha D_m + \lambda y P_0 e^{-\lambda t} \qquad (13.6)$$

where the degassing coefficient α is the same for both S and D and is assumed constant.

For the stated initial conditions the solutions to eqs. (13.5) and (13.6) are:

$$S_m = S_0 e^{-\alpha t} \qquad (13.7)$$

$$S_a = S_0(1 - e^{-\alpha t}) \qquad (13.8)$$

$$D_m = D_0 e^{-\alpha t} + \frac{\lambda}{\alpha - \lambda} (e^{-\lambda t} - e^{-\alpha t}) y P_0 \qquad (13.9)$$

$$D_a = D_0(1 - e^{-\alpha t}) + \left[\frac{\alpha}{\alpha - \lambda} (1 - e^{-\lambda t}) - \frac{\lambda}{\alpha - \lambda} (1 - e^{-\alpha t}) \right] y P_0 \qquad (13.10)$$

The behavior of S and D are uncoupled, other than by common α. It is, however, often convenient to restate these solutions in terms of isotopic ratios:

$$(D/S)_m = (D/S)_0 + \frac{\lambda}{\alpha - \lambda} (e^{(\alpha - \lambda)t} - 1) y P_0 / S_0 \qquad (13.11)$$

$$(D/S)_a = (D/S)_0 + \left[\frac{\alpha}{\alpha - \lambda} \frac{1 - e^{-\lambda t}}{1 - e^{-\alpha t}} - \frac{\lambda}{\alpha - \lambda} \right] y P_0 / S_0 \qquad (13.12)$$

The advantage, of course, is that it may be possible to characterize a reservoir better by an isotopic (or elemental) ratio than by an absolute abundance.

In the limiting case of fast degassing ($\alpha T \gg 1$) only a small fraction, $e^{-\alpha t} \ll 1$, of the original volatile inventory remains in the mantle and $S_m/S_a \ll 1$. The atmosphere forms quickly, with a characteristic degassing time $1/\alpha \ll T$. Most of the radiogenic daughter ever produced is also in the atmosphere: if $\alpha \gg \lambda$, only the daughter produced in the last degassing time $1/\alpha$ remains in the mantle; if $\lambda \gg \alpha$, the parent is now extinct and the daughter resembles a primordial species in that only the very small fraction $e^{-\alpha T}$ of the total production remains in the mantle. Since most of the radiogenic daughter is already in the atmosphere, the radiogenic enhancement of D/S in the atmosphere is close to its maximum limit, yP_0/S_0 if $\lambda T \gg 1$ or $(1-e^{-\lambda T})yP_0/S_0$ if not. If $\alpha \gg \lambda$, the mantle composition is much more radiogenic than atmospheric composition; if, however, $\lambda \gg \alpha$, the daughter again acts like a primordial isotope and mantle and atmospheric compositions are nearly identical.

In the opposite limiting case of slow degassing, $\alpha T \ll 1$, most of both the primordial and the radiogenic species are still in the mantle. Only a small fraction $\alpha T \ll 1$ of the primordial species is in the atmosphere, and the growth of the atmosphere is linear in time. Mantle and atmospheric compositions will be nearly identical unless $\lambda T \ll 1$ as well.

For all practical purposes this is a two-parameter model. One of the parameters is the degassing constant α; the other depends on the system considered. For the primordial isotope S, eqs. (13.7)–(13.8), the second parameter is the initial abundance S_0. For the radiogenic isotope D, eqs. (13.9)–(13.10), the second parameter is the parent abundance P_0, with the usual auxiliary assumption that D_0 is either negligible or can be inferred from S_0. For the ratio system, eqs. (13.11)–(13.12), the second parameter is P_0/S_0, with the equivalent auxiliary assumption that $(D/S)_0$ is known.

This two-parameter model needs two constraints to define it completely. One constraint is readily at hand: the present atmospheric abundances S_a, D_a, or their ratio. This leaves a system with one degree of freedom. Fig. 13.1 illustrates relationships among important variables in the ^{36}Ar–^{40}Ar–^{40}K system, and atmospheric evolution curves for Ar are illustrated in Fig. 13.2 and for Xe in Figs. 13.6–13.7. Various approaches to providing the second constraint are described below. Ideally, the goal is to find observational data which will overconstrain the system, and so provide a means of testing the validity of the model; unfortunately, this goal remains quite remote.

One approach to providing the second constraint is through S_0, the total inventory of the primordial species. As described in Sections 5.2–5.3, regularities in noble gas abundance patterns in solar system materials suggest the possibility of predicting elemental compositions for the total noble gas inventory but not of predicting absolute abundances with any useful precision. An alternative route is possible through comparison of mantle and atmospheric abundances. If, as suggested in Sections 9.2 and 11.8, the mantle inventory is small compared

Fig. 13.1. Illustration of parameter relationships in two-reservoir degassing model for Ar. The model has two free parameters (see text); constraining the model to account for the present atmosphere leaves one degree of freedom, with all relevant parameters functionally related as shown here. Slow degassing, $\alpha < 10^{-10}$ yr^{-1} corresponds to high K abundance (chondritic abundance is indicated at 880 ppm) and low $^{40}Ar/^{36}Ar$ in the mantle. The initial $^{40}K/^{36}Ar$ ratio varies only narrowly. The model variant with extensive early catastrophic degassing followed by continuous degassing does not affect the relation between K and α (panel a), predicts very high mantle $^{40}Ar/^{36}Ar$ for any α (cf. panel b), and changes the relation between initial $^{40}K/^{36}Ar$ and α to that shown by the dashed line in panel c. Reproduced from Bernatowicz & Podosek (1978).

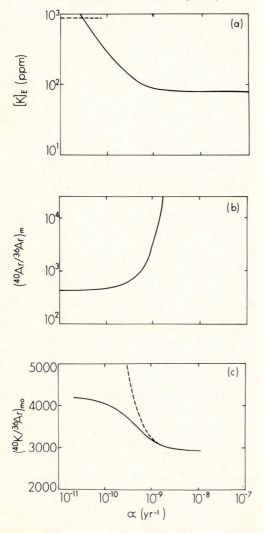

to the atmospheric inventory, i.e. if $S_m/S_a \ll 1$ at present, then degassing is rapid. As an example, $S_m/S_a = 0.1$ at present corresponds to $\alpha T = 2.4$ and generation of the atmosphere on a time scale of $1/\alpha = 1.9\,\text{Ga}$. Even this tentative inference must be highly qualified, however, given the substantial difficulties involved in trying to estimate any sort of representative mantle concentration from the igneous rock data.

The same approach through the radiogenic isotope amounts to a prediction of P_0. Atmospheric ^{40}Ar, for example, requires a minimum $K = 77$ ppm for the whole earth. Postulated values close to this correspond to extensive and therefore rapid outgassing, while values much higher correspond to little and therefore slow outgassing. A noteworthy early example of this approach is that of Turekian (1959), who assumed chondritic K abundance and therefore concluded slow degassing; $K = 880$ ppm, a representative chondritic value, for currently accepted geochronological parameters corresponds to $\alpha T = 0.13$ and only a small fraction, $1 - e^{-\alpha T} = 12\%$, of the earth's volatiles yet in the atmosphere. The Morgan & Anders (1980) value $K = 135$ ppm corresponds to fairly rapid degassing: $\alpha T = 1.3$ and a large fraction $1 - e^{-\alpha T} = 73\%$ of the earth's volatiles in the atmosphere. As with the primordial isotope, the same approach can also be effected by comparison of D_m and D_a. If, for example, a representative whole mantle ^{40}Ar concentration is $10^{-6}\,\text{cm}^3\,\text{STP/g}$, then $D_m/D_a = 0.11$ and we obtain fast degassing with $\alpha T = 4.4$ and a small residuum, $e^{-\alpha T} = 1.2\%$, of undegassed primordial volatiles.

A second approach consists of examining model predictions for the evolution of atmospheric isotopic composition (Figs. 13.2, 13.6 and 13.7) and trying to constrain the model by measurement of atmospheric composition at suitable times. The Xe compositions illustrated in Figs. 13.6 and 13.7, for example, clearly indicate, if they are real, a very rapid degassing. The methodology, that of examining trapped gases in old sedimentary rocks (Section 8.5), has met with rather little success so far, however. The atmospheric $^{40}\text{Ar}/^{36}\text{Ar}$ ratio should show large changes, but the experimental problem of distinguishing trapped Ar from *in situ* ^{40}Ar is formidable. The Xe isotopic measurements do not suffer from this problem, and the implications of the data illustrated in Figs. 13.6 and 13.7 are clear except for the ambiguity that since the compositions are indistinguishable from modern air they may in fact be nothing but modern air (Section 8.5). Actually, as pointed out by Alexander (1975), the range of isotopic evolution trajectories accessible by the simple first-order two-reservoir model is rather narrow and, in some regions, mutliple-valued. Thus, even a reliable paleoatmospheric composition would not necessarily force a definitive choice between fast and slow degassing even within the confines of the model. Such a measurement would, however, be very valuable in suggesting whether the simple model is even feasible according to whether it fell in the narrow field or outside it, in the region accessible only to more generalized models.

A third approach, the importance of which was first emphasized by Ozima & Kudo (1972), is measurement of present mantle composition $(D/S)_m$. High values, $(D/S)_m \gg (D/S)_a$, correspond to rapid and extensive outgassing; conversely, low values, $(D/S)_m \approx (D/S)_a$, correspond to slow degassing (unless $\lambda \gg \alpha$). The constraints for ^{36}Ar-^{40}Ar-^{40}K are illustrated in Fig. 13.1. The limit $\alpha T \to 0$ corresponds to $(^{40}Ar/^{36}Ar)_m = 428$, the minimum value possible in this model, and mantle compositions near this figure correspond to slow degassing. A high value such as $(^{40}Ar/^{36}Ar)_m = 10^4$ yields $\alpha T = 6.3$ and a very low residuum of undegassed primordial volatiles, $e^{-\alpha T} = 0.18\%$.

Fig. 13.2. Atmospheric composition evolutions for Ar possible in two-reservoir degassing. Trajectories for limiting cases of fast and slow degassing are indicated. Ar composition at any given time is not single-valued in α: the minimum value occurs for α intermediate between fast and slow extremes. The variant including extensive early catastrophic degassing gives the same evolution for $\alpha \gg \lambda$ but slow continuous degassing broadens the range of possible trajectories as indicated by dashed line. Open circle at upper right is paleoatmospheric composition reported by Cadogan (1977), a datum whose validity we consider questionable. Reproduced from Bernatowicz & Podosek (1978).

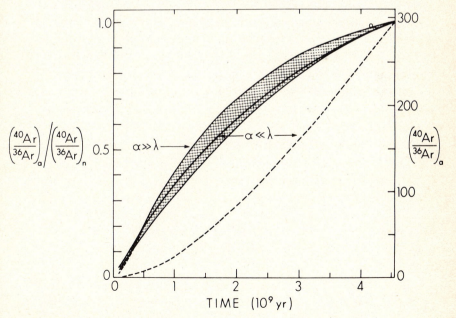

The observations of highly radiogenic 'excess' ^{40}Ar in many volcanic rocks (Section 9.4), interpreted as samples of 'mantle' Ar, indeed indicate (^{40}Ar/^{36}Ar)$_m$ > 10^4 (cf. Section 11.3 and Figs. 11.1, 11.2, and 11.8). This observation is the clearest and most unambiguous constraint on degassing models now available: the inference is, as noted above, quite rapid degassing. An approximate mantle ^{40}Ar abundance of some 10^{-6} cm^3 STP/g is also reasonably consistent with this level of degassing (see above).

It is, however, not clear that so high a degree of degassing as is inferred by (^{40}Ar/^{36}Ar)$_m$ > 10^4 is consistent with the absolute abundance levels of primordial species found in igneous rocks, which are characteristically higher than the nominal figure of an 0.18% undegassed residuum (cf. Fig. 9.1). This may reflect only atmospheric contamination (Section 9.5), or partitioning such that gases in magma are significantly higher than gases in the source rocks. It is suggestive of real information on degassing history, however. One possibility is different values of the degassing coefficient α for different gases. An especially interesting possibility is that this is a manifestation of the breakdown of the two-reservoir model and an indication that some more complicated model is needed. An obvious suggestion is slow leakage of gas from undepleted mantle into depleted mantle (cf. Sections 11.2 and 13.4).

In any case, the value of a model constraint through (^{40}Ar/^{36}Ar)$_m$ clearly illustrates the suggested dichotomy between depleted and undepleted mantle. The same reasoning applied to low, nearly atmospheric ^{40}Ar/^{36}Ar ratios believed measured in some mantle samples (Section 11.3) leads to just the opposite conclusion, namely slow and only partial degassing from the relevant reservoir.

The two-reservoir model is easily extended to include early catastrophic degassing. 'Early' here means at a time t_d such that $\lambda t_d \ll 1$ for whatever radioactive parent is concerned. For ^{40}Ar and ^{4}He, catastrophic degassing at any $t_d \lesssim 10^8$ yr would be considered early. This is not necessarily true for ^{129}Xe and ^{136}Xe, however, which have shorter-lived parents which would require catastrophic degassing within about 10^6–10^7 yr after formation of the earth to be considered early. If a catastrophic event causes degassing of a fraction f of primordial volatiles, then eqs. (13.7) and (13.8) are modified to

$$S_m = S_0(1-f)e^{-\alpha t} \tag{13.13}$$

$$S_a = S_0[1-(1-f)e^{-\alpha t}] \tag{13.14}$$

and, as long as the degassing is early, eqs. (13.9) and (13.10) become

$$D_m = D_0(1-f)e^{-\alpha t} + \frac{\lambda}{\alpha-\lambda}(e^{-\lambda t} - e^{-\alpha t})yP_0 \tag{13.15}$$

$$D_a = D_0(1-(1-f)e^{-\alpha t}) + \left[\frac{\alpha}{\alpha-\lambda}(1-e^{-\lambda t}) - \frac{\lambda}{\alpha-\lambda}(1-e^{-\alpha t})\right]yP_0 \tag{13.16}$$

Ratios are then given by

$$(D/S)_m = (D/S)_0 + \frac{\lambda}{\alpha - \lambda} (e^{(\alpha-\lambda)t} - 1)yP_0/S_0(1 - f) \qquad (13.17)$$

$$(D/S)_a = (D/S)_0 + \left[\frac{\alpha}{\alpha - \lambda} \frac{1 - e^{-\lambda t}}{1 - (1 - f)e^{-\alpha t}} \right.$$

$$\left. - \frac{\lambda}{\alpha - \lambda} \frac{1 - e^{-\alpha t}}{1 - (1 - f)e^{-\alpha t}} \right] yP_0/S_0 \qquad (13.18)$$

If subsequent continuous degassing is fast, degassing is nearly catastrophic anyway, and formal stipulation of early catastrophic degassing does not make very much difference. The case is thus principally interesting when subsequent continuous degassing is slow. In this case the principal effects of the early cata-strophic event are instantaneous generation of most of the atmosphere, decrease in S_m (to zero in the extreme $f = 1$), corresponding increase by the same factor in the radiogenic enhancement of $(D/S)_m$ (to infinity in the extreme $f = 1$), and a slower growth of $(D/S)_a$ (illustrated in Figs. 13.2, 13.6, and 13.7 for $f = 1$).

13.6   Argon

The basic qualitative features of Ar evolution in the simplest two-reservoir model, and plausible observational constraints, have been described in the previous section. Since most of the more sophisticated models have been developed for application to Ar, we will here describe the features of such general models. The discussion below follows that of Hamano & Ozima (1978).

The model considers three reservoirs, mantle, crust, and atmosphere, desig-nated by subscripts m, c, and a, respectively. Material transport is assumed to follow the first-order rate law. Mantle evolution is then governed by

$$\frac{d}{dt} {}^{40}K_m = -\lambda {}^{40}K_m - \gamma {}^{40}K_m \qquad (13.19)$$

$$\frac{d}{dt} {}^{40}Ar_m = y\lambda {}^{40}K_m - \alpha {}^{40}Ar_m \qquad (13.20)$$

$$\frac{d}{dt} {}^{36}Ar_m = -\alpha {}^{36}Ar_m \qquad (13.21)$$

The mantle is thus assumed to lose Ar to the atmosphere (not to the crust) with degassing coefficient α, equivalent to the α in the two-reservoir model; it also loses K to the crust with transport coefficient γ. Transport is assumed undirec-tional, with no return to the mantle (cf. Section 13.4).

Crustal evolution is governed by

$$\frac{d}{dt} {}^{40}K_c = -\lambda {}^{40}K_c + \gamma {}^{40}K_m \qquad (13.22)$$

$$\frac{d}{dt}\,{}^{40}\text{Ar}_c = y\lambda {}^{40}\text{K}_c - \beta {}^{40}\text{Ar}_c \tag{13.23}$$

The crust is thus assumed to contain no primordial gases, to accumulate K from the mantle, and to lose radiogenic ${}^{40}\text{Ar}$ to the atmosphere with degassing coefficient β.

Atmospheric evolution is then governed by

$$\frac{d}{dt}\,{}^{40}\text{Ar}_a = \alpha {}^{40}\text{Ar}_m + \beta {}^{40}\text{Ar}_c \tag{13.24}$$

$$\frac{d}{dt}\,{}^{36}\text{Ar}_a = \alpha {}^{36}\text{Ar}_m \tag{13.25}$$

Assumed boundary conditions are that initial abundances ${}^{40}\text{K}_0$ and ${}^{36}\text{Ar}_0$ are present entirely in the mantle (i.e. no crust, no atmosphere at $t = 0$) and that ${}^{40}\text{Ar}_0 = 0$. Hamano & Ozima (1978) consider the specific model that for $0 \leqslant t < t_d$, there is no transport of either K or Ar, so that eqs. (13.19)–(13.25) apply for $t_d < t \leqslant T$. They further allow that at time $t = t_d$ there was a catastrophic degassing event which removed a fraction f of mantle gases into the atmosphere (but no K to the crust), after which continuous degassing and crustal formation begin. Catastrophic degassing at t_d is not necessarily early; indeed, the allowance for catastrophic degassing of some radiogenic ${}^{40}\text{Ar}$ is what permits a possible constraint on t_d.

With these conditions the solution to the system (13.13)–(13.25) is

$$ {}^{40}\text{K}_m / {}^{40}\text{K}_0 = e^{-\lambda t_d} e^{-(\lambda + \gamma)(t - t_d)} \tag{13.26}$$

Fig. 13.3. Schematic illustration of three-reservoir (mantle, crust, and atmosphere) degassing model described in text. Reproduced from Hamano & Ozima (1978).

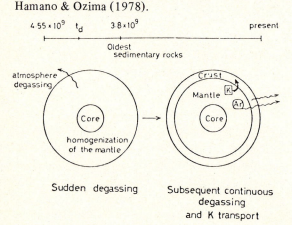

$$^{40}\mathrm{Ar_m}/y^{40}\mathrm{K_0} = (1-f)(1-e^{-\lambda t_\mathrm{d}})e^{-\alpha(t-t_\mathrm{d})}$$

$$+ \frac{\lambda}{\lambda+\gamma-\alpha}\, e^{-\lambda t_\mathrm{d}}e^{-\alpha(t-t_\mathrm{d})}(1-e^{-(\lambda+\gamma-\alpha)(t-t_\mathrm{d})}) \quad (13.27)$$

$$^{36}\mathrm{Ar_m}/^{36}\mathrm{Ar_0} = (1-f)e^{-\alpha(t-t_\mathrm{d})} \quad (13.28)$$

$$^{40}\mathrm{K_c}/^{40}\mathrm{K_0} = e^{-\lambda t}(1-e^{-\gamma(t-t_\mathrm{d})}) \quad (13.29)$$

$$^{40}\mathrm{Ar_c}/y^{40}\mathrm{K_0} = \frac{\lambda}{\beta-\lambda}\, e^{-\lambda t}(1-e^{-(\beta-\lambda)(t-t_\mathrm{d})})$$

$$+ \frac{\lambda}{\beta-\lambda-\gamma}e^{-\lambda t_\mathrm{d}}e^{-\beta(t-t_\mathrm{d})}(1-e^{-(\lambda+\gamma-\beta)(t-t_\mathrm{d})}) \quad (13.30)$$

and, for the atmosphere,

$$^{40}\mathrm{Ar_a}/y^{40}\mathrm{K_0} = (1-e^{-\lambda t_\mathrm{d}})[1-(1-f)e^{-\alpha(t-t_\mathrm{d})}]$$

$$+ \frac{\lambda(\beta-\alpha)}{(\alpha-\lambda-\gamma)(\beta-\lambda-\gamma)}\, e^{-\lambda t_\mathrm{d}}(1-e^{-(\lambda+\gamma)(t-t_\mathrm{d})})$$

$$- \frac{\lambda}{\alpha-\lambda-\gamma}e^{-\lambda t_\mathrm{d}}(1-e^{-\alpha(t-t_\mathrm{d})}) + \frac{\beta}{\beta-\lambda}\,(e^{-\lambda t_\mathrm{d}}-e^{-\lambda t})$$

$$+ \frac{\lambda\gamma}{(\beta-\lambda)(\beta-\lambda-\gamma)}\, e^{-\lambda t_\mathrm{d}}(1-e^{-\beta(t-t_\mathrm{d})}) \quad (13.31)$$

$$^{36}\mathrm{Ar_a}/^{36}\mathrm{Ar_0} = 1-(1-f)e^{-\alpha(t-t_\mathrm{d})} \quad (13.32)$$

This model reduces to the two-reservoir model for $\gamma=0$ (no crust). With the further simplification $t_\mathrm{d}=0$ it reduces to eqs. (13.13)–(13.16) and simplifying still further by setting $f=0$ yields eqs. (13.7)–(13.10).

Development of this model illustrates the rapid growth in number of parameters attendant on even modest improvement of generality. There are three transport coefficients (α, β, γ), two parameters associated with the catastrophic degassing (f, t_d), and two initial abundances ($^{40}\mathrm{K_0}$ and $^{36}\mathrm{Ar_0}$; the third, $^{40}\mathrm{Ar_0}$, was already assumed to be nil), or seven unknown parameters in all. Equations (13.26)–(13.32) specify seven present-day and thus potentially observable quantities, just enough to completely define the model. Additional constraints potentially providing enough data to overconstrain the model, could, in principle, be obtained from historical information, i.e. measuring any of the quantities at any time significantly different from the present; in practice, this is not now possible. Two firm constraints are available in terms of known present atmospheric abundances, $^{40}\mathrm{Ar_a}$ and $^{36}\mathrm{Ar_a}$, so the model has five free parameters. As with the two-reservoir model, the formalism can be cast in terms of isotopic composition, i.e. $^{40}\mathrm{Ar}/^{36}\mathrm{Ar}$ instead of $^{40}\mathrm{Ar}$ and $^{36}\mathrm{Ar}$ separately, and it is indeed convenient to do so. This reduces by one the number of parameters, $(^{40}K/^{36}Ar)_0$ instead of $^{40}K_0$ and $^{36}Ar_0$, but also the number of constraints, $(^{40}Ar/^{36}Ar)_a$ instead of $^{40}Ar_a$ and $^{36}Ar_a$, so the number of unknown parameters remains five.

Hamano & Ozima assumed a further constraint by taking $K_c = 3.86 \times 10^{23}$ g, from an average crustal K content of 1.91% (Holland & Lambert, 1972) and mass 2.02×10^{25} g; this crustal K contributes 65 ppm to the global inventory. They also assumed one more constraint by interpreting average K-Ar ages shorter than Rb-Sr ages for crustal rocks in terms of first-order loss of radiogenic ^{40}Ar, approximating thereby the crustal degassing coefficient β at 3.71×10^{-10} yr^{-1} ($\beta T = 1.69$). Since atmospheric ^{40}Ar could be produced by only 77 ppm K, and this value of β indicates extensive degassing of the crust, it can be seen that for these values a significant fraction of atmospheric ^{40}Ar enters the atmosphere through the crust rather than directly from the mantle. The importance of including the crust in degassing models is evident.

Fig. 13.4. Parameter relations in mantle–crust–atmosphere model (cf. Fig. 13.3). With constraints described in text, this model has three free parameters, taken to be α, f and t_d. Setting any model quantity constant constraints α, f, and t_d to a surface; intersections in the f-t_d plane at three values of α are shown for various values of mantle K content, mantle $^{40}Ar/^{36}Ar$ ratio, and for $\alpha = \gamma$. Shaded areas are solutions permitted for 100 ppm $\leqslant K_m \leqslant$ 400 ppm, for $(^{40}Ar/^{36}Ar)_m > 5000$, and for $\alpha > \gamma$. Reproduced from Hamano & Ozima (1978).

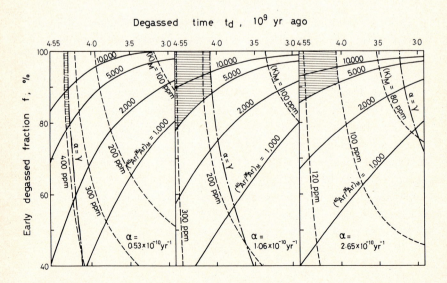

With these two additional constraints, the number of degrees of freedom is
reduced to three. Hamano & Ozima took the approach of considering α, f, and
t_d to be independent parameters. The locus obtained by setting any other model
quantity to be constant is then a surface in α-f-t_d space. Relevant loci are
illustrated in Fig. 13.4 as intersections of these surfaces with the f-t_d plane
(α = constant) at three values of α, corresponding to $\alpha T = 0.24$ (slow degassing),
$\alpha T = 0.48$ (transition), and $\alpha T = 1.21$ (approaching rapid degassing).

Hamano & Ozima then imposed further constraints in the form of inequalities.
They assumed that (i) $(^{40}Ar/^{36}Ar)_m > 5000$; (ii) mantle K concentration lies
between 100 ppm and 400 ppm; (iii) $\alpha > \gamma$. Parameter choices satisfying all three
inequalities are shown as shaded areas in Fig. 13.4. The field of possible solutions
satisfying these inequalities for any value of α is indicated in Fig. 13.5. This
analysis does not provide a strong constraint on α but it does lead to a definite
conclusion about catastrophic degassing; there indeed was a catastrophic degassing
and it was both extensive ($f > 77\%$) and relatively early ($t_d < 0.9$ Ga).

The requirement $\alpha > \gamma$ reflects the assumption that Ar is more mobile than
K. This certainly seems reasonable but is not well-founded experimentally and is
conceivably incorrect, e.g. if Ar partitions into a melt less strongly than K. The
requirement $\alpha > \gamma$ is not critical to the basic conclusion stated above, however:
if it is lifted, extensive catastrophic degassing is still required ($f > 70\%$), although
the limit on its timing is widened considerably ($t_d < 2.0$ Ga).

Any model, even one with the correct structure, is of course only as good as
its assumptions. The key condition here, as in the two-reservoir model, is the

Fig. 13.5. Synopsis of total field of permitted solutions as illustrated
in Fig. 13.4 for all values of α. Reproduced from Hamano & Ozima
(1978).

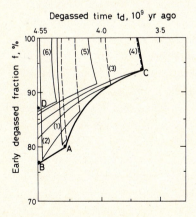

high $(^{40}Ar/^{36}Ar)_m$. In any model, $(^{40}Ar/^{36}Ar)_m \gg (^{40}Ar/^{40}Ar)_a$ implies extensive degassing. In the two-reservoir model this can be accommodated by either (or both) $f \approx 1$ or $\alpha T \gg 1$. In the present model the constraint on K abundances requires extensive catastrophic degassing; specifically, according to Hamano & Ozima, $(^{40}Ar/^{36}Ar)_m > 5000$ and $K_m > 100$ ppm eliminate any solution for $f = 0$ regardless of the value of α. Conversely, for $(^{40}Ar/^{36}Ar)_m < 500$, only $f \approx 0$ solutions are possible. We must thus repeat the ubiquitous qualification that if there exists an undepleted mantle reservoir, uncoupled from the depleted mantle considered above, and its $^{40}Ar/^{36}Ar$ ratio is indeed low, these same considerations lead to the conclusion that it has not been extensively degassed, either catastrophically or continuously.

13.7 Helium

Formally, a model for He degassing and evolution can be set up in the same way as an Ar model. There are a few differences, however. While ^{36}Ar is pure primordial and ^{40}Ar pure radiogenic, the primordial isotope ^3He is also produced in the radiogenic component, and while the radiogenic contribution to ^4He is dominant, the primordial contribution is not always negligible (Section 6.6). Also, ^4He has three important parents, ^{238}U, ^{235}U, and ^{232}Th, rather than just one. Radiogenic ^3He is evidently negligible in the mantle, however, and primordial ^4He in the mantle can be estimated reasonably well on the basis of meteorite analogy. Similarly, the importance of three parents does not really introduce any new degrees of freedom, since ^{238}U and ^{235}U are the same element and a Th/U ratio can apparently be assumed with reasonable reliability. These features thus complicate the mathematics but do not really change the nature of the problem.

Exospheric loss of He does change the nature of the problem. Since the atmosphere does not integrate the release of He from the solid earth, the vital boundary condition that the modern atmosphere provides to Ar degassing models is not available to He degassing models. Also, the constraints which potentially could have been provided by historical information, i.e. atmospheric $^3He/^4He$ in the past, are inapplicable. There are compensating advantages, however. Since atmospheric He concentration is very much lower than it would be without exospheric loss, contamination of juvenile He in volcanic rocks with atmospheric He is not much of a problem, and it is thus possible to access mantle He composition with much less ambiguity than in the case of Ar. For the same reason, atmospheric He dissolved in seawater is low enough that juvenile He is detectable and it is thus possible to estimate the flux of juvenile He from the mantle. This feature, unique to He, can be applied to provide a model constraint in place of the unavailable present atmosphere constraint.

We will first consider application of the simple two-reservoir model for a mantle–atmosphere system. Equation (13.7) becomes

$$^3He_m = {}^3He_0 e^{-\alpha T} \tag{13.33}$$

and eq. (13.9) generalizes to

$$^4He_m = {}^4He_0 e^{-\alpha t} + \sum \frac{\lambda_i}{\alpha - \lambda_i} (e^{-\lambda_i t} - e^{-\alpha t}) y_i P_{0i} \tag{13.34}$$

The summation is over three species, ^{238}U, ^{235}U, and ^{232}Th. Combining,

$$(^4He/^3He)_m = (^4He/^3He)_0 + \sum \frac{\lambda_i}{\alpha - \lambda_i} (e^{(\alpha-\lambda_i)t} - 1) y_i P_{0i}/^3He_0 \tag{13.35}$$

One constraint on the problem can be applied through eq. (13.35). We take $(^4He/^3He)_m = (1.2 \times 10^{-5})^{-1}$ (cf. Section 11.4). Also, we assume Helium A composition $(^4He/^3He)_0 = (1.43 \times 10^{-4})^{-1}$ (cf. Section 5.4); alternative possibilities of a lower $(^4He/^3He)_0$ ratio, either Helium B or a slight spallation contribution (Section 6.3) should not be ignored but would not make much of a difference.

A second constraint can be obtained by differentiating eq. (13.33):

$$-\frac{d}{dt} {}^3He_m = \alpha^3 He_0 e^{-\alpha t} \tag{13.36}$$

For this we will take $d^3He_m/dt = 0.108$ atoms g^{-1} yr^{-1}, which corresponds to the Craig *et al.* (1975) value of 4 atoms cm^{-2} sec^{-1} for juvenile ^3He release into the oceans; the mass normalization is for the whole earth.

Atmospheric accumulation provides two constraints on the three-parameter abundance system, eqs. (13.7)–(13.10), or, equivalently, one constraint on the two-parameter ratio system, eqs. (13.11)–(13.12). Without atmospheric accumulation the latter analysis cannot be applied to He, so we still need one more constraint.

For a third constraint we may assume the P_{0i}. We will take present $U = 14$ ppb (Table 12.3) and present $Th/U = 3.3$ by weight, both values referring to the earth as a whole. With this assumption, and those noted above, solution of eqs. (13.35)–(13.36) for α yields $\alpha = 1.37 \times 10^{-11}$ yr^{-1}, rather slow degassing $(\alpha T = 0.062)$ for which only $1 - e^{-\alpha T} = 6.0\%$ of the primordial species is yet degassed. This solution also corresponds to an initial endowment $^3He_0 = 3.1 \times 10^{-10}$ cm^3 STP/g. The 6.0% which, by this model, has already been degassed, and would be in the atmosphere now except for loss, corresponds to $(^3He/^{20}Ne)_a = 0.0017$, rather low in comparison with the planetary (Tables 5.3 and 5.4) value $^3He/^{20}Ne = 0.032$. This solution also yields present $^4He_m = 2.4 \times 10^{-5}$ cm^3STP/g, at least an order of magnitude higher than the concentration in most volcanic rocks. Increasing the radiogenic enhancement in $(^4He/^3He)_m$ by a factor of two, increasing the juvenile flux of ^{3}He by a factor of two, or decreasing the U content by a factor of two would all have the same effect on α, yielding $\alpha = 2.85 \times 10^{-11}\,\text{yr}^{-1}$ ($\alpha T = 0.13$), $1 - e^{-\alpha T} = 12\%$. Early catastrophic degassing would imply even slower continuous degassing. These results are not consistent with expectations for primordial ^{3}He/^{20}Ne, with absolute He concentrations observed in igneous rocks, and in particular not consistent with the conclusions reached by application of the same model to Ar.

In principle, the sense of model conclusions noted above could be changed by suitable adjustments in the values of the constraints. It seems more likely, however, that there is a significant problem in the model structure. If nonthermal mechanisms dominate exospheric loss, the atmospheric ^{4}He lifetime may not be much longer than the ^{3}He lifetime. Since nongeological sources are much more important for ^{3}He than for ^{4}He (Table 12.2), we might then expect that atmospheric ^{3}He/^{4}He should be comparable to or perhaps even higher than the juvenile (mantle) ratio. The fact that it is nearly an order of magnitude lower suggests that the major source term of ^{4}He is not the mantle flux, a reasonable proposition anyway (cf. Section 12.5). This other source must have very low ^{3}He/^{4}He ratio and is presumably radiogenic He from the crust rather than the mantle. As with Ar (Section 13.6), this indicates that transport of parent elements into the crust is an important effect in noble gas evolution and its neglect probably the most serious limitation on the applicability of the two-reservoir model.

Another approach (Bernatowicz & Podosek, 1978) to He degassing in the two-reservoir model is comparison of ^{3}He not with ^{4}He but with another gas species which does accumulate in the atmosphere, e.g. ^{20}Ne or ^{36}Ar. Of the three constraints needed, one can be provided by accumulation of this second species, and a second by the juvenile flux of ^{3}He; the third must be an assumption of the ratio of initial composition. It is also necessary to assume that the two species degas at the same rate α. Then, for ^{20}Ne,

$$^{20}\text{Ne}_a = {}^{20}\text{Ne}_0(1 - e^{-\alpha t}) \tag{13.37}$$

Combining with eq. (13.36), imposing the same juvenile flux of ^{3}He as above, and assuming $(^3\text{He}/^{20}\text{Ne})_0 = 0.032$ (the planetary value) yields $\alpha = 9.7 \times 10^{-10}\,\text{yr}^{-1}$, corresponding to rather rapid ($\alpha T = 4.4$) and extensive ($e^{-\alpha T} = 1.2\%$) degassing. The initial endowment for this case is $^3\text{He}_0 = 3.5 \times 10^{-10}\,\text{cm}^3\,\text{STP/g}$; this is somewhat larger than that for the first set of constraints described above, but since only $^3\text{He}_m = 4.1 \times 10^{-12}\,\text{cm}^3\,\text{STP/g}$ remains at present, the corresponding $^4\text{He}_m = 3.5 \times 10^{-7}$ is quite consistent with the igneous rock range (Fig. 9.2), with reasonable allowance for a He distribution coefficient favoring melt over solid. This value of α is also reasonably close to the value obtained in the two-reservoir model for Ar ($\alpha T \gtrsim 6.3$) in Section 13.5. The principal difference between this set of results and that obtained with the first set of constraints lies in the U abundance: insertion of $\alpha = 9.7 \times 10^{-10}$ yr^{-1} (and the corresponding ^{3}He$_0$) into eq. (13.35) requires a present $U_m = 1.2$ ppb.

Repeating this calculation for ^{36}Ar in place of ^{20}Ne and assuming ^{3}He/^{36}Ar = 0.0088 (Table 5.3) yields a slightly different result, $\alpha = 8.0 \times 10^{-10}$ yr^{-1} ($\alpha T = 3.6$), and e$^{-\alpha T} = 2.7\%$. An early catastrophic degassing lowers the degassing coefficient α calculated by this approach, as would also result if the actual juvenile ^{3}He flux is higher than the value assumed.

An interesting feature of this approach is that it employs only primordial gas data, which is possible only by relating a gas which accumulates in the atmosphere to one whose juvenile flux is observable (^{3}He is the only case). Two of the three input constraints, the initial elemental ratio and the juvenile flux, are rather uncertain of course, but as long as the actual values do not differ from those assumed by more than about a factor of two, the results will be little changed. It is particularly noteworthy that since no radiogenic isotopes are involved in specifying the model, transport of parent elements into the crust is irrelevant. As usual, however, it is necessary to express the qualification that these considerations probably apply only to the depleted mantle.

It is also possible to formulate a two-gas approach based on only the radiogenic isotopes. Ignoring or correcting for the approximately 10% primordial contribution to mantle ^{4}He, combining eq. (13.34) with eq. (13.9) applied to ^{40}Ar gives

$$(^4\mathrm{He}/^{40}\mathrm{Ar})_m = \frac{\sum \dfrac{\lambda_i}{\alpha - \lambda_i}(e^{-\lambda_i t} - e^{-\alpha t})y_i P_{0i}}{\dfrac{\lambda_{40}}{\alpha - \lambda_{40}}(e^{-\lambda_{40} t} - e^{-\alpha t})y_{40}\,^{40}K_0} \tag{13.38}$$

With the usual assumptions (e.g. present Th/U = 3.3, K/U = 10^4), eq. (13.38) gives present $(^4\mathrm{He}/^{40}\mathrm{Ar})_m$ as a function of α, so measurement of $(^4\mathrm{He}/^{40}\mathrm{Ar})_m$ can be inverted to determine α. Equivalent formulations can be made for a more complicated model. For the two-reservoir model with the values cited, the limiting value for fast degassing is $(^4\mathrm{He}/^{40}\mathrm{Ar})_m = 5.5$ and for slow degassing $(^4\mathrm{He}/^{40}\mathrm{Ar})_m = 2.1$ (cf. Fig. 11.7). This approach has the advantage of employing only ratios, of one radiogenic gas to another and one parent to another, rather than absolute abundances of anything. It has the further advantage that only radiogenic gases are needed, and these can usually be resolved in spite of atmospheric contamination. It has the disadvantage of requiring a ratio of noble gas elemental (rather than isotopic) abundances in the mantle, however, and in practice a reliable representative value for the mantle has not yet been obtained (Section 11.5).

13.8 Xenon

The attempt to relate terrestrial (atmospheric) Xe composition to meteoritic isotopic compositions has been an integral part of xenology since it first became known that compositional variations exist (see Section 5.6). This has included study of the radiogenic isotopes ^{129}Xe and ^{136}Xe (and other heavy isotopes produced in fission) and how they might help fit the formation of the earth into early solar system chronology. In contrast to Ar and He, however, Xe received rather scant attention in models for degassing and atmospheric evolution until relatively recently (cf. Bernatowicz & Podosek, 1978).

In this regard Xe differs from Ar and He in a number of interesting ways. One is that the radiogenic enrichments are much smaller, but while uncertainties about initial nonradiogenic composition are not trivial the magnitudes of expected effects are well within experimental limits. Another is that Xe contains two prominent radiogenic components, a feature which can often be turned to advantage. The most important difference, however, is that both of the radioactive parents involved, ^{129}I and ^{244}Pu, have half-lives very much shorter than the parents of ^{40}Ar and ^{4}He; Xe degassing is thus very sensitive to events very early in earth history.

Most undifferentiated meteorites 'formed' (in the sense of Xe isotopic closure) within a relatively brief span of the order of 10 Ma some 4.55 Ga ago, when Xe parent abundances were characterized by $(^{129}I/^{127}I)_u = 10^{-4}$ and $(^{244}Pu/^{238}U)_u = 0.007$ (cf. Hohenberg *et al.*, 1967; Hudson *et al.*, 1983). It seems reasonable to suppose that the earth too formed (in the same sense of Xe closure) at about this same time and thus with these parent abundances. If so, we would have to conclude that the earth is but very little degassed, since only a small fraction of the radiogenic ^{136}Xe and a very small fraction of the radiogenic ^{129}Xe are in the atmosphere now (Section 12.7). This conclusion conflicts with the same inventory analysis for ^{40}Ar (even if the earth's K content is as high as chondritic, at least 9% of the ^{40}Ar, mostly produced much later than the ^{129}Xe and ^{136}Xe, is now in the atmosphere), and with the conclusions of Ar and He degassing models described earlier (Sections 13.5–13.7). Furthermore, the inferred degrees of degassing separately inferred from ^{129}Xe and ^{136}Xe are rather dissimilar.

An alternative view, which has been gaining increasing acceptance, is that the earth did not form at the same time as meteorites but significantly later. If we use δt to represent the delay between formation of meteorites (specifically the time at which $^{129}I/^{127}I = (^{129}I/^{127}I)_u$ and $^{244}Pu/^{238}U = (^{244}Pu/^{238}U)_u$) and formation of the earth, both in the sense of Xe closure, and ^{129}Xe* and ^{136}Xe* to represent radiogenic contributions, then

$$^{129}\text{Xe*}/^{130}\text{Xe} = (^{129}I/^{127}I)_u (^{127}I/^{130}\text{Xe})_0 e^{-\lambda_{129}\delta t} \tag{13.39}$$

and

$$^{136}\text{Xe}^*/^{130}\text{Xe} = (^{244}\text{Pu}/^{238}\text{U})_u (^{238}\text{U}/^{130}\text{Xe})_0 y_{136} e^{-\lambda_{244}\delta t} \qquad (13.40)$$

In these equations the quantities are global inventories, subscript 0 has the same meaning as in Sections 13.5–13.7, i.e., value at formation of the earth, and $y_{136} = 7 \times 10^{-5}$ (Table 3.1); to be precise, eq. (13.40) requires a small adjustment for decay of ^{238}U in δt and for fission Xe produced by ^{238}U. These equations illustrate how sensitively Xe reflects early chronology: delayed formation of the order of $\delta t \approx 100$ Ma would have very little effect on either Ar or He but a very pronounced effect on Xe because of the rapid decay of ^{129}I and ^{244}Pu. Even if the present global abundances of I and U were known precisely, there would still be large uncertainty in the initial terrestrial abundances of ^{129}I and ^{244}Pu.

If it is assumed that the apparent discrepancies noted above are in fact due to delayed formation of the earth, eqs. (13.39)–(13.40) can be used to estimate δt. With appropriate assumptions, each can be solved for δt separately. Because both of the daughters are isotopes of the same element, however, many of the assumptions are obviated if eqs. (13.39) and (13.40) are combined (Wetherill, 1975):

$$^{136}\text{Xe}^*/^{129}\text{Xe}^* \approx \frac{(^{244}\text{Pu}/^{238}\text{U})_u}{(^{129}\text{I}/^{127}\text{I})_u} (^{238}\text{U}/^{127}\text{I})_0 y_{136} e^{(\lambda_{129} - \lambda_{244})\delta t} \qquad (13.41)$$

(this equation is still approximate: λ_{244} should be replaced by $\lambda_{244} - \lambda_{238}$ and $^{136}\text{Xe}^*$ should be corrected for the ^{238}U contribution.) The principal geochemical parameter needed is global U/I. If it is further assumed that both $^{129}\text{Xe}^*$ and $^{136}\text{Xe}^*$ have been degassed with the same efficiency, their ratio in eq. (13.41) is the atmospheric ratio. With any reasonable parameters, δt is indeed of the order of 10^8 yr. For the specific solution $\delta t = 121$ Ma (Section 12.7) in the time interval δt the abundance of ^{129}I decreases by a factor of 138 and the abundance of ^{244}Pu by a factor of 2.8. If the values cited above are correct, then for the absolute abundances listed in Table 12.3 the amounts of $^{129}\text{Xe}^*$ and $^{136}\text{Xe}^*$ now in air are about 53% of total radiogenic production.

While this interpretation is plausible, it is by no means assured. It is thus particularly treacherous to base arguments on assumed initial abundances of ^{129}I or ^{244}Pu or their ratio to primordial Xe. It is worth explicit note that none of the arguments in the rest of this section do so: in effect, all dependences on initial abundances are normalized to radiogenic enhancements in modern air.

Atmospheric evolution trajectories in the simple two-reservoir model, eqs. (13.12) and (13.18), are illustrated in Figs. 13.6 and 13.7. These are of particular interest for Xe for two reasons. One is that atmospheric compositional history can be considered a function of only one parameter, the degassing rate α; the other parameter, P_0/S_0, is fixed (as a function of α; cf. Fig. 13.1) by the constraint that the model produce modern atmospheric composition. This permits circumvention of the large uncertainty associated with initial abundances of short-lived ^{129}I and ^{244}Pu. The second is that, for reasons noted in Section 8.6, there actually appears to be a better chance of constraining models by measurement of paleoatmospheric composition for Xe than for Ar (or He). In fact, we may already have measurements of paleoatmospheric Xe compositions from sedimentary rock analyses (Figs. 13.6 and 13.7). All such compositions are indistinguishable from modern atmospheric composition (cf. Sections 8.3 and 8.6). It must be immediately suspected, of course, that this means nothing and provides no historical information because only actual modern air was analyzed, e.g. massive air contamination prior to laboratory analysis or rapid exchange of

Fig. 13.6. Atmospheric evolution trajectories for ^{129}Xe in the two-reservoir first-order degassing model. Atmospheric composition is shown on ordinate scale at right; because of uncertainty in initial ^{129}Xe/^{130}Xe, left ordinate scale illustrates evolution as ratio of radiogenic enhancement at any time t to present radiogenic enhancement. The shaded area is the range possible without catastrophic degassing; its narrowness reflects the rapid decay of ^{129}I; with complete catastrophic degassing ($f = 1$) at $t = 0$, evolution is essentially the same if $\alpha \gg \lambda$ and for lower α produces trajectories in the triangular region bounded by the dashed line. Open circles represent sedimentary rock analyses by Phinney (1972) and Frick & Chang (1977). Reproduced from Bernatowicz & Podosek (1978).

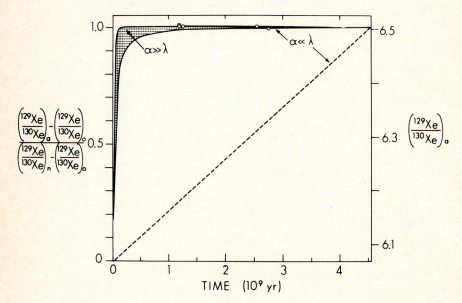

'trapped' gases and atmospheric gases in the field. Still, this is not necessarily the case, and essentially constant atmospheric Xe composition throughout most of geological time is a plausible evolutionary scenario. The observations may be valid and if so provide important information about degassing history.

If the paleoatmospheric ^{129}Xe data (Fig. 13.6) are valid, they do not really provide much information about α in two-reservoir degassing because even at the age of the oldest sediments the range of $(^{129}\text{Xe}/^{130}\text{Xe})_a$ corresponding to the full range of α is very narrow. They do, however, exclude the scenario of early catastrophic degassing followed by slow continuous degassing. This last conclusion also follows from the $(^{136}\text{Xe}/^{130}\text{Xe})_a$ data (Fig. 13.7). The ^{136}Xe data do, how-ever, provide a clear indication of fast rather than slow degassing. This would be very fast degassing indeed: $\alpha > \lambda_{244}$ corresponds to $\alpha T > 38$. This is far too fast to be realistic: it not only conflicts with the results of ^{40}Ar and ^{4}He degassing models, it predicts mantle abundances of both primordial and radiogenic species far too low to be observable.

The alternative approach, which has been profitable for Ar and He (Sections 13.5–13.7), is to attempt to constrain models by means of present mantle composition. This is somewhat more difficult for Xe because of ambiguity in assigning

Fig. 13.7. Atmospheric evolution trajectories for ^{136}Xe in the two-reservoir first-order degassing model; cf. Fig. 13.6. The dominant contribution is from ^{244}Pu but the model also includes the contribution from spontaneous fission of ^{238}U. Reproduced from Bernatowicz & Podosek (1978).

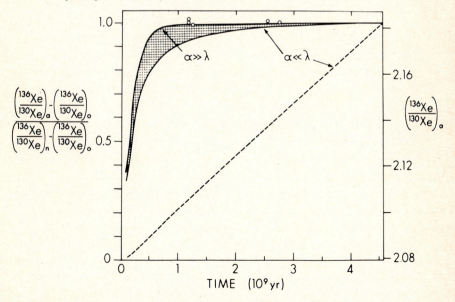

an appropriate mantle value. As noted in Chapters 9 and 11, most Xe compositions of suggested mantle provenance, in particular those associated with $^3He/^4He = 1.2 \times 10^{-5}$ and $(^{40}Ar/^{36}Ar)_m \gg (^{40}Ar/^{36}Ar)_a$, are indistinguishable from atmospheric composition. In this case the clear indication is 'slow' degassing. This is not necessarily in conflict with the fast degassing which seems to be the most plausible interpretation of the Ar and He models, however. For $\lambda t \gg 1$, which is certainly valid for both ^{129}Xe and ^{136}Xe, the two-reservoir degassing model (eqs. (13.11)–(13.12)) predicts very nearly equal $(D^*/S)_m$ and $(D^*/S)_a$ for 'slow' degassing, but the criterion for 'slow' is $\alpha \ll \lambda$ and not necessarily $\alpha T \ll 1$. Since even the relatively fast degassing suggested by Ar and He models satisfies $\alpha \ll \lambda$ (but not $\alpha T \ll 1$), there is no conflict between Ar and He models and Xe models on this account.

While most observations of apparent mantle Xe compositions show near identity with atmospheric composition, and none are highly radiogenic (in the sense that mantle He is highly radiogenic in comparison with primordial He, and mantle Ar is highly radiogenic in comparison with both primordial Ar and atmospheric Ar), it is nevertheless well known that some compositions are indeed radiogenic, i.e. that some samples exhibit excess ^{129}Xe relative to the atmosphere (Section 6.9). No matter what model for the formation of the earth or the evolution of the atmosphere is entertained, the existence of such excess ^{129}Xe requires that the earth has not been thoroughly mixed (isotopically equilibrated) since very early in its history, no later after its formation than a few half-lives of ^{129}I, i.e. less than about 10^8 yr. The especial relevance of this observation is that it is the strongest statement of its kind which can be made because ^{129}I is the shortest-lived radionuclide whose daughter is known to be inhomogeneously distributed in the earth. More detailed interpretations are evasive, however, since it has proved rather difficult to establish the provenance of excess ^{129}Xe. Many of the occurrences are suggestive of undepleted mantle provenance; if so, and if excess ^{129}Xe is restricted to undepleted mantle, we may conclude an early isolation of undepleted mantle from depleted mantle but not, on these grounds alone, anything about the timing of the degassing of the depleted mantle. Recently, however, it has been suggested that excess ^{129}Xe also occurs in depleted mantle (Section 11.7); if so, then as above the inference is that the atmosphere has long been isolated from its source, i.e. it formed early, by fast or catastrophic degassing. The first and still the best-known and most extreme enrichment of ^{129}Xe is in New Mexico CO_2 wells (Section 10.3), which remains particularly mysterious in terms of provenance.

Estimates for primordial terrestrial Xe compositions suggest that about 6.8% of atmospheric ^{129}Xe is radiogenic, i.e. $(^{129}Xe^*/^{130}Xe)_a = 0.44$. The most extreme mantle compositions yet known indicate ^{129}Xe enrichments, relative to air, of about 8%. If we *assume* that this represents a mantle composition, then $(^{129}Xe*/^{130}Xe)_m = 0.96$. In the simple two-reservoir model, inserting these values into eqs. (13.12) and (13.11), respectively, and taking their ratio permits elimination of P_0/S_0 and yields $\alpha = 22 \times 10^{-9}$ yr^{-1} ($\alpha T = 100$). This is, of course, impossibly fast. It is, however, a case where it is assumed that the source of the atmosphere retains some excess ^{129}Xe, and as such illustrates the generalization stated above about the rapid isotopic isolation required by the existence of excess ^{129}Xe: in this case the atmosphere is formed rapidly, in a timescale $1/\alpha = 45$ Ma, within a few (2.7) half-lives of ^{129}I.

Any reservoir which contains excess ^{129}Xe should also contain excess ^{136}Xe, and it is instructive to compare their abundances. The comparison is analagous to that between ^4He and ^{40}Ar (Section 13.7) except that in this case both daughters are isotopes of the same element and the question of elemental fractionation does not arise. If the radiogenic contribution to atmospheric ^{136}Xe is 4.7% (Section 6.7), then $(^{136}Xe*/^{129}Xe*)_a = 0.23$. As already noted, if $\alpha \ll \lambda$ for both λ_{244} as well as λ_{129}, then $(^{136}Xe*/^{129}Xe*)_m$ should be very nearly equal to $(^{136}Xe*/^{129}Xe*)_a$. If, however, α is sufficiently great as to produce an enrichment in $(^{129}Xe*/^{130}Xe)_m$ it will produce an even greater enrichment in $(^{136}Xe*/^{130}Xe)_m$. Thus, $\alpha = 22 \times 10^{-9}$ yr^{-1}, as above, by the same calculation yields $(^{136}Xe*/^{130}Xe)_m/(^{136}Xe*/^{130}Xe)_a = 4 \times 10^{26}$. The example is absurd in its extremity but it illustrates the qualitative point that excess $^{129}Xe*$ should be accompanied by $^{136}Xe*$ in proportions at least as great as that in air and perhaps very much greater.

The discussion above, based on the simple first-order two-reservoir model, reveals a number of conflicts and inconsistencies. The rapid growth of the atmosphere radiogenic Xe isotopes, and the existence of excess ^{129}Xe, both suggest extremely rapid degassing, which not only conflicts with Ar and He models but is self-contradictory in that there should be hardly any mantle gas at all left to observe. The equality of most mantle Xe compositions with atmospheric composition, and the absence of great enrichments of ^{136}Xe in company of excess ^{129}Xe, on the other hand, indicate slow degassing; these observations do not conflict with Ar and He models, since 'slow' means only slow in comparison with radiogenic daughter growth ($\alpha \ll \lambda$), and not necessarily slow in the sense $\alpha T \ll 1$. To some extent, the seeming conflicts can be dismissed as only apparent: the validity of the paleoatmospheric data can be challenged, for example, and even if they really are paleoatmospheric compositions there may be important experimental uncertainties (at the age of the oldest analyzed sediments the entire spread between fast and slow degassing in Fig. 13.7 is less than 0.4% in $^{136}Xe/^{130}Xe$). Nevertheless it seems evident that, in any of a number of ways (Section 13.4), there is an important failure of the basic model.

It is instructive to consider a simple and plausible interpretation which reconciles some of these conflicts. We may suppose that early in earth history degassing was very (catastrophically?) fast, that this fast degassing occurred at or until an age of the order of 10^8 yr, and has subsequently been much slower. Qualitatively, the important feature is that it is possible to find a time which is long as far as Xe is concerned, so that there is little subsequent change in mantle or atmospheric Xe composition, but which is short as far as Ar and He are concerned, so that most of their evolution occurs during the subsequent relatively slow degassing.

Formally, this scenario can be approximated even within the two-reservoir model by modest generalization to catastrophic degassing at time $t_d > 0$ (cf. Section 13.6). For simplicity, we ignore subsequent continuous degassing: by hypothesis, this has little effect on either mantle or atmospheric composition. If, at time t_d after the formation of the earth a fraction f of extant gas is injected into the atmosphere, then atmospheric composition is given by

$$(^{129}Xe*/^{130}Xe)_a = (^{129}I/^{130}Xe)_0 (1 - e^{-\lambda_{129}t_d}) \tag{13.42}$$

where $(^{129}I/^{130}Xe)_0$ is the ratio at $t = 0$, the 'formation' (Xe isotopic closure) of the earth. This ratio involves not only the global I/Xe ratio but also ^{129}I decay in any formation delay δt. Mantle composition is given by

$$(^{129}Xe*/^{130}Xe)_m = (^{129}I/^{130}Xe)_0 \left(1 + \frac{f}{1-f} e^{-\lambda_{129}t_d}\right) \tag{13.43}$$

Combining eliminates $(^{129}I/^{130}Xe)_0$:

$$\frac{(^{129}Xe*/^{130}Xe)_m}{(^{129}Xe*/^{130}Xe)_a} = 1 + \frac{e^{-\lambda_{129}t_d}}{(1-f)(1 - e^{-\lambda_{129}t_d})} \tag{13.44}$$

As above, we will approximate the left-hand side of eq. (13.44) by 0.96/0.44 = 2.2. By eq. (13.44) t_d can be taken arbitrarily long in the limit as f approaches unity. Realistically, a really complete catastrophic degassing is precluded by models for Ar and He as well as by the requirement that some gas remain for observation. For $f < 1$, the time t_d is narrowly restricted. For a plausible value such as $f = 0.9$, eq. (13.44) yields $t_d = 55$ Ma; in the limit of low f, the degassing time t_d approaches a lower limit of 15 Ma. Any subsequent continuous degassing will not affect $(^{129}Xe/^{130}Xe)_a$ as long as f is not small. Thus, in this scenario, Xe compositional evolution is essentially complete by t_d, before Ar and He have hardly started their evolution. Like any other, this model is unrealistically simple, but it does again illustrate the point that preservation of excess ^{129}Xe in the earth's interior requires formation of the atmosphere on a very rapid timescale (t_d in this case, $1/\alpha$ in continuous degassing). It also illustrates the important converse point which is typically less well appreciated: formation of the atmosphere cannot occur too fast, and the characteristic time for formation of the atmosphere must be no less than the order of the half-life of ^{129}Xe, lest there be produced interior Xe compositions which are much more radiogenic than the atmosphere.

Applying the same model to the growth of ^{136}Xe from ^{244}Pu, the analog of eq. (13.44) is

$$\frac{(^{136}\text{Xe*}/^{130}\text{Xe})_m}{(^{136}\text{Xe*}/^{130}\text{Xe})_a} = 1 + \frac{e^{-\lambda_{244}t_d}}{(1-f)(1-e^{-\lambda_{244}t_d})} \tag{13.45}$$

Because the lifetime of ^{244}Pu is longer than that of ^{129}I, the radiogenic enhancement of ^{136}Xe in the mantle is proportionately greater than that of ^{129}Xe, the more so the greater t_d. Thus, even at the minimum $t_d = 15$ Ma, $(^{136}\text{Xe*}/^{129}\text{Xe*})_m/(^{136}\text{Xe*}/^{129}\text{Xe*})_a = 3.8$, and this solution is not realistic since it violates ($f = 0$) the assumption that continuous degassing after t_d does not much change atmospheric composition. For the more realistic solution $t_d = 55$ Ma with $f = 0.9$, $(^{136}\text{Xe*}/^{129}\text{Xe*})_m/(^{136}\text{Xe*}/^{129}\text{Xe*})_a = 8.2$. For the nominal value $(^{136}\text{Xe*}/^{129}\text{Xe*})_a = 0.23$, this corresponds to $(^{136}\text{Xe*}/^{129}\text{Xe*})_m = 1.9$ and the mantle composition to be associated with $(^{129}\text{Xe}/^{130}\text{Xe})_m = 7.02$ (8% above the air value) is $(^{136}\text{Xe}/^{130}\text{Xe})_m = 3.9$.

In both the continuous and catastrophic models described above, generation of a mantle composition with a modest radiogenic enhancement of ^{129}Xe (relative to air) also generates a rather more substantial enhancement of ^{136}Xe. Qualitatively, this result is not specific to any particular model. If mantle Xe is to be radiogenic but not too highly radiogenic in ^{129}Xe, the formation of the atmosphere and subsequent effective isolation from the mantle must be rather narrowly constrained in time irrespective of model details; it must occur quickly enough to precede full decay of ^{129}I but slowly enough to permit substantial decay. These limits correspond to only relatively minor decay of ^{244}Pu, and the result is that mantle enrichment in ^{136}Xe must be substantially greater than the ^{129}Xe enrichment, both as compared to the atmosphere. As we have noted (Section 6.9), the predicted enrichment of excess ^{136}Xe which should accompany excess ^{129}Xe* is definitely not observed. While there may in fact be excesses of ^{136}Xe* which correlate with excess ^{129}Xe* (Section 11.7), the ^{136}Xe*/^{129}Xe* ratio is too low to be considered consistent with these model predictions. If such assessments are valid, the inference is a fundamental inadequacy of the basic model. We describe below two possible resolutions of this problem.

One possible viewpoint is simply to abandon the assumption that ^{244}Pu makes an important fission contribution. There will still be a fission effect, a smaller one, from ^{238}U. It is interesting to consider such a model simply for the reason that ^{238}U certainly does exist, and there certainly are excess ^{129}Xe effects in terrestrial samples, but the evidence for ^{244}Pu is less direct, and there is no known instance of a terrestrial Xe composition which unambiguously demands the existence of live ^{244}Pu in the earth. Further impetus arises from the observation that the most prominent occurrences of excess ^{129}Xe, in the CO_2 well gas (Section 10.3), is indeed accompanied by excess fission Xe, and the fissioning parent is demonstrably ^{238}U, not ^{244}Pu. It is usually thought that the fission Xe in this gas is crustal, but the He as well as the excess ^{129}Xe is evidently of mantle origin, so perhaps the fission Xe is too. Nevertheless, it should be appreciated that this is a somewhat radical approach. It is not sufficient to postulate that the ^{244}Pu was removed from the mantle very early, e.g. by differentiation into the crust; if that happened, and ^{244}Pu eventually decayed and made a significant contribution to atmospheric Xe, this contribution would be absent in mantle Xe, which is not the case. It is necessary to postulate that primordial terrestrial Xe composition was essentially identical to modern atmospheric Xe composition, except at ^{129}Xe, and that ^{244}Pu was absent or so scarce as to leave no trace in any known terrestrial samples. This, of course, rather complicates still further the problem of accounting for the origin of terrestrial Xe in general (Sections 5.6 and 6.7). This is perhaps excusable on the grounds that heavy Xe isotope cosmochemical interpretations are not nearly as well-founded as the inference that air contains radiogenic ^{129}Xe, but such considerations aside there *should* have been enough ^{244}Pu in the earth to make a globally prominent contribution to ^{136}Xe (Sections 6.2 and 12.7). Postulating the absence of ^{244}Pu in the earth thus has a variety of interesting cosmochemical ramifications.

Assuming no ^{244}Pu, the catastrophic model which yields eqs. (13.42)–(13.43) also yields

$$({}^{136}\text{Xe}^*/{}^{130}\text{Xe})_a = y({}^{238}\text{U}/{}^{130}\text{Xe})_0(1 - e^{-\lambda_{238}t_d}) \tag{13.46}$$

$$({}^{136}\text{Xe}^*/{}^{130}\text{Xe})_m = y({}^{238}\text{U}/{}^{130}\text{Xe})_0 \left(1 + \frac{fe^{-\lambda_{238}t_d} - e^{-\lambda_{238}t}}{1-f}\right) \tag{13.47}$$

where now $y = 3.5 \times 10^{-8}$ (Table 3.1). In practical terms, ^{238}U can account for no more than about 1% of atmospheric ^{136}Xe (Section 6.2) and very little of it will have been produced before $t_d \approx 10^8$ yr; $({}^{136}\text{Xe}^*/{}^{130}\text{Xe})_a$ will then be very small and $({}^{136}\text{Xe}/{}^{130}\text{Xe})_a \approx ({}^{136}\text{Xe}/{}^{130}\text{Xe})_0$. Also, there would be no independent cosmochemical framework for estimating $({}^{136}\text{Xe}/{}^{130}\text{Xe})_0$. For both reasons it is impractical to use the ratio of radiogenic enhancements in mantle and atmosphere to eliminate the elemental abundances, as was done in eqs. (13.44) and (13.45); it is thus necessary to work in a formalism retaining the elemental abundance ratios. Adding $({}^{136}\text{Xe}/{}^{130}\text{Xe})_0$ to both eqs. (13.46) and (13.47) and subtracting the first from the second yields

$$({}^{136}\text{Xe}/{}^{130}\text{Xe})_m - ({}^{136}\text{Xe}/{}^{130}\text{Xe})_a = y\,({}^{238}\text{U}/{}^{130}\text{Xe})_0 \frac{e^{-\lambda_{238}t_d} - e^{-\lambda_{238}t}}{1-f}$$

$$(13.48)$$

Doing the same for eqs. (13.42) and (13.43) and dividing into eq. (13.48) yields

$$\frac{({}^{136}\text{Xe}/{}^{130}\text{Xe})_m - ({}^{136}\text{Xe}/{}^{130}\text{Xe})_a}{({}^{129}\text{Xe}/{}^{130}\text{Xe})_m - ({}^{129}\text{Xe}/{}^{130}\text{Xe})_a} = y\,\frac{({}^{238}\text{U}/{}^{127}\text{I})_0}{({}^{129}\text{I}/{}^{127}\text{I})_0}\,\frac{e^{-\lambda_{238}t_d} - e^{-\lambda_{238}t}}{e^{-\lambda_{129}t_d}}$$

$$(13.49)$$

If there is indeed a fission excess which correlates with excess ^{129}Xe, (Section 11.7) the left-hand side of eq. (13.49) is a potentially observable quantity; for the sake of illustration we will estimate its value to be ≈ 0.6. Since decay of ^{238}U before t_d is minor, the present U/I ratio in the mantle is actually the more appropriate value for the right-hand side of eq. (13.49); again for illustration we will estimate a present I/U ≈ 2 (Becker *et al.*, 1968). As noted earlier, $({}^{129}\text{I}/{}^{127}\text{I})_0$ is quite uncertain, but this problem can be averted by noticing that the denominator on the right-hand side of eq. (13.49) is just $({}^{129}\text{I}/{}^{127}\text{I})_u e^{-\lambda_{128}(\delta t + t_d)}$. Combining with the estimated values cited above, solution of eq. (13.49) yields $\delta t + t_d = 2.1 \times 10^8$ yr. Since decay of ^{129}I is so fast, this result is insensitive to uncertainties in the putative observational constraints. This illustrates yet again the rapid timescale demanded by any excess ^{129}Xe, no matter what the model.

In considering this model for ^{244}Pu (eq. 13.45) it was legitimate to ignore continuous degassing after t_d as long as it was relatively slow ($\alpha \ll \lambda_{244}$ even if $\alpha T \gg 1$) and f was not small, in which case the rest of the ^{244}Pu (and ^{129}I) decays before continuous degassing has much effect. It can be objected that this is *not* legitimate for ^{238}U in place of ^{244}Pu unless $\alpha \ll \lambda_{238}$; in other words, because ^{238}U decays so slowly, rapid continuous degassing after t_d would create precisely the problem this model was constructed to avoid, namely the prediction of very high fission ^{136}Xe* wherever excess ^{129}Xe* is observed. The approach of ignoring ^{244}Pu and using ^{238}U instead to produce fission Xe effects is thus viable only if it is further assumed that continuous degassing for Xe, unlike Ar and He, has been slow in the absolute sense $\alpha T \ll 1$, or that U is extracted from mantle to the crust (cf. Section 13.6) sufficiently rapidly to avert growth of a large mantle fission Xe component.

The second possible resolution to the fission Xe problem is to retain the ^{244}Pu and relax the assumption that only two reservoirs, mantle and atmosphere, are involved. The basis of this approach is recognition that the key to the fission Xe problem is the assumption that excess ^{129}Xe is found in the same reservoir which generated the atmosphere; if the atmosphere also has a ^{244}Pu contribution, it is nearly impossible not to have a large fission ^{136}Xe in excess in the original reservoir as well, since the lifetime of ^{244}Pu is greater than that of ^{129}I. Once this assumption is dropped, the problem disappears. Thus, in this approach we assume that mantle (now specifically depleted mantle) Xe composition actually is very nearly the same as atmospheric Xe composition. This will follow from any history in which the major growth of the atmosphere is not completed until after decay of ^{244}Pu (and thus ^{129}I) is substantially complete. If degassing is continuous, the requirement is $\alpha \ll \lambda_{244}$; if catastrophic, $t_d \gg 1/\lambda_{244}$. This need not conflict with the analogous inferences for ^{40}Ar (or ^{4}He); the requisite timescale is of the order of 10^8 yr, which is slow and/or late for ^{136}Xe (and ^{129}Xe) but still fast and/or early for ^{40}Ar (and ^{4}He). As before, if the putative paleoatmospheric Xe compositions are valid (Fig. 13.7), then only the catastrophic alternative is viable.

The occurrences of excess ^{129}Xe must then be assigned to some other reservoir. This is the same approach as that taken for occurrences of low ^{40}Ar/^{36}Ar (Section 11.3) and high ^{3}He/^{4}He (Section 11.4) and the candidate reservoir is equally obvious: undepleted mantle. The principal difficulty with such an interpretation is seemingly that of ambiguous provenance: it is not clear that excess ^{129}Xe can be related to low ^{40}Ar/^{36}Ar in the same way as high ^{3}He/^{4}He, and the tectonic associations, particularly for the CO_2 well gas, are still less clear. If normal ocean floor basalts actually do contain excess ^{129}Xe (along with 'normal' mantle He and Ar compositions) the assignment of excesses to only undepleted mantle fails.

The relationship between ^{136}Xe and ^{129}Xe is important. If only ^{129}Xe were included, it would be possible to adopt an extreme position and argue, for Xe as for Ar, that there is no second mantle component, that mantle Xe (like Ar) is indeed highly radiogenic compared to air, and that observed compositional variations reflect dilution (overwhelming dilution in the case of Xe) of radiogenic mantle Xe with atmospheric (not juvenile) Xe introduced by contamination of tectonic recycling. The same arguments, however, lead to mantle Xe which is even more radiogenic at ^{136}Xe than at ^{129}Xe, and this prediction is contrary to observation.

Association of more highly radiogenic Xe with less radiogenic Ar and He in the undepleted mantle reservoir is not necessarily a problem for this model. As long as the reservoir has not been extensively degassed, present composition is mainly a reflection of initial conditions: the initial I/Xe, Pu/Xe, K/Ar, and U/He ratios. The similarity of undepleted mantle and atmospheric ^{136}Xe/^{130}Xe ratios would thus correspond to similar initial Pu/Xe ratios in undepleted and now-depleted mantle, while the higher ^{129}Xe/^{130}Xe would correspond to higher I/Xe in undepleted mantle. The controlling factor might be geochemical, but since the relevant initial condition is not ^{127}I/Xe but ^{129}I/Xe, it might also be temporal, i.e. in a heterogeneous accretion model the planetesimals which now constitute undepleted mantle may have formed (in the sense of Xe closure) somewhat earlier (around 10^7 yr) than those which are now depleted mantle.

13.9 Conclusions

While it would be naive to suppose that any of the formal models described in this chapter – or any formal models at all – are rigorously applicable to the earth, it may not be overly optimistic to hope that the basic qualitative features which emerge from these considerations approximate the actual history of the earth's noble gases. By extension, we may further hope that this history characterizes other volatiles and the atmosphere in general. In some cases, believable observational constraints and plausible assumptions are enough to determine the relevant free parameters and so define the models. In general, however, we sorely lack enough constraints to overdetermine the models and thus either support their validity or reject them and seek others. Nevertheless, by examining different systems for consistency and by integrating noble gas systematics into models of broader geological scope, it is possible to arrive at some plausible generalizations.

There is considerable evidence that atmospheric evolution has been rapid, in the absolute sense that most of the present atmospheric gases have been in the atmosphere since early in earth history. The process of transporting volatiles from the mantle to the atmosphere is ongoing today and has presumably been continuous throughout most of geological history; its rate has evidently been sufficient to 'turn over' the mantle several times in the age of the earth (formally, $\alpha T \approx 5$), so that the characteristic time for the growth of the atmosphere by continuous degassing is about 10^9 yr. The clearest and least arguable evidence that continuous degassing has been fast is the observation that $^{40}Ar/^{36}Ar$ in the mantle is much higher than $^{40}Ar/^{36}Ar$ in the atmosphere. With somewhat less firmness, the actual value $\alpha T \approx 5$ follows principally from comparison of the absolute abundance of ^{40}Ar in the mantle with the abundance in the atmosphere, and from the rate at which ^3He appears to be flowing from the mantle.

Additional evidence, consistent with that cited above but also complementary, further suggests that for a brief period very early in earth history, the rate of degassing was even higher, i.e. that the earth experienced early and extensive catastrophic degassing. This conclusion emerges from more sophisticated consideration of Ar evolution which includes transport of K from mantle to crust and, more directly, if mantle Xe composition actually is nearly indistinguishable from atmospheric Xe composition (and has been so throughout most of earth history). The probable timescale for early catastrophic degassing is about 10^8 yr: it may have begun very early, perhaps simultaneously with the accretion of the earth, but the Ar constraint indicates that it lasted not much more, and the Xe constraint not much less, than 10^8 yr beyond the formation of the earth.

The corollary of rapid degassing is extensive degassing. The models suggest, and observational data are consistent with, the appraisal that most of the relevant original volatiles are already in the atmosphere with a residue of no more than a few per cent remaining in the mantle.

There is, however, another body of evidence which denies this relatively simple picture. Isotopic compositions in some mantle samples are different from the majority of compositions referred to above, specifically in terms of less radiogenic Ar and He and more radiogenic Xe. The mere observation that different samples have different compositions indicates the inadequacy of treating the interior of the earth as a single entity, a single reservoir from which the atmosphere was derived. Among many possible detailed interpretations, the most plausible seems to be that these variations in isotopic composition reflect a fundamental dichotomy in the structure of the mantle: two kinds of mantle whose geochemical characteristics are distinct and whose evolutions have been mostly uncoupled for a long time. The basic geochemical distinction is that one mantle reservoir, the depleted mantle, has experienced substantial geochemical evolution involving differentiation and removal of incompatible elements, including volatiles, while the other, the undepleted mantle, has not, or at least not so extensively. The existence of such a dichotomy has long been suggested by several lines of evidence unrelated to noble gases and is suggested by noble gas geochemistry as well.

If this basic view of mantle structure is qualitatively correct, then the obvious association is that undepleted mantle is characterized by less radiogenic He, less radiogenic Ar, and more radiogenic Xe. This is a plausible and appealing view, and there is no evidence to contradict it, but the qualification should be expressed that the evidence which can be marshalled in its support is at present far from overwhelming. There are only relatively few observations which indicate an association of high ^{3}He/^{4}He and low ^{40}Ar/^{36}Ar with each other or of either with the tectonic setting and/or petrochemical and isotopic characteristics adduced for undepleted mantle, and the association of high ^{129}Xe/^{130}Xe with any such features is hardly more than hypothetical.

If this interpretation is correct, the generalization noted above – probable extensive and early catastrophic degassing and relatively rapid continuous degassing – refers only to the depleted mantle, not to undepleted mantle. The difference in He and Xe compositions between undepleted and depleted mantle compositions may partly reflect their different evolutions and partly their different initial conditions, i.e. initial ratio of parent to daughter elements. The same is true for Ar, but for Ar the difference in present compositions is so large as to suggest strongly that it arises mostly in different evolutions: specifically, that undepleted mantle has never been extensively outgassed, neither continuously nor catastrophically. The persistence of different Xe compositions indicates that the two end-member kinds of mantle have been effectively isolated from each other since the effective extinction of ^{129}I, less than 10^8 yr after the formation of the earth; of the radionuclides for which an equivalent statement may be made, ^{129}I has the shortest lifetime, so this is the strongest such limit which can be imposed.

If this basic picture is correct, then presumably the dominant source of the present atmosphere is depleted mantle; the contribution from undepleted mantle is evidently small, since it is inferred not to be extensively differentiated and degassed. The undepleted mantle is obviously making some juvenile inputs at present, however, and so has made some contribution to the present atmospheric inventory, and it is difficult to put a limit to how large this contribution may be. By the same considerations, it is difficult to pursue the degassing rate of undepleted mantle beyond the statement that it appears to be slow: in all the quantitative models it is essential to compare accumulated atmospheric characteristics with those of the reservoir from which it originated, and, if the present atmosphere indeed originated mostly from what is now depleted mantle, this integral constraint on undepleted mantle evolution is unavailable.

The inference that the present atmosphere was generated by the rapid and nearly complete degassing of the gases in its source (except for ^{40}Ar and ^{4}He, which are produced by long-lived parents) does not necessarily imply that the earth as a whole is nearly completely degassed, since undepleted mantle is *not* extensively degassed. It is not easy to determine how important this is to the global inventory. In the absence of the atmospheric constraint just noted, it is not possible to use evolution models to estimate absolute abundance levels for the undepleted mantle. The only other approach is comparison of atmospheric inventory with an independently predicted global inventory. For primordial species, both the noble gases and the major volatiles, the only basis for such an independent prediction is the analogy with extraterrestrial materials. The composition of the present atmosphere is, roughly, consistent with the predictions such analogy makes, but unfortunately the analogy cannot be used to predict global inventories narrowly enough to be useful in distinguishing full from minor degassing. Prospects appear brighter for the radiogenic gases, since estimates of the parent inventories can be guided not just by the analogy with extraterrestrial materials but also by the earth's present energy budget. The principal heat sources, U and Th, are not themselves directly useful in this regard, since their daughter ^{4}He does not accumulate in the atmosphere, but they are important indices for estimation of the initial global abundances of K, I, and Pu, whose daughters do accumulate. For estimated K concentration, it appears that about half the ^{40}Ar ever produced is now in the atmosphere; the degassed fraction of primordial volatiles presumably should be larger. Reasonable allowance for uncertainty in K content, however, would probably permit either conclusion – nearly complete or only minor degassing. The corresponding estimates for initial ^{129}I and ^{244}Pu are even more uncertain than estimates based on I and Pu chemistry, since these rapidly decaying radionuclides may have been significantly depleted during the time it took to accrete the earth. If the earth formed with the relatively high ^{129}I/^{127}I and ^{244}Pu/^{238}U abundances characteristic of primitive meteorites, the degree of global degassing is very low indeed; with allowance for decay during a formation interval of about 10^8 yr, chosen to match the ratio of radiogenic ^{129}Xe and ^{136}Xe in the atmosphere, the suggested degree of degassing is about 50%. Within this range of predicted degrees of degassing, the global degassing efficiency is comparable to the equally uncertain fraction of mantle mass assigned to the depleted mantle. Apparently, the initial volatile concentrations of what are now depleted and undepleted mantle were not drastically different.

The rather simple picture here sketched certainly cannot be supported with the rigor which all the mathematical development described in this chapter might suggest. Nevertheless, it has the virtue of plausibility as well as simplicity in that it is self-consistent as well as consistent with a broad range of geological information. Probably the most reasonable appraisal of this depleted/undepleted mantle model is that it merits the status of best working hypothesis, albeit one to be entertained with an active skepticism and an eye for its alternatives.

References

Alaerts, L., Lewis, R. S. & Anders, E. (1977). Primordial noble gases in chondrites: the abundance pattern was established in the solar nebula. *Science*, **198**, 927–30.

Alaerts, L., Lewis, R. S., Matsuda, J. & Anders, E. (1980). Isotopic anomalies of noble gases in meteorites and their origins – VI. Presolar components in the Murchison C2 chondrites. *Geochimica et Cosmochimica Acta*, **44**, 189–209.

Albarède, F. (1978). The recovery of spatial isotope distributions from stepwise degassing data. *Earth and Planetary Science Letters*, **39**, 387–97.

Aldrich, L. T. & Nier, A. O. (1948). The occurrence of He^3 in natural sources of helium. *Physical Review*, **74**, 1590–4.

Alexander, Jr, E. C. (1975). ^{40}Ar–^{39}Ar studies of Precambrian cherts; an unsuccessful attempt to measure the time evolution of the atmospheric ^{40}Ar/^{36}Ar ratio. *Precambrian Research*, **2**, 329–44.

Alexander, Jr, E. C., Lewis, R. S., Reynolds, J. H. & Michel, M. (1971). Plutonium-244: confirmation as an extinct radioactivity. *Science*, **172**, 837–40.

Altemose, V. O. (1961). Helium diffusion through glass. *Journal of Applied Physics*, **32**, 1309–16.

Anders, E. (1981). Noble gases in meteorites: evidence for presolar matter and superheavy elements. *Proceedings of the Royal Society, London*, A374, 207–38.

Anders, E. & Heymann, D. (1969). Elements 112 to 119: were they present in meteorites? *Science*, **164**, 821–3.

Anders, E., Heymann, D. & Mazor, E. (1970). Isotopic composition of primordial helium in carbonaceous chondrites. *Geochimica et Cosmochimica Acta*, **34**, 127–31.

Anders, E. & Owen, T. (1977). Mars and earth: origin and abundance of volatiles. *Science*, **198**, 453–65.

Anderson, D. L. (1981). Isotopic evolution of the mantle: the role of magma mixing. *Earth and Planetary Science Letters*, **57**, 1–12.

Armstrong, R. L. (1968). A model for evolution of strontium and lead isotopes in a dynamic earth. *Review of Geophysics*, **6**, 175–99.

Arrhenius, G. & Alfvén, H. (1971). Fractionation and condensation in Space. *Earth and Planetary Science Letters*, **10**, 253–67.

Aston, F. W. (1919). A positive ray spectrograph. *Philosophical Magazine VI*, **38**, 707–14.

Audouze, J., Bibring, J. P., Dran, J. C., Maurette, M. & Walker, R. M. (1976). Heavily irradiated grains and neon isotopic anomalies in carbonaceous chondrites. *Astrophysical Journal*, **206**, L185.

Austin, S. R. & Droullard, R. F. (1978). *Radon Emanation from Domestic Uranium Ores Determined by Modifications of the Closed Can, Gamma-only Assay Method.* Report of Investigations 8264, Denver: US Bureau of Mines.

Axford, W. E. (1968). The polar wind and the terrestrial helium budget. *Journal of Geophysical Research*, **73**, 6855–9.

Baadsgaard, H., Lipson, J. & Folinsbee, R. E. (1961). The leakage of radiogenic argon from sanidine. *Geochimica et Cosmochimica Acta*, 25, 147–57.

Banks, P. M. & Holzer, T. E. (1968). The polar wind. *Journal of Geophysical Research*, 73, 6846–54.

Baretto, P. M. C. (1975). Radon-222 emanation characteristics of rocks and minerals. In *Radon in Uranium Mining*, pp. 129–48. Vienna: International Atomic Energy Agency.

Barrer, R. M. & Edge, A. J. V. (1967). Gas hydrates containing argon, krypton, and xenon: kinetics and energetics of formation and equilibria. *Proceedings of the Royal Society, London*, A300, 1–24.

Basford, J. R., Dragon, J. C., Pepin, R. O., Coscio, Jr, M. R. & Murthy, V. R. (1973). Krypton and xenon in lunar fines. *Proceedings of the Fourth Lunar Science Conference*, 2, 1915–55.

Batiza, R., Bernatowicz, T. J., Hohenberg, C. M. & Podosek, F. A. (1979). Relations of noble gas abundances to petrogenesis and magmatic evolution of some oceanic basalts and related differentiated volcanic rocks. *Contributions to Mineralogy and Petrology*, 69, 301–14.

Becker, V., Bernet, J. H. & Manuel, O. K. (1968). Iodine and uranium in ultrabasic rocks and carbonatites, *Earth and Planetary Science Letters*, 4, 357–62.

Begemann, F. (1980). Isotopic anomalies in meteorites. *Reports on Progress in Physics*, 43, 1309–56.

Begemann, F., Weber, H. W. & Hintenberger, H. (1976). On the primordial abundance of argon-40. *Astrophysical Journal*, 203, L155–L157.

Behrmann, C. J., Drozd, R. J. & Hohenberg, C. M. (1973). Extinct lunar radioactivities: xenon from ^{244}Pu and ^{129}I in Apollo 14 breccias. *Earth and Planetary Science Letters*, 17, 446–55.

Benson, B. B. (1973). Noble gas concentration ratios as palaeotemperature indicators. *Geochimica et Cosmochimica Acta*, 37, 1391–5.

Benson, B. B. & Krause, D. J. (1976). Empirical laws for dilute aqueous solutions of non-polar gases. *Journal of Chemical Physics*, 64, 689–709.

Bernatowicz, T. J. (1981). Noble gases in ultramafic xenoliths from San Carlos, Arizona. *Contributions to Mineralogy and Petrology*, 76, 84–91.

Bernatowicz, T. J., Goettel, K. A., Hohenberg, C. M. & Podosek, F. A. (1979). Anomalous noble gases in josephinite and related rocks? *Earth and Planetary Science Letters*, 43, 368–84.

Bernatowicz, T. J., Hohenberg, C. M., Hudson, G. B., Kennedy, B. M. & Podosek, F. A. (1978). Excess fission xenon at Apollo 16. *Proceedings of the Ninth Lunar and Planetary Science Conference*, 1571–97.

Bernatowicz, T. J., Kramer, F. E., Podosek, F. A. & Honda, M. (1983). Adsorption and excess fission Xe: adsorption of Xe on vacuum crushed minerals. *Proceedings of Lunar Planetary Science Conference 13th*, in press.

Bernatowicz, T. J. & Podosek, F. A. (1978). Nuclear components in the atmosphere. In *Terrestrial rare gases*, ed. E. C. Alexander, Jr & M. Ozima, pp. 99–135. Tokyo: Japan Scientific Society Press.

Bieri, R. H. (1971). Dissolved noble gases in marine waters. *Earth and Planetary Science Letters*, 10, 329–33.

Bieri, R. H. & Koide, M. (1972). Dissolved noble gases in the east equatorial and southeast Pacific. *Journal of Geophysical Research*, 77, 1667–76.

Bieri, R. H., Koide, M. & Goldberg, E. D. (1964). Noble gases in sea water. *Science*, 146, 1035–7.

Bieri, R. H., Koide, M. & Goldberg, E. D. (1966). The noble gas contents of Pacific seawaters. *Journal of Geophysical Research*, 71, 5243–65.

Bieri, R. H., Koide, M. & Goldberg, E. D. (1967). Geophysical implications of the excess helium found in Pacific waters. *Journal of Geophysical Research*, 72, 2497–511.

Bieri, R. H., Koide, M. & Goldberg, E. D. (1968). Noble gas contents of marine waters. *Earth and Planetary Science Letters*, 4, 329–40.

Black, D. C. (1970). Trapped helium–neon isotopic correlations in gas-rich meteorites and carbonaceous chondrites. *Geochimica et Cosmochimica Acta*, 34, 132–40.

Black, D. C. (1971). Trapped neon–argon isotopic correlations in gas-rich meteorites and carbonaceous chondrites. *Geochimica et Cosmochimica Acta*, 35, 230–5.

Black, D. C. (1972a). On the origins of trapped helium, neon, and argon isotopic variations in meteorites – I. Gas-rich meteorites, lunar soil and breccia. *Geochimica et Cosmochimica Acta*, 36, 347–75.

Black, D. C. (1972b). On the origins of trapped helium, neon and argon isotopic variations in meteorites – II. Carbonaceous meteorites. *Geochimica et Cosmochimica Acta*, 36, 377–94.

Black, D. C. & Pepin, R. O. (1969). Trapped neon in meteorites – II. *Earth and Planetary Science Letters*, 6, 395–405.

Blander, M., Grimes, W. R., Smith, N. V. & Watson, G. N. (1959). Solubility of the noble gases in molten fluorides. II. In the LiF–NaF–KF eutectic mixture. *Journal of Physical Chemistry*, 63, 1164–7.

Bochsler, P., Stettler, A., Bird, J. M. & Weathers, M. S. (1978). Excess ^{3}He and ^{21}Ne in josephinite. *Earth and Planetary Science Letters*, 39, 67–74.

Bogard, D. D. & Gibson, Jr, E. K. (1978). The origin and relative abundance of C, N, and the noble gases of the terrestrial planets and in meteorites. *Nature*, 271, 150–3.

Bogard, D. D., Rowe, M. W., Manuel, O. K. & Kuroda, P. K. (1965). Noble gas anomalies in the mineral thucholite. *Journal of Geophysical Research*, 70, 703–8.

Boulos, M. S. & Manuel, O. K. (1971). The xenon record of extinct radioactivities in the earth. *Science*, 174, 1334–6.

Broecker, W. S. (1974). *Chemical Oceanography*. New York: Harcourt Brace Jovanovich, Inc.

Brown, H. (1952). Rare gases and formation of the earth's atmosphere. In *The Atmospheres of the Earth and Planets*, 2nd edn, ed. G. P. Kuiper, pp. 258–66. Chicago: University of Chicago Press.

Brunauer, S., Emmett, P. H. & Teller, E. (1938). Adsorption of gases in multimolecular layers. *Journal of the American Chemical Society*, 60, 309–19.

Bulter, W. A., Jeffery, P. M., Reynolds, J. H. & Wasserburg, G. J. (1963). Isotopic variations in terrestrial xenon. *Journal of Geophysical Research*, 68, 3283–91.

Burbidge, E. M., Burbidge, G. R., Fowler, W. A. & Hoyle, F. (1957). Synthesis of the elements in stars. *Reviews of Modern Physics*, 29, 547–650.

Cadogan, P. H. (1977). Paleoatmospheric argon in Rynie cherts. *Nature*, 268, 38–40.

Cameron, A. G. W. (1973). Abundances of the elements in the solar system. *Space Science Reviews*, 15, 121–46.

Canalas, R. A., Alexander, Jr, E. C. & Manuel, O. K. (1968). Terrestrial abundance of noble gases. *Journal of Geophysical Research*, 73, 3331–4.

Carlson, R. W., MacDougall, J. D. & Lugmair, G. W. (1978). Differential Sm/Nd evolution in oceanic basalts. *Geophysical Research Letters*, 5, 229–32.

Carslaw, H. S. & Jaeger, J. C. (1959). *Conduction of Heat in Solids*. Oxford: Clarendon Press.

Civetta, L., Cortini, M. & Gasparini, P. (1973). Interpretation of a discordant K-Ar age pattern (Capo Vaticano, Calabria). *Earth and Planetary Science Letters*, 20, 113–18.

Claasen, H. H. (1966). *The Noble Gases*. Boston: D. C. Heath & Co.

Clarke, W. B., Beg, M. A. & Craig, H. (1969). Excess ^{3}He in the sea: evidence for terrestrial primordial helium. *Earth and Planetary Science Letters*, 6, 213–20.

Clarke, W. B., Beg, M. A. & Craig, H. (1970). Excess helium 3 at the North Pacific Geosecs station. *Journal of Geophysical Research*, 75, 7676–8.

Clayton, D. D. (1968). *Principles of Stellar Evolution and Nucleosynthesis.* New York: McGraw-Hill.

Clayton, R. N. (1978). Isotopic anomalies in the early solar system. *Annual Review of Nuclear and Particle Science*, **28**, 501–22.

Clayton, R. N., Grossman, L. & Mayeda, T. K. (1973). A component of primitive nuclear composition in carbonaceous meteorites. *Science*, **182**, 485–8.

Clayton, R. N., Onuma, N. & Mayeda, T. K. (1976). A classification of meteorites based on oxygen isotopes. *Earth and Planetary Science Letters*, **30**, 10–18.

Cook, G. A. (ed.) (1961). *Argon, Helium and the Rare Gases: The Elements of the Helium Group.* 2 volumes. New York: Interscience Publishing Co.

Crabb, J. & Anders, E. (1981). Noble gases in E-chondrites. *Geochimica et Cosmochimica Acta*, **28**, 595–607.

Craig, H. (1963). The isotopic geochemistry of water and carbon in geothermal areas. In *Nuclear Geology on Geothermal Areas*, (Sept. 9–13, 1963), pp. 17–53. Pisa: Consiglio Nagionale delle Ricerche, Laboratoirio di Geologia Nucleare.

Craig, H. (1966). Isotopic composition and origin of the Red Sea and Salton Sea geothermal brines. *Science*, **154**, 1544–8.

Craig, H. (1969). Geochemistry and origin of the Red Sea brines. In *Hot Brines and Recent Heavy Metal Deposits*, ed. E. T. Degens & D. A. Ross, pp. 208–42. Holland: Springer-Verlag.

Craig, H. & Clarke, W. B. (1970). Oceanic ^{3}He: contribution from cosmogenic tritium. *Earth and Planetary Science Letters*, **9**, 45–8.

Craig, H., Clarke, W. B. & Beg, M. A. (1975). Excess ^{3}He in deep water on the East Pacific Rise. *Earth and Planetary Science Letters*, **26**, 125–32.

Craig, H. & Lal, D. (1961). The production rate of natural tritium. *Tellus*, **13**, 85–105.

Craig, H. & Lupton, J. E. (1976). Primordial neon, helium, and hydrogen in oceanic basalts. *Earth and Planetary Science Letters*, **31**, 369–85.

Craig, H., Lupton, J. E. & Horibe, Y. (1978a). A mantle helium component in circum-Pacific volcanic gases: Hakone, the Marianas, and Mt. Lassen. In *Terrestrial Rare Gases*, ed. E. C. Alexander, Jr & M. Ozima, pp. 3–16. Tokyo: Japan Scientific Society Press.

Craig, H., Lupton, J. E., Welhan, J. A. & Poreda, R. (1978b). Helium isotope ratios in Yellowstone and Lassen Park volcanic gases. *Geophysical Research Letters*, **5**, 897–900.

Craig, H., Poreda, R., Lupton, J. E., Marti, K. & Regnier, S. (1979). Rare gases and hydrogen in josephinite (abstract). *EOS, Transactions, American Geophysical Union*, **60**, 970.

Craig, H. & Weiss, R. F. (1968). Argon concentrations in the ocean: a discussion. *Earth and Planetary Science Letters*, **5**, 175–82.

Craig, H. & Weiss, R. F. (1971). Dissolved gas saturation anomalies and excess helium in the ocean. *Earth and Planetary Science Letters*, **10**, 289–96.

Craig, H., Weiss, R. F. & Clarke, W. B. (1967). Dissolved gases in the equatorial and south Pacific Ocean. *Journal of Geophysical Research*, **72**, 6165–81.

Crank, J. (1975). *The Mathematics of Diffusion*, 2nd edn. Oxford: Clarendon Press.

Dalrymple, G. B. (1969). ^{40}Ar/^{36}Ar analyses of historic lava flows. *Earth and Planetary Science Letters*, **6**, 47–55.

Dalrymple, G. B. & Moore, J. G. (1968). Argon 40: excess in submarine pillow basalts from Kilauea volcano, Hawaii. *Science*, **161**, 1132–5.

Damon, P. E. & Kulp, J. L. (1958a). Excess helium and argon in beryl and other minerals. *American Mineralogist*, **43**, 433–59.

Damon, P. E. & Kulp, J. L. (1958b). Inert gases and the evolution of the atmosphere. *Geochimica et Cosmochimica Acta*, **13**, 280–92.

Dempster, A. J. (1918). A new method of positive ray analysis. *Physical Review*, **11**, 316–25.

Dietz, R. S. (1964). Sudbury structure as an astrobleme. *Journal of Geology*, **72**, 412–34.

Donahue, T. M., Hoffman, J. H. & Hodges, Jr, R. R. (1981). Krypton and xenon in atmosphere of Venus. *Geophysical Research Letters*, 8, 513–16.

Donahue, T. M. & Pollack, J. B. (1983). Origin and evolution of the atmosphere of Venus. In *Proceedings of the International Conference on the Venus Environment*, in press.

Downing, R. G., Hennecke, E. W. & Manuel, O. K. (1977). Josephinite: a terrestrial alloy with radiogenic xenon-129 and the noble gas imprint of iron meteorites. *Geochemical Journal*, 11, 219–29.

Drozd, R. J., Hohenberg, C. M. & Morgan, C. J. (1974). Heavy rare gases from Rabbit Lake (Canada) and the Oklo Mine (Gabon): natural spontaneous chain reactions in old uranium deposits. *Earth and Planetary Science Letters*, 23, 28–33.

Drozd, R. J., Kennedy, B. M., Morgan, C. J., Podosek, F. A. & Taylor, G. J. (1976). The excess fission xenon problem in lunar samples. *Proceedings of Lunar Scientific Conference 7th*, 599–623.

Drozd, R. J., Morgan, C. J., Podosek, F. A., Poupeau, G., Shirck, R. J. & Taylor, G. J. (1977). Plutonium-244 in the early solar system? *Astrophysical Journal*, 212, 567–80.

Drozd, R. J. & Podosek, F. A. (1976). Primordial ^{129}Xe in meteorites. *Earth and Planetary Science Letters*, 31, 15–30.

Dubey, V. S. & Holmes, A. (1929). Estimates of the ages of the Whin Sill and the Cleveland Dyke by the helium method. *Nature*, 123, 794–5.

Dymond, J. (1970). Excess argon in submarine basalt pillows. *Geological Society of America, Bulletin*, 81, 1229–32.

Dymond, J. & Hogan, L. (1973). Noble gas abundance patterns in deep-sea basalts – primordial gases from the mantle. *Earth and Planetary Science Letters*, 20, 131–9.

Dymond, J. & Hogan, L. (1978). Factors controlling the noble gas abundance patterns of deep-sea basalts. *Earth and Planetary Science Letters*, 38, 117–28.

Eberhardt, P. (1974). A Neon-E rich phase in the Orgueil carbonaceous chondrite. *Earth and Planetary Science Letters*, 24, 182–7.

Eberhardt, P. (1978). A Neon-E rich phase in Orgueil: results of stepwise heating experiments. *Proceedings of Lunar and Planetary Scientific Conference 9th*, 1027–51.

Eberhardt, P., Eugster, O. & Marti, K. (1965). A redetermination of the isotopic composition of atmospheric neon. *Zeitschrift für Naturforschung*, 20a, 623–4.

Eberhardt, P., Geiss, J., Graf, H., Grögler, N., Krähenbühl, U., Schwaller, H., Schwarzmüller, J. & Stettler, A. (1970). Trapped solar wind noble gases, exposure age and K/Ar age in Apollo 11 lunar fine material. *Proceedings of the Apollo 11 Lunar Science Conference*, 1037–70.

Eberhardt, P., Geiss, J., Graf, H., Grögler, N., Mendia, M. D., Mörgeli, M., Schwaller, H. & Stettler, A. (1972). Trapped solar wind gases in Apollo 12 lunar fines 12001 and Apollo 11 breccia 10046. *Proceedings of Third Lunar Science Conference*, 2, 1821–56.

Emerson, D. E., Stroud, L. & Meyer, T. O. (1966). The isotopic abundance of neon from helium-bearing natural gases. *Geochimica et Cosmochimica Acta*, 30, 847–54.

Eugster, O., Eberhardt, P. & Geiss, J. (1967). Krypton and xenon isotopic composition in three carbonaceous chondrites. *Earth and Planetary Science Letters*, 3, 249–57.

Fanale, F. P. (1971). A case for catastrophic early degassing of the earth. *Chemical Geology*, 8, 79–105.

Fanale, F. P. (1976). Martian volatiles: their degassing history and geochemical fate. *Icarus*, 28, 179–202.

Fanale, F. P. & Cannon, W. A. (1971a). Physical adsorption of rare gas on terrigenous sediments. *Earth and Planetary Science Letters*, 11, 362–86.

Fanale, F. P. & Cannon, W. A. (1971b). Adsorption on the Martian regolith. *Nature*, **230**, 502–4.

Fanale, F. P. & Cannon, W. A. (1972). Origin of planetary primordial rare gas: the possible role of adsorption. *Geochimica et Cosmochimica Acta*, **36**, 319–28.

Fanale, F. P., Cannon, W. A. & Owen, T. (1978). Mars: regolith adsorption and the relative concentrations of atmospheric rare gases. *Geophysical Research Letters*, **5**, 77–80.

Farrar, H., Fickel, H. R. & Tomlinson, R. H. (1962). Cumulative yields of light fragments in ^{235}U thermal neutron fission. *Canadian Journal of Physics*, **40**, 1017–26.

Farrar, H. & Tomlinson, R. H. (1962). Cumulative yields of the heavy fragments in U^{235} thermal neutron fission. *Nuclear Physics*, **34**, 367–81.

Fields, P. R., Friedman, A. M., Milsted, J., Lerner, J., Stevens, C. M., Metta, D. & Sabine, W. K. (1966). Decay properties of plutonium-244, and comments on its existence in nature. *Nature*, **212**, 131–4.

Fisher, D. E. (1970). Heavy rare gases in a Pacific seamount. *Earth and Planetary Science Letters*, **9**, 331–5.

Fisher, D. E. (1971). Excess rare gases in a subaerial basalt from Nigeria. *Nature*, **232**, 60–1.

Fisher, D. E. (1974). The planetary primordial component of rare gases in the deep earth. *Geophysical Research Letters*, **1**, 161–4.

Fisher, D. E. (1975). Trapped helium and argon and the formation of the atmosphere by degassing. *Nature*, **256**, 113–14.

Fisher, D. E. (1979). Helium and xenon in deep-sea basalts as a measure of magmatic differentiation. *Nature*, **282**, 825–7.

Fisher, D. E. (1981). A search for primordial atmospheric-like argon in an iron meteorite. *Geochimica et Cosmochimica Acta*, **45**, 245–9.

Fisher, D. E., Bonatti, E., Joensun, O. & Funkhouser, J. (1968). Ages of Pacific deep-sea basalts, and spreading of the sea floor. *Science*, **160**, 1106–7.

Fleming, W. H. & Thode, H. G. (1953). Argon 38 in pitchblende minerals and nuclear processes in nature. *Physical Review*, **90**, 857–8.

Folland, K. A. (1974). ^{40}Ar diffusion in homogeneous orthoclase and an interpretation of Ar diffusion in K-feldspars. *Geochimica et Cosmochimica Acta*, **38**, 151–66.

Francis, P. W., Moorbath, S. & Thorpe, R. S. (1977). Strontium isotope data for recent andesites in Equador and North Chile. *Earth and Planetary Science Letters*, **37**, 197–202.

Frank, R. C., Swets, D. E. & Lee, R. W. (1961). Diffusion of neon isotopes in fused quartz. *Journal of Chemical Physics*, **35**, 1451–9.

Frick, U. (1977). Anomalous krypton in Allende meteorite. *Proceedings of the Eighth Lunar Science Conference*, **1**, 273–92.

Frick, U. & Chang, S. (1977). Ancient carbon and noble gas fractionation. *Proceedings of the Eighth Lunar Science Conference*, **1**, 263–72.

Frick, U., Mack, R. & Chang, S. (1979). Noble gas trapping and fractionation during synthesis of carbonaceous matter. *Proceedings of the Tenth Lunar and Planetary Science Conference*, **2**, 1961–73.

Frick, U. & Moniot, R. K. (1977). Planetary gas components in Orgueil. *Proceedings of the Eighth Lunar Science Conference*, **1**, 229–61.

Frischat, G. H. & Oel, H. J. (1967). Diffusion of neon in glass melt. *Physics and Chemistry of Glasses*, **8**, 92–5.

Funk, H., Podosek, F. A. & Rowe, M. W. (1967). Fissiogenic xenon in the Renazzo and Murray meteorites. *Geochimica et Cosmochimica Acta*, **31**, 1721–32.

Funkhouser, J. G., Fisher, D. E. & Bonatti, E. (1968). Excess argon in deep-sea rocks. *Earth and Planetary Science Letters*, 5, 95–100.

Funkhouser, J. G. & Naughton, J. J. (1968). Radiogenic helium and argon in ultramafic inclusions from Hawaii. *Journal of Geophysical Research*, 73, 4601–8.

Ganapathy, R. & Anders, E. (1974). Bulk compositions of the moon and earth, estimated from meteorites. *Proceedings of Lunar Scientific Conference 5th*, 1181–1206.

Garrels, R. M. & Mackenzie, F. T. (1972). A quantitative model for the sedimentary rock cycle. *Marine Chemistry*, 1, 27–41.

Geiss, J., Buehler, F., Cerutti, H., Eberhardt, P. & Filleaux, C. H. (1972). Solar wind composition experiments. *Apollo 15 Preliminary Scientific Report*, Ch. 15.

Gerling, E. K. & Levskii, L. K. (1956). On the origin of the rare gases in stony meteorites. *Doklady Akademiya Nauk SSSR*, 10, 750.

Gerling, E. K., Mamyrin, B. A., Tolstikhin, I. N. & Yakovlevka, S. S. (1971). Helium isotope composition in some rocks. *Geochemical International*, 8, 755–61.

Gerling, E. K., Morozova, I. M. & Sprintsson, V. D. (1968). On the nature of the excess argon in some minerals. *23rd International Geological Congress*, 6, 9–15.

GHCSBHP (The Group of Hydro-Chemistry, The Seismological Brigade of Hebei Province) (1975). Studies on forecasting earthquakes in light of the abnormal variations of Rn concentration in groundwater. *Acta Geophysica Cino*, 18 (4), 279–83.

Giletti, B. J. & Tullis, J. (1977). Studies in diffusion – IV. Pressure dependence of Ar diffusion in phlogopite. *Earth and Planetary Science Letters*, 35, 180–3.

Ginzburg, A. I. & Panteleyev, A. I. (1971). Excess argon in beryl. *Geokhimiya*, 12, 1218–27.

Göbel, R., Ott, U. & Begemann, F. (1978). On trapped noble gases in ureilites. *Journal of Geophysical Research*, 83, 855–67.

Goettel, K. A. (1976). Models for the origin and composition of the earth, and the hypothesis of potassium in the earth's core. *Geophysical Surveys*, 2, 369–97.

Goldreich, P. & Ward, W. R. (1973). The formation of planetesimals. *Astrophysical Journal*, 183, 1051–61.

Gooding, J. L., Keil, K., Mayeda, T. K., Clayton, R. N., Fukuoka, T. & Schmitt, R. A. (1980). Oxygen isotopic compositions of petrologically characterized chondrules from unequilibrated chondrites (abstract). *Meteoritics*, 15, 295.

Gramlich, J. W. & Naughton, J. J. (1972). Nature of source material for ultramafic minerals from Salt Lake Crater, Hawaii, from measurement of helium and argon diffusion. *Journal of Geophysical Research*, 77, 3032–42.

Green II, H. W. (1972). A CO_2 charged asthenosphere. *Science*, 238, 2–5.

Grimes, W. R., Smith, N. V. & Watson, G. M. (1958). Solubility of noble gases in molten fluorides. – I. In mixtures of NaF–ZrF_4 (53–47 mole %) and NaF–ZrF_4–UF_4 (50–46–4 mole %). *Journal of Physical Chemistry*, 62, 862–6.

Hall, H. T. & Murthy, V. R. (1971). The early chemical history of the earth: some critical elemental fractionations. *Earth and Planetary Science Letters*, 11, 239–44.

Halpern, I. (1971). Three fragment fission. *Annual Review of Nuclear Science*, 21, 245–94.

Hamano, Y. & Ozima, M. (1978). Earth-atmosphere evolution model based on Ar isotopic data. In *Terrestrial Rare Gases*, ed. E. C. Alexander Jr & M. Ozima, pp. 155–71. Tokyo: Japan Scientific Society Press.

Hamilton, E. I. & Dears, T. (1963). Isotopic composition of strontium in some African carbonatites and limestones and in strontium minerals. *Nature*, 198, 776–7.

Harper, C. T. & Schamel, S. (1971). Note on the isotopic composition of argon in quartz veins. *Earth and Planetary Science Letters*, 12, 129–33.

Hart, S. R. & Dodd, R. T. (1962). Excess radiogenic argon in pyroxenes, *Journal of Geophysical Research*, 61, 2998–9.

Hart, S. R., Dymond, J. & Hogan, L. (1979). Preferential formation of the atmosphere-sialic crust system from the earth. *Nature*, 278, 156–9.

Hawksworth, C. J., O'Nions, R. K. & Pankhurst, R. J. (1977). A geochemical study of island-arc and back-arc tholeiites from the Scotia Sea. *Earth and Planetary Science Letters*, 36, 253–62.

Hayashi, C. (1972). Origin of the solar system (in Japanese). *Report of the 5th Symposium on the Moon and Planets*, pp. 13–18. The Space and Aeronautical Research Institute, University of Tokyo.

Hayashi, C., Nakazawa, K. & Mizuno, H. (1979). Earth's melting due to the blanketing effect of the primordial dense atmosphere. *Earth and Planetary Science Letters*, 43, 22–8.

Hayatsu, A. & Palmer, H. C. (1975). K–Ar isochron study of the Tudor gabbro, Grenville Province, Ontario. *Earth and Planetary Science Letters*, 25, 208–12.

Hayatsu, A. & Waboso, C. E. (1982). The solubility of rare gases in rock melts: implications for K–Ar dating and earth degassing. *Abstracts: Fifth International Conference on Geochronology, Cosmochronology, and Isotope Geology, Nikko, Japan*, 139–40.

Hebeda, E. H., Boelrijk, N. A. I. M., Priem, H. N. A., Verdurmen, E. A. Th., Verschure, R. H. & Simon, O. J. (1980). Excess radiogenic Ar and undisturbed Rb–Sr systems in basic intrusives subjected to alpine metamorphism in southeastern Spain. *Earth and Planetary Science Letters*, 47, 81–90.

Hennecke, E. W. & Manuel, O. K. (1971). Mass fractionation and the isotopic anomalies of xenon and krypton in ordinary chondrites. *Zeitschrift für Naturforschung*, 26a, 1980–5.

Hennecke, E. W. & Manuel, O. K. (1975a). Noble gases in lava rock from Mount Capulin, New Mexico. *Nature*, 256, 284–7.

Hennecke, E. W. & Manuel, O. K. (1975b). Noble gases in an Hawaiian xenolith. *Nature*, 257, 778–80.

Hennecke, E. W. & Manuel, O. K. (1975c). Noble gases in CO_2 well gas, Harding County, New Mexico. *Earth and Planetary Science Letters*, 27, 356–55.

Hennecke, E. W. & Manuel, O. K. (1977). Argon, krypton and xenon in iron meteorites. *Earth and Planetary Science Letters*, 36, 29–43.

Heymann, D., Dziczkaniec, M. & Palma, R. (1976). Limits for the accretion time of the earth from cosmogenic ^{21}Ne produced in planetesimals. *Proceedings of the Seventh Lunar Science Conference*, 3, 3411–19.

Hiyagon, H. (1981). Preliminary studies on partition of rare gases between crystals and melts. MA Thesis, University of Tokyo.

Hiyagon, H. & Ozima, M. (1982). Noble gas distribution between basalt melt and crystals. *Earth and Planetary Science Letters*, 58, 255–64.

Hoffman, J. H., Hodges, R. R., Donahue, T. M. & McElroy, M. B. (1980). Composition of the Venus lower atmosphere from the Pioneer Venus mass spectrometer. *Journal of Geophysical Research*, 85, 7882–90.

Hohenberg, C. M., Davis, P. K., Kaiser, W. A., Lewis, R. S. & Reynolds, J. H. (1970). Trapped and cosmogenic rare gases from stepwise heating of Apollo 11 samples. *Proceedings of the Apollo 11 Lunar Science Conference*, 2, 1283–1309.

Hohenberg, C. M., Podosek, F. A. & Reynolds, J. H. (1967). Xenon–iodine dating: sharp isochronism in chondrites. *Science*, 156, 202–6.

Holland, J. G. & Lambert, R. St J. (1972). Major elemental and chemical composition of shields and the continental crust. *Geochemica et Cosmochimica Acta*, 36, 673–83.

Holloway, J. R. (1966). Fluids in the evolution of granitic magmas: consequences of finite CO_2 solubility. *Geological Society of America, Bulletin*, 87, 1513–18.

Holub, R. F. & Brady, B. T. (1981). The effect of stress on radon emanation from rock. *Journal of Geophysical Research*, 86, 1776–85.

Honda, M., Kurita, K., Hamano, Y. & Ozima, M. (1982). Experimental studies of He and Ar degassing during rock fracturing. *Earth and Planetary Science Letters*, 59, 429–36.

Honda, M., Ozima, M., Nakada, Y. & Onaka, T. (1979). Trapping of rare gases during the condensation of solids. *Earth and Planetary Science Letters*, 43, 197–200.

Horn, M. K. & Adams, J. A. S. (1966). Computer-derived geochemical balances and element abundances. *Geochimica et Cosmochimica Acta*, 30, 279–97.

Hostetler, C. J. (1981). A possible common origin for the rare gases on Venus, Earth, and Mars. *Proceedings of the Twelfth Lunar and Planetary Science Conference*, 12B, 1387–93.

Hudson, G. B., Flynn, G. J., Fraundorf, P., Hohenberg, C. M. & Shirck, J. (1981). Noble gases in stratospheric dust particles: confirmation of extraterrestrial origin. *Science*, 211, 383–6.

Hudson, G. B., Kennedy, B. M., Podosek, F. A. & Hohenberg, C. M. (1983). The early solar system abundance of ^{244}Pu as inferred from the St. Severin chondrite. *Journal of Geophysical Research*, submitted.

Hughes, D. W. (1978). Meteors. In *Cosmic Dust*, ed. J. A. M. McDonnell, pp. 123–85, New York: Wiley.

Huneke, J. C. & Smith, S. P. (1976). The realities of recoil: ^{39}Ar recoil out of small grains and anomalous age patterns in ^{39}Ar-^{40}Ar dating. *Proceedings of the Lunar Science Conference 7th*, 1987–2008.

Hurley, P. M., Hughes, H., Faure, G., Fairbain, H. W. & Pinson, W. H. (1962). Radiogenic strontium-87 model of continent formation. *Journal of Geophysical Research*, 67, 5315–34.

Hyde, E. K. (1964). *The Nuclear Properties of the Heavy Elements III, Fission Phenomena*. Englewood Cliffs (NJ): Prentice-Hall.

Hyman, H. H. (ed.) (1963). *Noble-gas Compounds*, Chicago: The University of Chicago Press.

Inghram, M. G. & Reynolds, J. H. (1950). Double beta-decay of Te^{130}. *Physical Review*, 78, 822–3.

Istomin, V. G., Grechnev, K. V. & Kochnev, V. A. (1980). Mass spectrometry measurements of the composition of the lower atmospheres of Venus. *NASA Translation Report*, TM-75477.

Jaeger, T. A. (1940). Magmatic gases. *American Journal of Science*, 278, 313–53.

Jambon, A. & Shelby, J. E. (1980). Helium diffusion and solubility in obsidians and basaltic glass in the range 200–300 °C. *Earth and Planetary Science Letters*, 51, 206–14.

Jeffrey, P. M. & Anders, E. (1970). Primordial noble gases in separated meteoritic minerals – I. *Geochimica et Cosmochimica Acta*, 34, 1175–98.

Jeffrey, P. M. & Reynolds, J. H. (1961). Origin of excess Xe^{129} in stone meteorites. *Journal of Geophysical Research*, 66, 3582–3.

Jenkins, W. J., Beg, M. A., Clarke, W. B., Wangersky, P. J. & Craig, H. (1972). Excess ^{3}He in the Atlantic Ocean. *Earth and Planetary Science Letters*, 16, 122–6.

Jenkins, W. J. & Clarke, W. B. (1976). The distribution of ^{3}He in the western Atlantic Ocean. *Deep Sea Research*, 23, 481–94.

Jenkins, W. J., Edmond, J. M. & Corliss, J. B. (1978). Excess ^{3}He and ^{4}He in Galapagos submarine hydrothermal waters. *Nature*, 272, 156–8.

Johnson, H. E. & Axford, W. I. (1969). Production and loss of He^3 in the earth's atmosphere. *Journal of Geophysical Research*, 74, 2433–8.

Jokipii, J. R. (1964). The distribution of gases in the protoplanetary nebula. *Icarus*, 3, 248–52.

Kaneoka, I. (1974). Investigation of excess argon in ultramafic rocks from the Kola Peninsula by the ^{40}Ar/^{39}Ar method. *Earth and Planetary Science Letters*, 22, 145–56.

Kaneoka, I. (1980). Rare gas isotopes and mass fractionation: an indicator of gas transport into or from a magma. *Earth and Planetary Science Letters*, 48, 284–92.

Kaneoka, I. & Takaoka, N. (1978). Excess ^{129}Xe and high ^{3}He/^{4}He ratios in olivine pheno-crysts of Kapuho lava and xenolithic dunites from Hawaii. *Earth and Planetary Science Letters*, 39, 382–6.

Kaneoka, I. & Takaoka, N. (1980). Rare gas isotopes in Hawaiian ultramafic nodules and volcanic rocks: constraint on genetic relationships. *Science*, 208, 1366–8.

Kaneoka, I., Takaoka, N. & Aoki, K. (1978). Rare gases in mantle-derived rocks and minerals. In *Terrestrial Rare Gases*, ed. E. C. Alexander, Jr & M. Ozima, pp. 71–83. Tokyo: Japan Scientific Society Press.

Kester, D. R. (1975). Dissolved gases other than CO_2. In *Chemical Oceanography*, 2nd edn vol. 1, ed. J. P. Riley & G. Skirro, pp. 498–556. New York: Academic Press.

King, C. Y. (1978). Episodic radon changes in subsurface soil gas along active faults and possible relation to earthquakes. *Journal of Geophysical Research*, 85, 3065–78.

Kirsten, T. (1968). Incorporation of rare gases in solidifying enstatite melts. *Journal of Geophysical Research*, 73, 2807–10.

Kirsten, T. & Gentner, W. (1966). K–Ar Alterbestimmungen an Ultrabasiten des Baltischen Schildes. *Zeitschrift für Naturforschung*, 21a, 119–26.

Kirsten, T. & Müller, O. (1967). Argon and potassium in mineral fractions of three ultra-mafic rocks from the Baltic Shield. In *Radioactive Dating and Methods of Low Level Counting*, pp. 438–97. Vienna: International Atomic Energy Agency.

Kirsten, T., Richter, H. & Storzer, D. (1981). Abundance patterns of rare gases in sub-marine basalt and glasses (abstract). *Meteoritics*, 16, 341.

Klots, C. E. & Benson, B. B. (1963). Isotopic effect in the solution of oxygen and nitrogen in distilled water. *Journal of Chemical Physics*, 38, 890–2.

Kockarts, G. (1973). Helium in the terrestrial atmosphere. *Space Science Reviews*, 14, 723–57.

Kockarts, G. & Nicolet, M. (1962). Le probléme aéronomique de l'hélium et de l'hydrogéne neutres. *Annales de Geophysique*, 18, 269.

Konig, H. (1963). Über die Löslichkeit der Edelgase in Meerwasser. *Zeitschrift für Natur-forschung*, 18, 363–7.

Kothari, B. K., Marti, K., Niemeyer, S., Regnier, S. & Stephens, J. R. (1979). Noble gas trapping during condensation: a laboratory study (abstract). In *Lunar and Planetary Science X*, pp. 682–84. Houston: Lunar and Planetary Institute.

Krummenacher, D. (1970). Isotopic composition of argon in modern surface volcanic rocks. *Earth and Planetary Science Letters*, 8, 109–17.

Krummenacher, D., Merrihue, C. M., Pepin, R. O. & Reynolds, J. H. (1962). Meteoritic krypton and barium versus the general isotopic anomalies in meteoritic xenon. *Geochimica et Cosmochimica Acta*, 26, 231–49.

Kuroda, P. K., Beck, J. N., Efurd, D. W. & Miller, D. K. (1974). Xenon isotope anoma-lies in the carbonaceous chondrite Murray. *Journal of Geophysical Research*, 79, 3981–92.

Kuroda, P. K. & Sherrill, R. D. (1977). Xenon and krypton isotope anomalies in the Besner Mine, Ontario, thucholite. *Geochemical Journal*, 11, 9–19.

Kuroda, P. K., Sherrill, R. D. & Jackson, K. C. (1977). Abundances and isotopic compositions of rare gases in granites. *Geochemical Journal*, 11, 75–90.

Kurz, M. D. & Jenkins, W. J. (1981). The distribution of helium in oceanic basalt glasses. *Earth and Planetary Science Letters*, 53, 41–54.

Kyser, T. K. & Rison, W. (1982). Systematics of rare gas isotopes in basic lavas and ultra-mafic xenoliths. *Journal of Geophysical Research*, 87, 5611–30.

Lancet, M. S. & Anders, E. (1973). Solubility of noble gases in magnetite: implications for planetary gases in meteorites. *Geochimica et Cosmochimica Acta*, 37, 1371–88.

Langmuir, I. (1918). The adsorption of gases on plane surfaces of glass, mica and platinum. *Journal of the American Chemical Society*, **40**, 1361-1403.

Larimer, J. W. & Anders, E. (1967). Chemical fractionation in meteorites - II. Abundance patterns and their interpretation. *Geochimica et Cosmochimica Acta*, **31**, 1239-70.

Lederer, C. M. & Shirley, V. S. (1978). *Table of Isotopes*, (7th edn.) New York: Wiley, 1523 pp.

Lewis, R. S. (1975). Rare gases in separated whitlockite from the St. Severin chondrite: xenon and krypton from fission of extinct ^{244}Pu. *Geochimica et Cosmochimica Acta*, **39**, 417-32.

Lewis, R. S. & Anders, E. (1981). Isotopically anomalous xenon in meteorites: a new clue to its origin. *Astrophysical Journal*, **247**, 1122-4.

Lewis, R. S., Hertogen, J., Alaerts, L. & Anders, E. (1979). Isotopic anomalies in meteorites and their origins - V. Search for fission fragment recoils in Allende sulfides. *Geochimica et Cosmochimica Acta*, **43**, 1743-52.

Lewis, R. S., Srinivasan, B. & Anders, E. (1975). Host phase of a strange xenon component in Allende. *Science*, **190**, 1251-62.

Li, Y. H. (1972). Geochemical mass balance among lithosphere, hydrosphere, and atmosphere. *American Journal of Science*, **72**, 119-37.

London, F. (1930). Zur Theorie und Systematik der Molekularkräfte. *Zeitschrift für Physik*, **63**, 245-79.

Lovering, J. F. & Richards, J. R. (1964). Potassium–argon age study of possible lower-crust and upper-mantle inclusions in deep-seated intrusions. *Journal of Geophysical Research*, **69**, 4895-901.

Lupton, J. E. (1976). The ^{3}He distribution in deep water over the Mid-Atlantic Ridge. *Earth and Planetary Science Letters*, **32**, 371-4.

Lupton, J. E. (1979). Helium-3 in the Guaymas Basin: evidence for injection of mantle volatiles in the Gulf of California. *Journal of Geophysical Research*, **84**, 7446-52.

Lupton, J. E. & Craig, H. (1975). Excess ^{3}He in oceanic basalts: evidence for terrestrial primordial helium. *Earth and Planetary Science Letters*, **26**, 133-9.

Lupton, J. E. & Craig, H. (1981). A major helium-3 source at 15°S on the East Pacific Rise. *Science*, **214**, 13-18.

Lupton, J. E., Weiss, R. F. & Craig, H. (1977a). Mantle helium in the Red Sea brines. *Nature*, **266**, 244-6.

Lupton, J. E., Weiss, R. F. & Craig, H. (1977b). Mantle helium in hydrothermal plumes in the Galapagos Rift. *Nature*, **267**, 603-4.

McDougall, I. (1977). Uranium in marine basalts: concentration, distribution, and implications. *Earth and Planetary Science Letters*, **35**, 65-70.

McDougall, I. & Green, D. H. (1964). Excess radiogenic argon in pyroxenes and isotopic ages on minerals from Norwegian eclogite. *Norsk Geologisk Tidsskrift*, **44**, 183-96.

McDougall, I., Polack, H. A. & Stipp, J. J. (1969). Excess radiogenic argon in young subareal basalts from the Auckland volcanic field, New Zealand. *Geochimica et Cosmochimica Acta*, **33**, 1485-1520.

Macedo, C. R., Costa, C. V., Ferreira, J. T. & Reynolds, J. H. (1977). Rare-gas dating, III. Evaluation of a double-spiking procedure for potassium–argon dating. *Earth and Planetary Science Letters*, **34**, 411-18.

McElroy, M. B., Yung, Y. L. & Nier, A. O. (1978). Isotopic composition of nitrogen: implications for the past history of Mars' atmosphere. *Science*, **194**, 70-2.

MacGregor, I. D. & Basu, A. R. (1974). Thermal structure of the lithosphere: a petrologic model. *Science*, **185**, 1007-11.

McSween, Jr, H. Y. (1979). Are carbonaceous chondrites primitive or processed? A review. *Review of Geophysics and Space Physics*, **17**, 1059-78.

Magaritz, M., Whitford, D. J. & James, D. E. (1978). Oxygen isotopes and the origin of high $^{87}Sr/^{86}Sr$ andesites. *Earth and Planetary Science Letters*, **40**, 220–30.

Mamyrin, B. A., Anufriyev, G. S., Kamenskii, I. L. & Tolstikhin, I. N. (1970). Determination of the isotopic composition of atmospheric helium. *Geochemistry International*, **7**, 498–505.

Manuel, O. K., Hennecke, E. W. & Sabu, D. D. (1972). Xenon in carbonaceous chondrites. *Nature*, **240**, 99–101.

Marti, K. (1967). Isotopic composition of trapped krypton and xenon in chondrites. *Earth and Planetary Science Letters*, **3**, 243–8.

Marti, K. (1969). Solar-type xenon: a new isotopic composition of xenon in the Pesyanoe meteorite. *Science*, **166**, 1263–5.

Mason, B. (1962). *Meteorites*. New York: Wiley, 274 pp.

Matsuda, J., Lewis, R. S., Takahashi, H. & Anders, E. (1980). Isotopic anomalies of noble gases in meteorites and their origins – VII. C3V carbonaceous chondrites. *Geochimica et Cosmochimica Acta*, **44**, 1861–74.

Matsuo, S. (1961). On the chemical nature of fumarolic gases of volcano Showa-Shinzan, Hokkaido, Japan. *Journal of Earth Science, Nagoya University*, **9**, 80–100.

Matsuo, S. (1979). Volcanic gases. In *Chikyu-Kagaku* (in Japanese) ed. I. Yokoyama, S. Aramaki & K. Nakamura, pp. 121–32. Tokyo: Iwanami-Shoten.

Matsuo, S. & Miyake, Y. (1966). Gas composition in ice samples from Antarctica. *Journal of Geophysical Research*, **71**, 5235–41.

Matsuo, S., Suzuki, M. & Mizutani, Y. (1978). Nitrogen to argon ratio in volcanic gases. In *Terrestrial Rare Gases*, ed. E. C. Alexander, Jr & M. Ozima, pp. 17–25. Tokyo: Japan Scientific Society Press.

Mazor, E. (1972). Paleotemperatures and other hydrological parameters deduced from noble gases dissolved in groundwaters; Jordan Rift Valley, Israel. *Geochimica et Cosmochimica Acta*, **36**, 1321–36.

Mazor, E. (1975). Atmospheric and radiogenic noble gases in thermal waters: their potential application to prospecting and steam production studies. *Proceedings 2nd UN Symposium on the Development and Use of Geothermal Resources. San Francisco*, pp. 793–802.

Mazor, E. & Fournier, R. O. (1973). More on noble gases in Yellowstone National Park hot waters. *Geochemica et Cosmochimica Acta*, **37**, 515–25.

Mazor, E., Heymann, D. & Anders, E. (1970). Noble gases in carbonaceous chondrites. *Geochimica et Cosmochimica Acta*, **34**, 781–824.

Mazor, E. & Wasserburg, G. J. (1965). Helium, neon, argon, krypton and xenon in gas emanations from Yellowstone and Lassen volcanic National Parks. *Geochimica et Cosmochimina Acta*, **29**, 443–54.

Mazor, E., Wasserburg, G. J. & Craig, H. (1964). Rare gases in Pacific Ocean water. *Deep-Sea Research*, **11**, 929–32.

Meier, F. O., Jungck, M. & Eberhardt, P. (1980). Evidence for pure neon-22 in Orgueil and Murchison. In *Lunar and Planetary Science XI*, pp. 723–5. Houston: Lunar and Planetary Institute.

Meijer, A. (1976). Rb and Sr isotopic data bearing on the origin of volcanic rocks from the Mariana island-arc system. *Geological Society of America, Bulletin*, **87**, 1358–69.

Melton, C. E. & Giardini, A. A. (1976). Experimental evidence that oxygen is the principal impurity in natural diamonds. *Nature*, **263**, 309–10.

Melton, C. E. & Giardini, A. A. (1980). The isotopic composition of argon included in an Arkansas diamond and its significance. *Geophysical Research Letters*, **7**, 461–4.

Merrihue, C. & Turner, G. (1966). Potassium–argon dating by activation with fast neutrons. *Journal of Geophysical Research*, **71**, 2852–7.

Mirtov, B. A. (1961). *Gaseous Composition of the Atmosphere and its Analysis*. Moscow, (in Russian).

Mizuno, H., Nakazawa, K. & Hayashi, C. (1980). Dissolution of the primordial rare gases into the molten earth's material. *Earth and Planetary Science Letters*, 50, 202–10.

Mogro-Campero, A. & Fleischer, R. L. (1977). Subterrestrial fluid convection: a hypothesis for long-distance migration of radon within the earth. *Earth and Planetary Science Letters*, 34, 321–5.

Moniot, R. K. (1980). Noble-gas-rich separates from ordinary chondrites. *Geochimica et Cosmochimica Acta*, 44, 253–71.

Moorbath, S. (1975). Evolution of Precambrian crust from strontium isotopic evidence. *Nature*, 254, 395–8.

Morgan, W. J. (1972). Plate motions and deep mantle convection. *Geological Society of America, Memoir*, 132, 7–22.

Morgan, J. W. & Anders, E. (1980). Chemical composition of Earth, Venus and Mercury. *Proceedings of the National Academy of Sciences, USA*, 77, 6973–7.

Morrison, P. & Pine, J. (1955). Radiogenic origin of the helium isotopes in rock. *Annals of the New York Academy of Sciences*, 62, 71–9.

Morrison, T. J. & Johnstone, N. B. (1954). Solubilities of the inert gases in water. *Journal of the Chemical Society*, 3, 3441–6.

Mulfinger, H. O. & Scholze, H. (1962). Löslichkeit und Diffusion von Helium in Glasschmelzen, II. Diffusion. *Glastechnische Berichte; Zeitschrift für Glaskunde*, 35, 459–67.

Mysen, B. O. & Boeftchen, A. L. (1975). Melting of a hydrous mantle: II. Geochemistry of crystals and liquids formed by anatexis of mantle peridotite at high pressures and high temperatures as a function of controlled activities of water, hydrogen and carbon dioxide. *Journal of Petrology*, 16, 549–93.

Nagao, K. (1979). Isotopic composition of terrestrial rare gases and application to earth science. PhD Thesis, Osaka University.

Nagao, K., Takaoka, N. & Matsubayashi, O. (1981). Rare gas isotopic compositions in natural gases from Japan. *Earth and Planetary Science Letters*, 53, 175–88.

Nakagawa, Y. (1978). Statistical behavior of planetesimals in the primitive solar system. *Progress in Theoretical Physics*, 59, 1834–51.

Niederer, F. R., Papanastassiou, D. A. & Wasserburg, G. J. (1980). Endemic isotopic anomalies in titanium. *Astrophysical Journal*, 240, L37–L77.

Nief, G. (1960). As reported in: 'Isotopic abundance ratios reported for reference samples stocked by the National Bureau of Standards' (ed. F. Mohler). *NBS Technical Note 51*.

Niemeyer, S. & Leich, D. A. (1976). Atmospheric rare gases in lunar rock 60015. *Proceedings of the Seventh Lunar Science Conference*, 1, 587–97.

Nier, A. O. (1950a). A redetermination of the relative abundances of the isotopes of carbon, nitrogen, oxygen, argon and potassium. *Physical Review*, 77, 789–93.

Nier, A. O. (1950b). A redetermination of the relative abundances of the isotopes of neon, krypton, rubidium, xenon and mercury. *Physical Review*, 79, 450–4.

Okabe, S. (1956). Time variation of the atmospheric radon-content near the ground surface with relation to some geophysical phenomena. *University of Kyoto, Memoir of College of Science*, 28, 99–115.

Okita, T. & Shimozuru, D. (1975). Remote sensing measurements of mass flow of sulfur dioxides. *Bulletin of the Volcanological Society of Japan*, 19, 151–7.

O'Nions, R. K., Evensen, N. M. & Hamilton, P. J. (1979). Geochemical modeling of mantle differentiation and crust growth. *Journal of Geophysical Research*, 84, 6091–101.

Owen, T., Biemann, K., Rushneck, D. R., Biller, J. E., Howarth, D. W. & Lafleur, A. L. (1977). The composition of the atmosphere at the surface of Mars. *Journal of Geophysical Research*, 82, 4635–9.

Owen, T., Crabb, J. & Anders, E. (1983). Venus: a chondritic source for atmospheric noble gases? *Icarus*, in press.

Oyama, V. I., Carle, G. C., Woeller, F., Pollack, J. B., Reynolds, R. T. & Craig, R. A. (1980). Pioneer Venus gas chromatography of the lower atmosphere of Venus. *Journal of Geophysical Research*, **85**, 7891–902.

Ozima, M. (1975). Ar isotopes and earth–atmosphere evolution models. *Geochimica et Cosmochimica Acta*, **39**, 1127–34.

Ozima, M. & Alexander, Jr, E. C. (1976). Rare gas fractionation patterns in terrestrial samples and the earth–atmosphere evolution model. *Review of Geophysics and Space Physics*, **14**, 385–90.

Ozima, M. & Kudo, K. (1972). Excess argon in submarine basalts and an earth–atmosphere evolution model. *Nature Physical Science*, **239**, 23–4.

Ozima, M. & Nakazawa, K. (1980). Origin of rare gases in the earth. *Nature*, **284**, 313–16.

Ozima, M. & Podosek, F. A. (1980). Mantle–crust material interaction from the rare gas viewpoint. *Proceedings of Japan Academy*, Series B, **56**, 260–2.

Ozima, M., Takaoka, N., Nito, O. & Zashu, S. (1982). Ar isotopic ratios and K, Na and other trace element contents in Premier and Finsch mine diamonds. In *Material Sciences of the Earth's Interior*, ed. I. Sunagawa, pp. 377–88. Tokyo: Terra Publishing Co.

Ozima, M. & Takigami, Y. (1980). Activation energy for thermal release of Ar from some DSDP submarine rocks. *Geochimica et Cosmochimica Acta*, **44**, 141–4.

Ozima, M. & Zashu, S. (1983a). Primitive He in diamonds. *Science*, **219**, 1067–8.

Ozima, M. & Zashu, S. (1983b). Noble gases in submarine pillow volcanic glasses. *Earth and Planetary Science Letters*, **62**, 24–40.

Pepin, R. O. (1967). Trapped neon in meteorites. *Earth and Planetary Science Letters*, **2**, 13–18.

Pepin, R. O. (1968). Neon and xenon in carbonaceous chondrites. In *Origin and Distribution of Elements*, ed. L. H. Ahrens, pp. 379–86. New York: Pergamon.

Pepin, R. O. & Phinney, D. (1983). Components of xenon in the solar system. *The Moon and Planets*, in press.

Pepin, R. O. & Signer, P. (1965). Primordial rare gases in meteorites. *Science*, **149**, 253–65.

Pereira, E. B. & Adams, J. A. S. (1982). Helium production in natural gas reservoirs. *Geophysical Research Letters*, **9**, 87–90.

Perkins, W. G. & Begal, D. R. (1971). Diffusion and permeation of He, Ne, Ar, Kr and D_2 through silicon oxide thin films. *Journal of Chemical Physics*, **54**, 1683–94.

Phinney, D. (1972). [36]Ar, Kr and Xe in terrestrial materials. *Earth and Planetary Science Letters*, **16**, 413–20.

Phinney, D., Tennyson, J. & Frick, U. (1978). Xenon in CO_2 well gas revisited. *Journal of Geophysical Research*, **83**, 2313–19.

Podosek, F. A. (1970a). Dating of meteorites by high-temperature release of iodine-correlated [129]Xe. *Geochimica et Cosmochimica Acta*, **34**, 341–65.

Podosek, F. A. (1970b). The abundance of [244]Pu in the early solar system. *Earth and Planetary Science Letters*, **8**, 183–7.

Podosek, F. A. (1972). Gas retention chronology of Petersburg and other meteorites. *Geochimica et Cosmochimica Acta*, **36**, 755–72.

Podosek, F. A. (1978). Isotopic structures in solar system materials. *Annual Review of Astronomy and Astrophysics*, **16**, 293–334.

Podosek, F. A., Bernatowicz, T. J. & Kramer, F. E. (1981). Adsorption of xenon and krypton on shales. *Geochimica et Cosmochimica Acta*, **45**, 2401–15.

Podosek, F. A., Honda, M. & Ozima, M. (1980). Sedimentary noble gases. *Geochimica et Cosmochimica Acta*, **44**, 1875–84.

Podosek, F. A., Huneke, J. C., Burnett, D. S. & Wasserburg, G. J. (1971). Isotopic composition of xenon and krypton in the lunar soil and in the solar wind. *Earth and Planetary Science Letters*, **10**, 199–216.

Polak, B. G., Kononov, V. I., Tolstikhin, I. N., Mamyrin, B. A. & Khabarin, L. V. (1975). The helium isotopes in thermal fluids. In *Thermal and Chemical Problems of Thermal Waters*, ed. A. I. Johnson, pp. 15–29. Grenoble: International Association of Hydrological Science, Publication No. 119.

Pollack, J. B. & Black, D. C. (1979). Implications of the gas compositional measurements of Pioneer Venus for the origin of planetary atmospheres. *Science*, **205**, 56–9.

Pollack, J. B. & Black, D. C. (1982). Noble gases in planetary atmospheres: implications for the origin and evolution of atmospheres. *Icarus*, **51**, 169–98.

Powell, J. L., Hurley, P. M. & Fairbain, H. W. (1962). Isotopic composition of strontium in carbonatites. *Nature*, **196**, 1085–6.

Press, F. & Siever, R. (1973). *Earth*. San Francisco: W. H. Freeman and Co.

Rayleigh, J. W. S. (1933). Beryllium and helium I: the helium contained in beryls of different geological age. *Proceedings of the Royal Society, London*, **A142**, 370–81.

Reeves, H. (1974). On the origin of the light elements. *Annual Review of Astronomy and Astrophysics*, **12**, 437–69.

Reimer, G. M. (1980). Use of soil-gas helium concentrations for earthquake prediction: limitations imposed by diurnal variation. *Journal of Geophysical Research*, **85**, 3107–14.

Revelle, R. & Suess, H. E. (1962). Interchange of properties between sea and atmosphere. In *The Sea*, vol. I, ed. M. N. Hill, New York: Wiley (Interscience).

Reynolds, J. H. (1956). High-sensitivity mass spectrometer for noble gas analysis. *Review of Scientific Instruments*, **27**, 928–34.

Reynolds, J. H. (1960). Isotopic composition of primordial xenon. *Physical Review Letters*, **4**, 351–4.

Reynolds, J. H. (1963). Xenology. *Journal of Geophysical Research*, **68**, 2939–56.

Reynolds, J. H. (1967). Isotopic abundance anomalies in the solar system. *Annual Review of Nuclear Science*, **17**, 253–316.

Reynolds, J. H., Frick, U., Niel, J. M. & Phinney, D. L. (1978). Fare-gas-rich separates from carbonaceous chondrites. *Geochimica et Cosmochimica Acta*, **42**, 1775–97.

Reynolds, J. H. & Turner, G. (1964). Rare gases in the chondrite Renazzo. *Journal of Geophysical Research*, **69**, 3263–81.

Ringwood, A. E. (1975). *Composition and petrology of the earth's mantle*. New York: McGraw-Hill.

Rison, W. (1980a). Isotopic studies of the rare gases in igneous rocks: implications for the mantle and atmosphere. PhD Thesis, University of California (Berkeley).

Rison, W. (1980b). Isotopic fractionation of argon during stepwise release from shungite. *Earth and Planetary Science Letters*, **47**, 383–90.

Roddick, J. C. & Farrar, E. (1971). High initial argon ratios in hornblendes. *Earth and Planetary Science Letters*, **12**, 208–14.

Ross, S. & Olivier, J. P. (1964). *On Physical Adsorption*. New York: Wiley (Interscience).

Rubey, W. W. (1951). Geologic history of sea water. *Geological Society of America, Bulletin*, **62**, 1111–48.

Russell, R. D. (1972). Evolutionary model for lead isotopes in conformable ores and in ocean volcanics. *Review of Geophysics and Space Physics*, **10**, 529–49.

Russell, R. D. & Birnie, D. J. (1974). A bi-directional model for lead isotope evolution. *Physics of the Earth and Planetary Interiors*, 8, 158–66.

Sabu, D. D. & Manuel, O. K. (1980a). Noble gas anomalies and synthesis of the chemical elements. *Meteoritics*, **15**, 117–38.

Sabu, D. D. & Manuel, O. K. (1980b). The neon alphabet game. *Proceedings of Lunar and Planetary Science Conference 11th*, 879–99.

Safronov, V. S. (1969). Evolution of the protoplanetary cloud and formation of the earth and planets (translated into English). *NASA*, TFF-677.

Saito, K., Alexander, Jr, E. C. & Fragon, J. C. (1983). Rare gases in cyclosilicates and cogenetic minerals. *Journal of Geophysical Research*, in press.

Saito, K., Basu, A. R. & Alexander, Jr, E. C. (1978). Planetary-type rare gases in an upper mantle-derived amphibole. *Earth and Planetary Science Letters*, **39**, 274–80.

Sams, J. R., Constabaris, G. & Halsey, G. D. (1960). Second virial coefficients of neon, argon, krypton and xenon with a graphitized carbon black. *Journal of Physical Chemistry*, **64**, 1689–96.

Schilling, J. G. (1973). Iceland mantle plume: geochemical study of Reykjanes Ridge. *Nature*, **242**, 565–71.

Schreyer, W., Yoder, H. S. & Aldrich, L. T. (1960). Synthesis of argon-containing cordierite. *Annual Report of Geophysical Laboratory, Carnegie Institution, Washington*, **59**, 94–6.

Schwartzman, D. W. (1973a). Argon degassing model of the earth. *Nature Physical Science*, **245**, 20–1.

Schwartzman, D. W. (1973b). Argon degassing and the origin of the sialic crust. *Geochimica et Cosmochimica Acta*, **37**, 2479–95.

Sclater, J. G., Jaupart, C. & Galson, D. (1980). The heat flow through oceanic and continental crust and the heat loss of the earth. *Review of Geophysics and Space Physics*, **18**, 269–311.

Segre, E. (1952). Spontaneous fission. *Physical Review*, **86**, 21–8.

Sekiya, M., Nakazawa, K. & Hayashi, C. (1980). Dissipation of the rare gases contained in the primordial earth's atmosphere. *Earth and Planetary Science Letters*, **50**, 197–201.

Shelby, J. E. (1971). Diffusion of helium isotopes in vitreous silica. *Physical Review*, **4**, 2681–6.

Shelby, J. E. (1973). Helium migration in alkali borate glasses. *Journal of Applied Physics*, **44**, 3880–8.

Shelby, J. E. (1974). Helium diffusion and solubility in K_2O–SiO_2 glasses. *Journal of the American Ceramic Society*, **57**, 260–3.

Shelby, J. E. & Eagan, R. J. (1976). Helium migration in sodium–aluminosilicate glasses. *Journal of the American Ceramic Society*, **59**, 420–5.

Shelby, J. E. & Wayen, R. C. (1974). Gas migration in vitreous B_2O_3. *Journal of Applied Physics*, **45**, 2536–9.

Sheldon, W. R. & Kern, J. W. (1972). Atmospheric helium and geomagnetic field reversals. *Journal of Geophysical Research*, 77, 6194–201.

Shepherd, E. S. (1938). The gases in rocks and some related problems, *American Journal of Science*, **35A**, 311–51.

Signer, P. & Suess, H. E. (1963). Rare gases in the Sun, in the atmosphere and in meteorites. In *Earth Science and Meteorites*, ed. J. Geiss, pp. 241–72. New York: Wiley.

Sigurgeirsson, Th. (1962). *Dating recent basalt by the potassium argon method* (in Icelandic). Report of Physical Laboratory of the University of Iceland, 9 pp.

Sill, G. & Wilkening, L. (1978). Ice clathrate as a possible source of the atmospheres of the terrestrial planets. *Icarus*, **33**, 13–22.

Sinha, A. K. & Hart, S. R. (1972). A geochemical test of the subduction hypothesis for general measurements and an absolute growth curve for single stage model. *Carnegie Institution, Washington, Yearbook*, **70**, 335–58.

Smith, J. V. & Schreyer, W. (1962). Location of argon and water in cordierite. *Mineralogical Magazine*, **33**, 226–36.

Smith, S. P. (1978). Noble gases in plutonic igneous rocks. *US Geological Survey Open-File Report 78-701*, pp. 400–2.

Smith, S. P., Huneke, J. C., Rajan, R. S. & Wasserburg, G. J. (1977). Neon and argon in the Allende meteorite. *Geochimica et Cosmochimica Acta*, **41**, 627–47.

Smith, S. P., Huneke, J. C. & Wasserburg, G. J. (1978). Neon in gas-rich samples of the carbonaceous chondrites Mokoida, Murchison, and Cold Bokkeveld. *Earth and Planetary Science Letters*, **39**, 1–13.

Smith, S. P. & Reynolds, J. H. (1981). Excess ^{129}Xe in a terrestrial sample as measured in a pristine system. *Earth and Planetary Science Letters*, **54**, 236–8.

Srinivasan, B. (1976). Barites: anomalous xenon from spallation and neutron-induced reactions. *Earth and Planetary Science Letters*, **31**, 129–41.

Srinivasan, B., Alexander, Jr, E. C. & Manuel, O. K. (1971). Iodine-129 in terrestrial ores. *Science*, **173**, 327–8.

Srinivasan, B. & Anders, E. (1978). Noble gases in the Murchison meteorite: possible relics of s-process nucleosynthesis. *Science*, **201**, 51–6.

Staudacher, Th. & Allègre, C. J. (1982). Terrestrial xenology, *Earth and Planetary Science Letters*, **60**, 389–406.

Steiger, R. H. & Jäger, E. (1977). Subcommission on Geochronology: convention on the use of decay constants in geo- and cosmochronology. *Earth and Planetary Science Letters*, **36**, 359–62.

Stoiber, R. E. & Jepsen, A. (1973). Sulphur dioxide contributions to the atmosphere by volcanoes. *Science*, **182**, 577–8.

Strutt, R. J. (1908). Helium and radioactivity in rare and common minerals. *Proceedings of the Royal Society (London)*, **A80**, 572–94.

Suess, H. E. (1949). Die Haufigkeit der Edelgase auf der Erde und im Kosmos. *Journal of Geology*, **57**, 600–7.

Suess, H. E. (1962). Thermodynamic data on the formation of solid carbon and organic compounds in primitive planetary atmospheres. *Journal of Geophysical Research*, **67**, 2029–34.

Suess, H. E., Wänke, H. & Wlotzka, F. (1964). On the origin of gas-rich meteorites. *Geochimica et Cosmochimica Acta*, **28**, 595–607.

Sugisaki, R. (1978). Changing He/Ar and N_2/Ar ratios of fault air may be earthquake precursors. *Nature*, **275**, 209–11.

Sun, S. S., Tatsumoto, M. & Schilling, J. G. (1975). Mantle mixing along the Reykjanes Ridge axis: lead isotopic evidence. *Science*, **190**, 143–7.

Swets, D. E., Lee, R. W. & Frank, R. C. (1961). Diffusion coefficient of helium in fused quartz. *Journal of Chemical Physics*, **34**, 17–22.

Takagi, J., Hampel, W. & Kirsten, T. (1974). Cosmic-ray muon-induced ^{129}I in tellurium ores. *Earth and Planetary Science Letters*, **24**, 141–50.

Takaoka, N. & Nagao, K. (1980). Rare-gas studies of Cretaceous deep-sea basalts. *Initial Report of the Deep Sea Drilling Project, LI, LII, LIII*, 1121–6.

Takaoka, N. & Ozima, M. (1978). Rare gas isotopic compositions in diamonds. *Nature*, **271**, 45–6.

Talwani, M., Windisch, C. C. & Langseth, Jr, M. G. (1971). Reykjanes Ridge crest: a detailed geophysical study. *Journal of Geophysical Research*, 76, 473–517.

Tanner, A. B. (1964). Radon migration in the ground: a review. In *The Natural Radiation Environment*, ed. J. A. S. Adams & W. M. Lowder, pp. 161–90. Chicago: University of Chicago Press.

Tanner, A. B. (1978). Radon migration in the ground: a supplementary review. *US Geological Survey Open-File Report.*

Tatsumoto, M. (1978). Isotopic composition of lead in oceanic basalt and its implication to mantle evolution. *Earth and Planetary Science Letters*, 38, 63–87.

Teitsma, A. & Clarke, W. B. (1978). Fission xenon isotope dating. *Journal of Geophysical Research*, 83, 5443–53.

Teng, T. L. (1980). Some recent studies on groundwater radon content as an earthquake precursor. *Journal of Geophysical Research*, 85, 3089–99.

Thompson, D. P., Basu, A. R., Hennecke, E. W. & Manuel, O. K. (1978). Noble gases in the earth's mantle. *Physics of the Earth and Planetary Interiors*, 17, 98–107.

Thomson, J. J. (1912). Further experiments on positive rays. *Philosophical Magazine*, 24, 209–53.

Tolstikhin, I. N. (1975). Helium isotopes in the earth's interior and in the atmosphere: a degassing model of the earth. *Earth and Planetary Science Letters*, 26, 88–96.

Tolstikhin, I. N. (1978). A review: some recent advances in isotope geochemistry of light rare gases. In *Terrestrial Rare Gases*, ed. E. C. Alexander, Jr & M. Ozima, pp. 27–62. Tokyo: Japan Scientific Society Press.

Tolstikhin, I. N., Mamyrin, B. A., Khabarin, L. B. & Erlich, E. N. (1974). Isotope composition of helium in ultrabasic xenoliths from volcanic rocks of Kamchatka. *Earth and Planetary Science Letters*, 22, 75–84.

Trimble, V. (1975). The origin and abundances of the chemical elements. *Review of Modern Physics*, 47, 877–976.

Turekian, K. K. (1959). The terrestrial economy of helium and argon. *Geochimica et Cosmochimica Acta*, 17, 37–43.

Turekian, K. K. (1964). Degassing of argon and helium from the earth. In *The Origin and Evolution of Atmospheres and Oceans*, ed. P. J. Brancazio & A. G. W. Cameron, pp. 74–82. New York: Wiley.

Turekian, K. K., Nozaki, Y. & Benninger, L. K. (1977). Geochemistry of atmospheric radon and radon products. *Annual Review of Earth and Planetary Sciences*, 5, 227–55.

Turner, G. (1972). ^{40}Ar-^{39}Ar age and cosmic ray irradiation history of the Apollo 15 anorthosite 15415. *Earth and Planetary Science Letters*, 14, 169–75.

Turner, G., Miller, J. A. & Grastry, R. L. (1966). Thermal history of the Bruderheim meteorite. *Earth and Planetary Science Letters*, 1, 155–62.

Uhlig, H. H. (1937). The solubilities of gases and surface tension. *Journal of Physical Chemistry*, 41, 1215–25.

Ulomov, V. I. & Mavashev, B. Z. (1971). Forerunner of the Tashkent earthquakes, *Izvestiya Akademiya Nauk Uzbek, USSR*, 188–200.

US Standard Atmosphere, 1962. Washington, DC: US Government Printing Office.

Veizer, J. & Jansen, S. L. (1979). Basement and sedimentary recycling and continental evolution. *Journal of Geology*, 87, 341–70.

Verniani, F. (1966). The total mass of the earth's atmosphere. *Journal of Geophysical Research*, 71, 385–91.

von Weizäcker, C. F. (1937). Über die Möglichkeit eines dualen β-Zerfalls von Kalium. *Physikalishe Zeitschrift*, 38, 623–4.

von Zahn, U., Fricke, K. H., Hunten, D. M., Krankowsky, D., Mauersberger, K. & Nier, A. O. (1980). The upper atmosphere of Venus during morning conditions. *Journal of Geophysical Research*, 85, 7829–40.

Wakita, H. (1978). Earthquake prediction and geothermal studies in China. In *Chinese Geophysics*, vol. 1, ed. T. L. Teng & W. H. K. Lee, pp. 443–57. Washington, DC: American Geophysical Union.

Wakita, H., Fujii, N., Matsuo, S., Nagao, K. & Takaoka, N. (1978). 'Helium Spots': caused by a diapiric magma from the upper mantle. *Science*, **200**, 430–2.

Wakita, H., Nakamura, Y., Notsu, K., Noguchi, M. & Asada, T. (1980). Radon anomaly: a possible precursor of the 1978 Izu-Oshima-Kinkai earthquake. *Science*, **207**, 882–3.

Walker, J. C. G. (1977). *Evolution of the Atmosphere*. New York: Macmillan.

Walker, W., Kirouac, G. J. & Rourke, F. M. (1977). *Chart of the Nuclides*. California: General Electric Company. Nuclear Energy Group.

Wasserburg, G. J. & Hayden, R. J. (1955). Ar^{40}-K^{40} dating. *Geochimica et Cosmochimica Acta*, **7**, 51–60.

Wasserburg, G. J., MacDonald, G. J. F., Hoyle, F. & Fowler, W. A. (1964). Relative contributions of uranium, thorium, and potassium to heat production in the earth. *Science*, **143**, 465–7.

Wasserburg, G. J. & Mazor, E. (1965). Spontaneous fission xenon in natural gases. In *Fluids in Subsurface Environments – A Symposium*. pp. 386–98. Tulsa: American Association of Petroleum Geologists, Memoir No. 4.

Wasserburg, G. J., Papanastassiou, D. A. & Lee, T. (1980). Isotopic heterogeneities in the solar system. In *Early Solar System Processes and the Present Solar System, LXXIII Corso*, Bologna: Soc. Italiana di Fisica.

Wasson, J. T. (1974). *Meteorites*. Berlin: Springer-Verlag.

Wedepohl, K. H. (1969). Composition and abundance of common sedimentary rocks. In *Handbook of Geochemistry*, ed. K. H. Wedepohl. Berlin: Springer-Verlag.

Weiss, R. F. (1970a). The solubility of nitrogen, oxygen and argon in water and seawater. *Deep Sea Research*, **17**, 721–35.

Weiss, R. F. (1970b). Helium isotope effect in solution in water and seawater. *Science*, **168**, 247–8.

Weiss, R. F. (1971a). Solubility of helium and neon in water and seawater. *Journal of Chemical and Engineering Data*, **16**, 235–41.

Weiss, R. F. (1971b). The effect of salinity on the solubility of argon in seawater. *Deep Sea Research*, **18**, 225–30.

Weiss, R. F., Lonsdale, P., Lupton, J. E., Bainbridge, A. E. & Craig, H. (1977). Hydrothermal plumes in the Galapagos Rift. *Nature*, **267**, 600–3.

Wescott, M. R. (1966). Loss of argon from biotite in a thermal metamorphism. *Nature*, **210**, 83–4.

Wetherill, G. W. (1953). Spontaneous fission yields from uranium and thorium. *Physical Review*, **92**, 907–12.

Wetherill, G. W. (1954). Variations in the isotopic abundances of neon and argon extracted from radioactive minerals. *Physical Review*, **96**, 679–83.

Wetherill, G. W. (1975). Radiometric chronology of the early solar system. *Annual Review of Nuclear Science*, **25**, 283–328.

Wetherill, G. W. (1981). Solar wind origin of ^{36}Ar on Venus, *Icarus*, **46**, 70–80.

White, D. E. & Waring, G. A. (1963). Volcanic emanation. In *US Geological Survey Professional Paper 440-K*, pp. 1–27.

Wilhelm, E., Battino, R. & Wilcock, R. J. (1977). Low-pressure solubility of gases in liquid water. *Chemical Reviews*, **77**, 219–62.

Wilkening, L. & Marti, K. (1976). Rare gases and fossil particle tracks in the Kenna ureilite. *Geochimica et Cosmochimica Acta*, **40**, 1465–73.

Wood, D. S. & Caputi, R. (1966). Solubilities of Kr and Xe in fresh and sea water. *United States Naval Radiological Defense Laboratory*-TR-988.

Wood, J. A. (1979). *The Solar System*. Englewood Cliffs (NJ): Prentice-Hall.

Yamanouchi, T. & Shimo, M. (1980). Variation of Rn content before and after earthquake in Mikawa Crustal Movement Observatory, 'Nagoya University' *Abstracts. The Seismological Society of Japan*, No. 2, p. 27.

Yang, J. & Anders, E. (1982a). Sorption of noble gases by solids with reference to meteorites. II. Chromite and carbon. *Geochimica et Cosmochimica Acta*, **46**, 861–75.

Yang, J. & Anders, E. (1982b). Sorption of noble gases by solids, with reference to meteorites. III. Sulfides, spinels, and other planetary gases. *Geochimica et Cosmochimica Acta*, **46**, 877–92.

Yang, J., Lewis, R. S. & Anders, E. (1982). Sorption of noble gases by solids, with reference to meteorites. I. Magnetite and carbon. *Geochimica et Cosmochimica Acta*, **46**, 841–60.

York, D. & Farquhar, R. M. (1972). *The Earth's Age and Geochronology*. New York: Pergamon Press.

Young, B. G. & Thode, H. G. (1960). Absolute yields of the xenon and krypton isotopes in U^{238} spontaneous fission. *Canadian Journal of Physics*, **38**, 1–9.

Young, D. M. & Crowell, A. D. (1962). *Physical Adsorption of Gases*. London: Butterworth.

Zähringer, J. (1968). Rare gases in stone meteorites. *Geochimica et Cosmochimica Acta*, **32**, 209–38.

Zaikowski, A. & Schaeffer, O. A. (1979). Solubility of noble gases in serpentine: implications for meteoritic noble gas abundances. *Earth and Planetary Science Letters*, **45**, 141–54.

Zartman, R. E., Wasserburg, G. J. & Reynolds, J. H. (1961). Helium, argon and carbon in some natural gases. *Journal of Geophysical Research*, **66**, 277–306.

Index

Note: (f) and (t) indicate figure and table respectively.

activation energy, 49, 51, 52, 58, 60
  mass dependence of, 64
  of noble gas diffusion, 49, 49(t)
adsorption, 24–35, 38, 42, 43, 44, 47, 120, 186
  chemical, 26
  energy, 34
  Kr, Xe on various substances, 27(t), 32(f), 33
  Langmuir model, 28–30
  multilayer, 30
  multiple, 120
  phenomenological descriptions of, 27, 28–30
  physical, 26
  time scale, 43
air, 11–14
  elemental abundance, 12(t)
  industrial standard of, 14
  isotopic composition, 13(t)
air contamination, 33, 42, 211–15
  in meteorites, 104
ambipolar diffusion, 121
Apollo missions, 74, 95, 274
Ar (argon), 5, 47, 48, 69, 72(t), 76, 78, 134–5, 253–7
  degassing, 315–20; see also under degassing of the earth
  diffusion, 50(f), 51, 52, 60, 60(f)
  discovery of, 1
  excess, see excess Ar
  in air, 11–14
  in iron meteorites, 150
^{40}Ar
  atmospheric inventory of, 286(t)
  in Venus and Mars, 293
^{40}Ar/^{36}Ar
  in mantle-derived materials, 253(f)
  in submarine glasses, 213, 216, 254, 254(f), 255
  in the mantle, 253–7, 314
  in xenoliths, 217
  lower than air ratio, 257
  primordial, 95

^{40}Ar/^{39}Ar (dating) method, 22, 54, 56, 57, 58, 59
Arrhenius equation, 49
Arrhenius plot, 38(f), 50(f), 60, 61(f)
atmosphere, 9
  see also planetary atmosphere
atmosphere (terrestrial)
  evolution of, 299–338
  external sources for, 273–7
  loss (or dissipation) of, 277
  primary, 273, 290–1
  principal elements in, 272(f)
  recycling of, 278
  secondary, 3, 273
  sinks for, 277–80
  solar wind contribution in, 274–5
  sources of, 276
  subduction of, 278
  volatiles in, 270(t), 271
AVCC (average carbonaceous chondrite), 99, 111
  Kr, 110
  Xe, 99

barite (BaSO$_4$), 184
basalts
  Kr and Xe adsorption, 27(t)
  Kr and Xe solubility, 37(t)
  submarine, see submarine basalts
beryl, 209, 224–7, 227(f)
BET
  area, 27(t), 30
  plot, 31(f)
  theory, 30
biotite, 49, 52, 53(f)
  excess Ar in, 210
bubbles
  in ices, 167, 168
  in seawater, 154, 163
Bunsen coefficients, 155

C (carbonaceous matter), 188
  noble gases in, 188
carbonatite, 238

CCF (carbonaceous chondrite fission), 103, 104
 hypothesis, 101, 105
 Kr, 112
 Xe, 99, 110, 112
chemisorption, 26, 34
chondrites
 carbonaceous, 75, 78, 80, 81; Ne in, 92, 96; rich in planetary gases, 81
 H, 78
 LL, 78
 noble gases in, 75, 79(f); abundance patterns, 76(f), 77(f), 79(f)
 ordinary, 77, 80; metamorphosed, 78
clathrates, 45-7, 123
 compositions of, 47
 heat of clathrate formation, 47
 ice-methane, 48, 124
 ice-noble gas, 47
clinopyroxene, 41
CO_2, 168, 229, 236, 269, 271
 in mantle xenoliths, 238
 in Mars, 292
 radiogenic ^{129}Xe in, 237-8
 well gas, New Mexico, 237-8, 264(f), 267; ^{4}He/^{40}Ar in, 237
comets, 48, 121
components (noble gas), 7-9, 127
 atmospheric, 9
 in situ, 8, 54, 57
 juvenile, 9
 nuclear, 8
 planetary, 9
 primordial, 8, 68
 radiogenic, 8
 separation of, 52, 54, 59
 solar, 9
 trapped, 8, 54, 70
cordierite, 209, 224, 226, 227(f)
cosmic abundances, 68, 69
cosmic ratios, 71, 73(t), 74
cosmic rays, 21, 133, 281, 283
cosmic rock, 72(t)
crust, 40, 131, 258, 307, 318
crystal-melt partitioning, *see* partitioning
cyclosilicates, 39, 224-7, 228
 excess He and Ar in, 209, 224-9; age effect of, 226
 noble gas concentrations in, 227(f)
 noble gas fractionation in, 226

degassed fraction of the earth
 viewed from Ar, 314; from Xe, 330
degassing coefficient, 303, 305, 309, 310, 314
 of Ar, 315, 316, 318, 319, 320
 of Xe, 327, 328, 329
degassing efficiency, 287

degassing models
 catastrophic, 300-3, 314, 335
 continuous, 301, 303-5
 ^{4}He/^{40}Ar, 262-3
 three reservoir, 307, 308, 315-20
 two gas (approaches), 322-3
 two reservoir, 306, 308-15, 321, 329
 see also degassing of the earth
degassing of Mars, 293, 296
degassing of the earth, 2, 82, 228, 288-90, 305
 characteristic degassing time, 310
 of ^{40}Ar, 131, 150, 287. 289, 315-20
 of ^{4}He, 320-3
 of Ne, 322-3
 of ^{129}Xe and ^{136}Xe, 287, 324-35
delta values (Δ or δ), 6-7, 67
 definition of, 6
 ^{3}He, in seawater, 169, 170, 171
 ^{4}He, in seawater, 170
 saturation anomaly, 154
desorption, 26, 28, 33, 34, 43
 time scale, 28, 29, 33, 43
diamonds, 202(f), 203(f), 204(f), 221-3, 228, 253, 257
 ^{3}He/^{4}He in, 221-2
 in ureilites, 116
 thermal release of noble gases, 223(f)
diffusion
 characteristic dimension, 49, 50-1; effects of exsolution on, 51
 coefficient, *see* diffusion coefficients
 equation, 48, 57
 fractionation in, 64
 grain boundary, 51
 of Ar, 50(f), 51, 52, 60, 61(f)
 of light noble gases in glasses, 50(f)
 single stage, 64
 surface, 51
 volume, 51, 56
diffusion coefficients, 48-52, 49(t), 50(f), 54, 60, 64, 218
 of cations, 217
 of He in the earth, 303
 of He, Ne, and Ar in glasses, 50(f)
 of noble gases, 43
 temperature dependence of, 49
dilatancy, 246, 247, 248(f)
diorites, excess He and Ar in, 209, 210
discrimination (isotopic), 7, 9-10, 14, 67
 see also isotopic fractionation
dispersion forces, 23, 26
dissociation energy, 64
distribution coefficients, 10, 39, 40, 41(t), 42, 44, 61, 118, 119, 120
 Ar, 43(t)
 He, 322
 see also partitioning

double spike technique, 10, 67

earthquake prediction, 239–46
  He/Ar ratio, 246
  helium emanation, 245–6
  radon emanation, 239–45
elemental abundance (noble gas)
  in volcanic rocks, 196–207, 198–9(t),
    200(f), 202–4(f)
  pattern, 70–81, 205(f), 212(f), 216,
    218(f), 219(t), 220(f)
  Type I, 182, 186
  Type II, 205, 206
  Type III, 182
elemental fractionation, 9, 45, 54, 74
  in trapping of noble gases, 45, 122(f),
    137(f)
emanation, 231–49
  argon, 234, 247–8; stress effects on,
    248–9
  helium, 231, 232, 234, 245–6, 280;
    stress effects on, 247–9
  in subduction zones, 232
  radon, 238–9
  volcanic, 231–5; juvenile gases in, 231
enstatite, 37(t), 46(t), 118, 212
excess Ar (or ^{40}Ar), 9, 208–11, 301, 314
  in beryl, 209, 225–6
  in metamorphism, 210–11
excess He (or ^{4}He), 168, 174, 208–11, 215,
  301
  in beryl, 209, 225–6
  in ocean, 281
excess ^{3}He, 208–11
  at the East Pacific Rise, 171, 172
  in Atlantic Ocean, 171, 172
  in brines, 173
  in deep-sea waters, 171
  in Pacific Ocean, 170(f), 172
  in plumes, 172
  origin of, 171
excess ^{21}Ne, 130, 151, 225
excess volatiles, 231, 269, 276
excess ^{129}Xe, *see under* ^{129}Xe
excess ^{136}Xe (and heavy Xe isotopes),
  see under ^{136}Xe
exospheric loss, 267, 277, 278, 292
  from Venusian atmosphere, 292
  see also Jeans escape
extinct radionuclides, 19(t)
extraterrestrial materials, 57, 71, 85, 90,
  134, 137, 276, 281, 337
  in the earth, 204
  infall rate of, 275
  Ne in, 152
  spallation ^{21}Ne in, 90

feldspars, 51, 52, 53(f), 227(f), 229

  excess Ar in, 210
  noble gas concentration in, 227(f)
Fick's Law, 57
Fig Tree Formation, 194
  Xe in, 194
first-order rate law (or process), 303,
  306, 315
fission
  ^{244}Pu spontaneous, 20(t)
  ^{235}U neutron-induced, 20(t), 22
  ^{238}U spontaneous, 20(t)
fission Xe (dating) method, 22, 54
formation interval, 288
formation of the earth, 288, 289
fractionation, 9–10
  elemental, *see* elemental fractionation
  isotopic, *see* isotopic fractionation
Freundlich isotherm, 30
  for Xe and Kr, 32(f)
FUN (fractionation and unknown nuclear
  effects), 88

Gabbros, 202(f), 203(f), 204(f)
  excess He and Ar in, 209
gas constant, 6(t)
geobarometer, 217
geochronology, 2, 54, 68
geothermal energy, 167, 168
geothermal waters, 166–8
  in Jordan Rift Valley, 167
  origin of, 166
granites, 199(t), 202(f), 203(f), 204(f),
  222, 224, 246, 247, 248(f)
  fission Xe in, 224
granodiorites, 52, 53(f), 60, 61(f)
graphitization, 221, 229

Hawaii, 217, 251, 255, 260, 267
  volcanic gases, 232, 233; ^{40}Ar/^{36}Ar in,
    232; ^{3}He/^{4}He in, 232
  volcanics, 217, 220, 229, 238, 256(f),
    261(f)
He (helium), 140–2, 257–61
  abundances in various substances, 72(t)
  auroral precipitation, 275, 283
  diffusion of, 50(f)
  discovery of, 1
  emanation, 245–6; along fault zones, 245
  escape from air, 84, 142, 215, 271, 273,
    281–3, 320
  excess, *see* excess He
  flux in the atmosphere, 282(t)
  He-A, 94, 97, 321
  He-B, 94, 97, 321
  He-C, 96
  He-D, 96
  in air, 11–14, 84, 280–3
  in the crust, 140

He (*cont.*)
  isotopic compositions of, 91(t)
    mantle, 141, 142, 208, 231, 258–60
    planetary, 94
    primordial, 141–3
    solar, 94, 208
^{3}He, 64, 81, 84
  concentration profile in seawater, 175(f)
  excess, *see* excess ^{3}He
  flux, 281, 290; vs. heat flow, 234–5
  in hydrothermal waters, 174(f)
  in sedimentary rocks, 184
  in the sun, 94
  primordial, 142, 170, 171, 216, 283, 320
  production of: in air, 281; in the crust,
    130
  solubility in water, 63
  spallation production of, 18
^{4}He
  atmospheric inventory of, 286
  atmospheric life time, 322
  flux, 274, 280–3
  in air, 282(t)
  solubility in water, 63
  trapped in sedimentary rocks, 183, 184
^{4}He/^{40}Ar, 131, 236, 261–4
  degassing model, 262–3
  in CO_2 well gas, 237
  in mantle-derived materials, 264(f)
  in natural gases, 236
  in radiogenic gases, 169
  in the mantle, 262–3, 323
  in Venus and Mars, 293
  vs. ^{40}Ar/^{36}Ar, 264(f)
^{3}He/^{4}He ratios
  at plate margins, 142
  distribution of, 258(f), 259(f)
  in diamonds, 231
  in Hawaii, 232, 260
  in hot spots, 260
  in hydrothermal plumes, 259
  in Japan, 233
  in Marianas, 233
  in Precambrian platforms, 235
  in Red Sea brines, 175
  in South Pacific seawater, 170
  in subduction regions, 232
  in submarine glasses, 216
  in the crust, 257–61
  in the mantle, 257–61, 321
  in volcanic gases, 232–3
  of juvenile He, 174
  secular change in air, 283
  vs. heat flow, 234–5
heat of adsorption, 24, 27, 30
  temperature dependence of, 38
heat evaporation, 47
heat of solution, 38, 39

helium A (He-A), 94
  in mantle He, 208
helium B (He-B), 94
helium escape from air, *see under* He
Henry constant, 27(t), 33, 34(f), 36, 39,
  52, 164
  mass dependence of, 63
  temperature dependence of, 38, 38(f)
Henry's Law, 28, 29, 30, 31, 35, 36, 47
  region, 33
H_2O, 47, 269, 271, 273
  in Mars, 292
hot spots, 217, 251, 252, 260
  ^{40}Ar/^{36}Ar in, 217, 260
  ^{3}He/^{4}He ratios in, 260
hot springs, 142, 164, 234
  ^{3}He/^{4}He ratios in, 234
hydrothermal circulations, 172, 173
hydrothermal vents, 231, 234

^{129}I, 19(t), 54, 56, 324
  initial abundance of, 325
^{129}I–^{129}Xe (dating) method, 54, 115
ice, 84, 167, 168
Iceland, 260
implantation of noble gases, 44
  solar wind, 44, 55
in situ components (noble gases), 8–9, 54,
  79, 80, 81
incompatible elements, 40
inert gases, 23
  see also noble gas
isotopic anomalies (or heterogeneity),
  85–90, 97
  in noble gases, 89; origin of, 88
  in O, 86–7
isotopic effects, 7, 66
  see also isotopic fractionation
isotopic fractionation, 54, 60–7, 86, 185
  in adsorption, 63
  in diffusion, 64, 65(f)
  in dissociation, 64
  in electric discharge, 45(f), 66
  in gravitational field, 63
  in mass spectrometry, 10, 67
  in nonequilibrium transfer, 64
  in sediments, 184
  in solution, 63
  of O, 87–8

Jeans (thermal) escape, 64, 84, 273, 282
josephinite, 151, 152
  noble gases in, 151–2
  origin of, 151
juvenile flux, 143, 215
juvenile gases, 9, 40, 84, 211, 214, 215,
  216, 231, 276–7, 278, 307

juvenile noble gases, 9
  ^{40}Ar, 169
  ^{3}He, 169–76, 277
  ^{4}He, 168–9, 174, 175, 231, 232; flux of, 168, 176
  ^{3}He/^{4}He ratio, 231
  Ne, 266
  Xe, 268

K (potassium), 40
  global inventory of, 131
K–Ar ages, 58
  sediments, 184
  whole rock, 58
K–Ca (dating) method, 5
kerogen, 43(t), 46(t), 66, 119, 181(f)
kersutite, 206, 222(f)
kimberlite, 217, 220
Kr (krypton), 47, 48, 69, 78, 79, 145–7, 266–8
  abundances in various substances, 72(t)
  air Kr, 11–14
  CCF, 112, 147
  discovery of, 2
  excess ^{86}Kr, 145, 147, 149; in air, 132
  in the mantle, 266–8
  isotopic compositions of, 110(t)
  planetary, 112, 113, 145, 147
  solar, 110(t), 111, 112, 113; isotopic compositions of, 111(f)
  terrestrial, 145, 147

Langmuir isotherm, 28–9
lattice defects (or vacancies), 36
lherzolites, 220, 222(f), 229
  noble gas distributions in, 228(t)
locations of noble gases, 228–9
  in diamonds, 221
  in meteorites, 117–18
lunar breccia, 55(f), 71
lunar rocks (or samples), 35, 60, 62(f), 85
  adsorption potential of Xe on, 35
  implantation of solar wind on, 44
lunar soil
  light noble gas isotopic compositions in, 91(t)
  Ne in, 95

magma, 209, 250
magnetite, 43(t), 44
  noble gas solubility in, 37(t)
  noble gases in, 78
mantle
  argon in, 314
  component (noble gas), 209
  depleted, 232, 251, 252, 260, 308, 314, 328, 336, 337

helium (mantle He), 208, 231, 258, 259, 260
  heterogeneity, 251–2
  Kr in, 266–8
  Ne in, 140
  primitive, 308
  undepleted, 197, 251, 252, 255, 257, 260, 305, 314, 320, 334, 336, 337
  upper and lower, *see* depleted and undepleted mantle
  xenon in, 268, 334
mantle dichotomy, 336
mantle evolution, 315
Mars, 291–8
  ^{40}Ar in, 293
  degassing of, 293, 296
  noble gases in, 294–5(t)
  primordial noble gases in, 293
  radiogenic ^{129}Xe in, 296
  volatiles in, 291–2, 294–5(t)
mass discrimination (or fractionation), 10, 11
  see also isotopic fractionation
mass spectrometers, 6, 43
  noble gas, 10
  Reynolds type, 3
  solid-source, 10
megacryst (olivine), 217
meteorite, 4, 70, 71
  Allende, 27(t), 105, 107(f), 109(f), 113(f), 115(f)
  Bjurböle, 56(f)
  Bruderheim, 58, 59, 79
  gas rich, 81
  Mokoia, 77(f), 78, 105, 107(f)
  Murchison, 108
  Murray, 56, 100(f), 102(t), 103(f), 107(f), 112(f)
  noble gas data for, 72–3(t)
  Orgueil, 112(f)
  parent bodies, 75, 117, 124
  primitive, 68, 75, 82, 99; Xe in, 99
  refractory inclusions in, 86, 88, 89
  Renazzo, 100(f), 101(f), 105, 107(f)
  solar-gas-rich, 101
  solar wind implantation on, 44
  Xe isotopic compositions in, 100(f)
meteorite analogy, 82, 84, 85, 137, 283, 284, 295, 320
meteorite impacts, 224
methane, 47, 123, 124, 235, 236, 273
  clathrate, 47, 124
microcline, 49(t)
muscovite, 210, 227(t)

N (nitrogen), 31(f), 69(t), 160, 291
N_2/Ar
  in the mantle, 233
  in volcanic gases, 233

natural gases, 235–8
Ne (neon), 69, 76, 78, 79, 135–40, 265–6
  abundances, 72(t)
  air, 11–14; isotopic composition of,
    14, 137
  discovery of, 2
  earth's original, 266
  enrichment, 182, 188, 189
  fractionation of air, 139–40
  gravitational escape of, 140
  in mantle-derived materials, 265(f)
  isotopic compositions, 91(t)
  juvenile, 266
  mantle, 140
  Ne-A, 92, 93(f), 95, 96, 97, 137, 138
  Ne-B, 92, 93(f), 95, 97
  Ne-C, 95
  Ne-D, 96, 97
  Ne-E, 91(t), 93(f), 96, 97, 137, 138
  origin of terrestrial, 137
  planetary, 96
  primordial, 90, 138, 140
  solar, 95, 96, 137, 138; $^{20}Ne/^{22}Ne$ in, 96
  solar flare, 95, 137
  solar wind, 96
  spallation, 90
^{21}Ne
  excess, 130, 151, 225
  production in minerals, 130, 265
  radiogenic, 129(t), 130, 139, 194;
    terrestrial inventory of, 139
^{22}Ne, radiogenic, production of, 130
$^{20}Ne/^{36}Ar$
  in meteorites, 76, 116
  in submarine glasses, 201
  in Venus, 297
nebula, 125, 273
noble gas
  abundance data, 72(t)
  atomic sizes, 24, 25(t)
  average concentration in the earth, 11
  components, 7–8
  compounds, 23
  detection limits, 4
  elemental abundance patterns, *see*
    under elemental abundance
  fractionation, 80
  gravitational escape, 117
  identification of, 15
  in air, 11–14
  in chondrites, 75
  in rain water, 166–8
  in solar system, 4
  interaction with other substances, 23
  ionization potentials, 24, 25(t)
  isotopic abundances, 15–17, 20, 91(t)
  locations in rocks, 53, 54, 228, 229
  nuclear properties, 15–22

physical and chemical properties of,
  25(t)
  primordial abundances, 69
  production rate of, 19(t)
  radiogenic, 126–32, 129(t)
  solar, 69, 77
  terrestrial inventory, 11, 18, 48; external
    sources for, 275; radiogenic, 131;
    solar wind contribution in, 274–5
  trapped, 81
noble gas compounds, 23
noble gas geochemistry, 7–10, 86
  nomenclature in, 7–10
nuclear components, 8
nuclear weapon tests, 21
nucleosynthesis, 69, 86, 94, 95, 108,
  117, 150
  ^{40}Ar production in, 95
  ^{3}He production in, 94

O (oxygen), 21(t), 271, 272, 291
obsidian, 49(t)
occlusion (of noble gases), 44
Oklo phenomenon, 21, 184, 266
olivine, 41, 228
orthoclase, 49(t)

paleoatmosphere
  $^{40}Ar-^{36}Ar$ in, 193–4
  $^{3}He/^{4}He$ in, 194
  ^{21}Ne in, 194
  noble gases in, 192–5, 312
  Xe in, 194, 326, 329
partial melting, 251
partitioning (of noble gas), 4, 39–41, 60,
  197, 304
pegmatite, 226
  see also distribution coefficients
peridotites, 217
permeation, 52
perthitization, 51
phlogopite, 49(t), 217
photosynthesis, 271
phyllosilicates, 43, 119
planetary atmosphere, 82–5
  noble gases in Mars, 83(f), 85, 123, 124
  noble gases in the earth, 82, 83(f), 123,
    124
  noble gases in Venus, 83(f), 85, 123,
    124
  origin of noble gases in, 116–26
planetary components, 9
  see also planetary noble gases
planetary helium, 94
  see also under He
planetary neon, 96
  see also under Ne

planetary noble gases, 73(t), 77, 78, 80, 81, 116
  in meteorites, 77, 114
  in the earth, 204
  origin of, 75, 79, 81, 116–26
planetary pattern (noble gas abundance), 71, 75, 83(f), 205(f)
planetesimals, 124, 125, 126
  noble gas captured by, 124–6, 125(f)
plumes, 172
  excess ^{3}He in, 172
plutonic rocks, 222–4
  see also granites
polarizability (electronic), 24, 25(t)
  of noble gases, 25(t)
potassium–argon (dating) method, 2, 5
  see also K–Ar ages
preaccretionary earth materials, 21(t), 133, 135
primeval noble gases, 81
primordial components, 8, 75, 81, 86, 142, 149, 201
  ^{3}He, 170, 171
  lunar, 85
  Xe, 104
^{244}Pu, 19, 54, 132, 267, 324
  in the earth, 332–4
  initial abundance of, 325
^{244}Pu–^{136}Xe (dating) method, 54
pyroxene, 43(t), 228

'Q' (noble gas carrier), 80
quartz, 227(f)

radon (Rn), 5, 11, 18, 235, 238–9, 240
  discovery of, 2
  emanation, 238–9, 246; continental flux, 239; earthquake precursory, 239–45; effect of water on, 247; seasonal variation in, 241; stress effects on, 246–9; transport mechanism, 244–5
  in hydrothermal plumes, 239
  in seawater, 239
  in spring water, 239
rain (or meteoric) water, 166–8
Rayleigh distillation, 65, 96
Red Sea brines, 168
regolith, 85, 292
Reynolds mass spectrometer, 3
Rn, *see* radon

sanidine, 49(t)
schungite (amorphous carbon), 49(t), 178(t), 183(f), 189(f)
seafloor spreading, 215
seawater, 160–6, 168, 212, 213
  air injection in, 154, 163, 164, 165
  juvenile (or excess) He in, 163, 164, 169–76

  N$_2$ in, 160
  noble gases in, 154, 160–6
  O$_2$ in, 160
  Pacific deep, 168; juvenile ^{4}He in, 168
  ^{222}Rn in, 239
  South Pacific, 170
secondary atmosphere, 273
sedimentary rocks (or sediments), 177–95
  ^{3}He in, 184
  noble gases in, 176–95, 178(t), 179(f); origin of, 185–9
  trapped ^{4}He in, 183, 184
  Xe/Ar in, 280
  Xe in, 120, 185, 189, 190, 191, 194
serpentine, 44, 119, 186, 212, 213
  noble gas solubility in, 37(t), 46(t), 119, 212, 213
Setchenow relation, 155
shales, 46, 178, 180, 181, 183, 185, 187, 188, 189, 190, 191
  Kr adsorption on, 27(t), 32(f), 33, 34(f)
  N$_2$ adsorption on, 31(f)
  Xe adsorption on, 27(t), 32(f), 33, 34(f), 84
shales hypothesis, 190, 191, 192
silicate melts, 37, 38, 39
  solubility of noble gases in, 37(t)
Skaergaard intrusion, 224, 225(f), 267
  excess ^{129}Xe in, 267
solar activity, 282, 283
solar components, 9
solar composition, 47, 68
  primordial, 68
solar flare, 69, 281
  Ne in, 95, 137
  spallation, 96, 134
solar helium, 94
  in the mantle, 208
solar nebula, 47, 81, 117, 118
  pressures in, 48, 118
  primitive, 47
  temperatures in, 118
solar neon, 92
solar noble gas, 69, 73(t), 77, 78, 80, 81, 83(f), 205(f)
  elemental abundance patterns of, 71, 74, 74(f)
  in meteorites, 77
  Kr in, 110, 110(t), 111(f)
  origin of, 71, 81
  Xe in, 101
solar pattern (noble gas abundance), 75
  definition of, 71
solar system, 68, 69, 70, 86, 88
  composition of, 68, 69(t)
  noble gas isotopic anomalies in, 89
solar wind, 44, 73(t), 74, 82, 94, 274, 281
  D/H in, 94
  D/^{4}He in, 94

solar wind (*cont.*)
   ^{3}He/^{4}He in, 94, 95
   implantation, 44
   in air, 275
   light noble gas isotopic compositions in, 91(t)
   measured by Apollo missions, 74
   Ne in, 96
solubility, 35-9, 40, 52, 211
   data for noble gases, 37(t)
   noble gas:in enstatite melt, 37(t), 46(t), 118, 212(f); in silicate melt, 36, 37(t)
   parameters, 157
   pressure dependence of, 36
   temperature dependence of, 38, 38(f)
   theories of, 39
solution (noble gases), 26, 35-9, 42, 44, 47
   heat of, 38, 39
soot, 43(t), 46(t), 119
spallation, 18, 21, 70, 90, 94, 95, 96, 98, 133-4
   gases in the earth, 79
   in preaccretionary matter, 133
   in terrestrial noble gas inventory, 134
   products in meteorites, 21
spike, in noble gas analyses, 14
spinel, 228, 228(t)
stepwise heating (*or* degassing), 52-60, 92, 229
   in Kr isotopic analyses, 112(f)
   in Ne isotopic analyses, 92
   in Xe isotopic analyses, 99, 101(f), 105, 106
subareal volcanics, 202(f), 203(f), 204(f), 264(f), 265(f)
   excess He and Ar in, 209
subduction, 84, 307
   of juvenile gases, 84, 278
   of water, 278
   of Xe, 278
subduction zones, 232
   gas emanation in, 232
   ^{3}He/^{4}He in, 232
submarine basalts, 61(f), 202(f), 203(f), 204(f), 251, 253, 264(f), 265(f)
   air contamination in, 214
   excess He and Ar in, 209
   glassy rims of: ^{40}Ar/^{36}Ar in, 213, 216; Ar concentration, 213; ^{3}He/^{4}He in, 216; ^{3}He/^{20}Ne in, 216; juvenile gases in, 215-16; noble gas abundance patterns in, 219(f)
   holocrystalline, 214
   noble gas abundance patterns in, 218(f)
   vesicles in, 229
subsolar components, 297
SUCOR (surface correlated components), 102(t), 103(f)

Xe in, 104
superheavy element, 104, 110

^{130}Te, 21
terrestrial planets, 4
Th (thorium), 40
^{232}Th, 19
thermal waters, in Jordan Rift Valley, 167
tholeiites, 216-17
thucholite, 183(f), 184
tourmaline, 209, 224, 225, 226, 227(f)
transport coefficient (of noble gases), 303, 304-5
   of He, 321, 322, 323
   of K, 315, 317, 319
   see also degassing coefficient
trapped components, 54
trapped gases, 42, 70, 71, 81
   in meteorites, 70, 71, 72(t), 73(t), 74, 75, 76, 77
trapping (of noble gases), 41-5, 46(t)
   by adsorption, 120
   by electric discharge, 44
   during crystal growth, 44
   elemental fractionation in, 45
   in sediments, 63

U (uranium)
   global inventory of, 131
   production of ^{4}He, 128
   Th/U ratio in rocks, 128
   U/K ratio in rocks, 131; in chondrites, 131
^{235}U, 19(t), 54
   fission of, 98, 128
^{238}U, 19(t), 54
   fission of, 98, 128, 267
uranium-helium age, 3

vacancies (lattice), 36, 39
van der Waals
   bonds, 63
   force, 25(t), 23, 26, 39
   interaction, 24
   type , 4, 46
van't Hoff equation, 155
Venus, 291-8
   ^{36}Ar in, 297
   ^{40}Ar in, 293
   ^{4}He/^{40}Ar in, 293
   noble gases in, 294-5(t)
   primordial noble gases in, 293
   surface temperature of, 291
   volatiles in, 291-2, 294-5(t)
volatiles, in planets, 294-5(t)
volcanic ash, Kr and Xe adsorption on, 27
volcanic glasses, 49(t)

water, 153–76
  juvenile, 84, 166, 278
  noble gas solubility in, 37(t); pressure
    effects, 154; salinity dependence,
    157, 159; temperature effects, 154
  solubility of O_2 and N_2 in, 155–9

Xe (xenon), 21(t), 47, 48, 69, 72(t), 78, 79,
    91(t), 143–5, 266–8
  air Xe, 143, 145
  CCF, 99, 110, 112; in planetary Xe, 103
  concentration in rocks, 4
  deficiency in the earth, 84, 190–2, 284
  degassing model, 324–35
  discovery of, 2
  enhancement (or enrichment), 206
  fission, 8, 21, 54, 55, 130, 144, 266,
    285; from ^{244}Pu in air, 145
  H, 105, 108, 110
  in air, 8, 11–14, 99, 145
  in AVCC, 99, 146(f)
  in Mars, 285
  in Precambrian atmosphere, 194
  in sedimentary rocks, 120, 185, 189,
    190, 191, 194
  in the mantle, 266–8
  in the sun, 108
  in Venus, 285
  in water, 165–6
  isotopic compositions, 100(f), 102(f),
    103(f), 106(f), 107(f)
  juvenile, 268
  L, 105, 108, 110
  mantle, 268, 328, 331
  planetary, 99, 103(f), 143
  primitive, 99, 104, 105, 108, 110, 112
  primordial, 104, 105, 147
  S, 108, 110, 114
  solar, 101, 103(f), 143, 144
  SUCOR, 102, 103(f), 104
  terrestrial, 103(f)
  trapped, 56
  U, 106, 108, 115, 145, 146(f)
  ^{238}U-fission, 8
  X-Xe, 105, 108, 100, 113
^{129}Xe
  atmospheric evolution of, 326(f), 327(f)
  atmospheric inventory of, 286(t)
  degassing of, 287
  excess, 147–9, 224, 225, 266–7, 331,
    333; in CO_2 well gases, 237–8, 268,
    332, 334; in Hawaii volcanics, 267;
    in mantle-derived samples, 266–7;
    in submarine glasses, 267; in the
    mantle, 268, 334
  in Mars, 296
  radiogenic, 194, 287, 324, 328–9
  trapped in meteorites, 115
^{136}Xe
  degassing of, 287
  excess ^{136}Xe, 147–9, 331, 333;
    ^{244}Pu origin, 148; ^{238}U origin, 149
  fission, 128, 129(t), 132, 194; in the
    earth, 288
  inventory of, 286(t)
  radiogenic, 194, 287, 288, 324; in the
    earth, 288
Xe–Xe age of the earth, 288
^{129}Xe–^{128}Xe (dating) method, 22
xenoliths (ultramafic), 41, 217–21, 253,
    264(f), 265(f)
  ^{40}Ar/^{36}Ar in, 217
  CO_2 inclusions in, 238
  excess He and Ar in, 209
  excess ^{129}Xe in, 238
  noble gas distribution in, 202–4(f)
xenology, 98, 101, 104, 105, 108